Pressure Vessels

About the Author

Robert F. Stricker, P.E., retired as Project Manager from an international engineering services firm, WorleyParsons, Abu Dhabi, UAE. His previous engagements included Manager of Engineering at the Houston, Texas pressure vessel fabrication companies, Wyatt Industries and General Welding Works. Mr. Stricker has B.S. and M.S. degrees in mechanical engineering from the University of Central Florida and has taught engineering classes at Lone Star College in Houston. Since 1985 Mr. Stricker has been providing consulting services to pressure vessel fabrication shops, with emphasis on the stress analyses of component geometries beyond the typical Code formulae.

Pressure Vessels

The ASME Code Simplified
Ninth Edition

Robert F. Stricker, P.E.
J. Phillip Ellenberger, P.E.

McGraw Hill

New York Chicago San Francisco
Athens London Madrid
Mexico City Milan New Delhi
Singapore Sydney Toronto

Library of Congress Cataloging-in-Publication Data

Names: Stricker, Robert F., author. | Ellenberger, J. Phillip, author.
Title: Pressure vessels : the ASME code simplified / Robert F. Stricker, P.E., J. Phillip Ellenberger, P.E.
Description: Ninth edition. | New York : McGraw Hill Education, 2021. | Includes index. | Summary: "New edition of a guide to the most recent ASME Boiler & Pressure Vessel Code (BPV2017 Edition) loaded with code commentary, practical information, examples, and data critical for code compliance"—Provided by publisher.
Identifiers: LCCN 2020025617 | ISBN 9781260455410 | ISBN 9781260455427 (ebook)
Subjects: LCSH: Pressure vessels—Standards—United States.
Classification: LCC TS283 .C53 2021 | DDC 681/.76041—dc23
LC record available at https://lccn.loc.gov/2020025617

Pressure Vessels: The ASME Code Simplified, Ninth Edition

1 2 3 4 5 6 7 8 9 CCD 24 23 22 21

ISBN 978-1-260-45541-0

MHID 1-260-45541-6

Sponsoring Editor
Robert Argentieri

Editorial Supervisor
Donna M. Martone

Acquisitions Coordinator
Elizabeth Houde

Project Manager
Poonam Bisht

Copy Editor
Kirti Sharma

Proofreader
Shweta Bharti Dayal

Indexer
Cheryl ATX Stitcher

Production Supervisor
Pamela A. Pelton

Composition
MPS Limited

Art Director, Cover
Jeff Weeks

A very special thanks goes to my good friends Minh Le and
Curtis Dahl. Their support and help back in the earlier days of
my career assisted in defining the path I have followed.
This book is but a small manifestation of that influence.
I wish you both were still here to help me celebrate.
Robert Stricker

Contents

Preface ... xi
Acknowledgments ... xiii

1 Origin, Development, and Jurisdiction of the ASME Code ... 1
History of the ASME Code ... 1
Additions to the Code ... 3
ASME Boiler and Pressure Vessel Committee ... 4
Procedure for Obtaining the Code Symbol and Certificate ... 6
The National Board of Boiler and Pressure Vessel Inspectors ... 6
Shop Reviews by the National Board ... 7
National Board Requirements ... 8
Code Case Interpretations ... 9
Canadian Pressure Vessel Requirements ... 11
Department of Transportation Regulations ... 15
Welded Repair or Alteration Procedure ... 16
Managerial Factors in Handling Code Work ... 19
Work-Flow Schedules ... 24
Mill Orders ... 27
Warehouse Orders ... 29
Shop Organization ... 31

2 Descriptive Guide to the ASME Code Section VIII, Division 1, Pressure Vessels ... 33
Requirements for Establishing Design Thickness Based on the Degree of Radiography (Code Pars. UW-11 and UW-12) ... 33
Use of Internal-Pressure Cylindrical Shell Thickness Tables ... 44
Use of Simplified External-Pressure Cylindrical Shell Thickness Charts ... 53
Use of Internal-Pressure Thickness Tables for Ellipsoidal and Torispherical Heads ... 57
Elements of Joint Design for Heads ... 70

Thickness of Flat Heads and Bolted Flat Cover Plates 73
Various Other Designs for Heads 75
Allowable Pressure, Pitch, and Thickness 75
Calculations for Reinforcement of Openings 75
Cryogenic and Low-Temperature Vessels 81
Impact-Test Exemptions 84
Service Restrictions and Welded Joint Category 93
Noncircular-Cross-Section-Type Pressure Vessels 95
Pressure Vessels Stamped with the UM Symbol 95
Marking and Nameplate Requirements 97

3 Design for Safety 101
Basic Causes of Pressure Vessel Accidents 101
Corrosion Failures 101
Stress Failures 101
Other Failures 102
Design Precautions 102
Inspection Openings 104
Quick-Opening Closures 106
Pressure Relief Devices 108

4 Guide to Quality Control Systems for ASME Code Vessels 111

5 Inspection and Quality Control of ASME Code Vessels 137
Shop Inspection 137
Tests 142
ASME Manufacturer's Data Reports 143
Field Assembly Inspection 144
Responsibility for Field Assembly 144
Inspection Procedure 145
Radiography 146
Conditions Requiring Radiography 146
Interpretation of Welding Radiographs 148
Sectioning 154
Hydrostatic Testing 155
Test Gage Requirements 157
Seven Simplified Steps to Efficient Weld Inspection 157

6 **Welding, Welding Procedure, and Operator Qualification** 161
Examination of Procedure and Operator 161
Record of Qualifications 162
Welding Procedures and Qualification Simplified. 162
Welding procedure specification forms. . . 164
Variables [Code Pars. QW-400, QW-250, and QW-350 and Tables QW-415 (WPS) and QW-416 (WPQ)] 167
Welder Qualifications 170
Welding Details and Symbols. 176
Joint Preparation and Fit-Up 180
Effects of Welding Heat. 183
Preheating 187
Postweld Heat Treatment 188
Conditions requiring postweld heat . . . . . treatment 190

7 **Nondestructive Examination.** 193
Leak Testing 200
Acoustic Emission Testing 203

8 **ASME Code Section VIII, Division 2, Alternative Rules, and Division 3, Alternative Rules for High Pressure Vessels** 205

9 **Power Boilers: Guide to the ASME Code Section I, Power Boilers** 211

10 **Nuclear Vessels and Required Quality Assurance Systems** 219
Power Plant Cycles. 220
Pressurized Water Reactor Plants. 220
Boiling Water Reactor Plants 221
Quality Assurance 225

11 **Section XII Rules for Construction and Continued Service of Transport Tanks** 235
Overview of Section XII 235
Part TG: General Requirements. 236
Part TM: Material Requirements 236
Part TD: Design Requirements. 236
Part TW: Requirements for Tanks Fabricated by Welding. 237
Part TF: Fabrication Requirements 237

Part TE: Examination Requirements 237
Part TT: Testing Requirements 237
Part TR: Pressure Relief Devices 237
Part TS: Stamping, Marking, Certification, Reports, and Records 238
Part TP: Requirements for Repair, Alteration, Testing, and Inspection for Continued Service 238
Modal Appendices 238
Modal Appendix 1: Cargo Tanks 238
Modal Appendix 2: Rail Tank Cars 242
Modal Appendix 3: Portable Tanks 242
Modal Appendix 4: Category 106A500-X, 106A800-X, 110A500-W, 110A600-W, 110A800-W, 110A1000-W, and 110A2000-W Ton Containers 243

Appendix A **245**

Appendix B **251**

Appendix C **263**

Appendix D **265**

Appendix E **273**

Appendix F **277**

Index **283**

Preface

This Ninth Edition of *Pressure Vessels: The ASME Code Simplified* has been brought up to date by Robert Stricker, who has around 40 years of experience in the field, along with valuable input by J. Philip Ellenberger. This edition includes expanded details and revised sections that outline recent changes to the *Boiler and Pressure Vessel Code* of the American Society of Mechanical Engineers (ASME).

Much of the material here is based on the content of the Eighth Edition written by J. Philip Ellenberger, with his more than 40 years of experience in the design, fabrication, and inspection of the ASME Code pressure vessels. The wealth of real-world knowledge possessed by the authors makes this book a reference tool that is accurate, practical, and easily understood.

Since the publication of the Eighth Edition various processes have been changed by the use of computers. What has not changed is the need for both theoretical and practical information to make a quality product.

The Code has been accepted for many years as the standard for the construction of safe boilers and pressure vessels. It is progressive and viable. Important changes and additions are made to it when required.

This book is geared to meet the needs of engineering, manufacturing, repair, and testing companies that have to comply with the Code requirements and make the Code work better, more efficiently, and more profitably.

The rules for cargo tanks have been shifted from the U.S. Department of Transportation to the newly issued (as of 2004) ASME Section XII. These changes are explained in rewritten Chap. 11.

Welding is one of the most important functions in building pressure vessels. This book is a practical reference and text that can be used in schools and industry. It explains the theory of the weld, welding metallurgy, effects of welding heat, types of welding tests, welding procedures, and qualification of procedures so they can be

CHAPTER 1

Origin, Development, and Jurisdiction of the ASME Code

History of the ASME Code

On March 20, 1905, a disastrous boiler explosion occurred in a shoe factory in Brockton, Massachusetts, killing 58 persons, injuring 117 others, and causing a quarter of a million dollars in property damage. For years prior to 1905, boiler explosions had been regarded as either an inevitable evil or "an act of God" (see Figs. 1.1 and 1.2). But this catastrophic accident had the effect of making the people of Massachusetts see the necessity and desirability of legislating rules and regulations for the construction of steam boilers in order to secure their maximum safety. After much debate and discussion, the state enacted the first legal code of rules for the construction of steam boilers in 1907. In 1908, the state of Ohio passed similar legislation—the Ohio Board of Boiler Rules adopting, with a few changes, the rules of the Massachusetts Board.

Subsequently, other states and cities in which explosions had taken place began to realize that accidents could be prevented by the proper design, construction, and inspection of boilers and pressure vessels and, therefore, began to formulate rules and regulations for this purpose. As regulations differed from state to state and often conflicted with one another, manufacturers began to find it difficult to construct vessels for use in one state that would be accepted in another. Because of this lack of uniformity, both manufacturers and users made an appeal in 1911 to the Council of the American Society of Mechanical Engineers to correct the situation. The Council answered the appeal by appointing a committee "to formulate standard specifications for the construction of steam boilers and other pressure vessels and for their care in service."

Figure 1.1 The Brockton, Massachusetts, shoe factory. (*Courtesy of The Hartford Steam Boiler Inspection and Insurance Company.*)

Figure 1.2 Shoe factory after the boiler explosion of March 20, 1905, which led to the adoption of many state boiler codes and the *ASME Boiler and Pressure Vessel Code*. (*Courtesy of The Hartford Steam Boiler Inspection and Insurance Company.*)

The first committee consisted of seven members, all experts in their respective fields—one boiler insurance engineer, one material manufacturer, two boiler manufacturers, two professors of engineering, and one consulting engineer. The committee was assisted by an advisory committee of 18 engineers representing the various phases of design, construction, installation, and operation of boilers.

Following a thorough study of the Massachusetts and Ohio rules and other useful data, the committee made its preliminary report in 1913 and sent 2000 copies of it to the professors of mechanical engineering, engineering departments of boiler insurance companies, chief inspectors of the boiler inspection departments of states and cities, manufacturers of steam boilers, editors of engineering journals, and others interested in the construction and operation of steam boilers, with a request for the suggestions of changes or additions to the proposed regulations.

After three years of countless meetings and public hearings, a final draft of the first *ASME Rules for the Construction of Stationary Boilers and for Allowable Working Pressures,* known as the 1914 edition, was adopted in the spring of 1915.

Additions to the Code

Since 1914, many changes have been made and new sections added to the Code as the need arose. The present sections are listed in the following order:

I. Rules for Construction of Power Boilers

II. Materials
- Part A: Ferrous Material Specifications
- Part B: Nonferrous Material Specifications
- Part C: Specifications for Welding Rods, Electrodes, and Filler Metals
- Part D: Properties (Customary)
- Part D: Properties (Metric)

III. Rules for Construction of Nuclear Facility Components
- Subsection NCA: General Requirements for Division 1 and Division 2
- Appendices
- Division 1
  - Subsection NB: Class 1 Components
  - Subsection NC: Class 2 Components
  - Subsection ND: Class 3 Components

  - Subsection NE: Class MC Components
  - Subsection NF: Supports
  - Subsection NG: Core Support Structures
- Division 2: Code for Concrete Containments
- Division 3: Containment Systems for Transportation and Storage of Spent Nuclear Fuel and High-Level Radioactive Material
- Division 5: High Temperature Reactors

IV. Rules for Construction of Heating Boilers

V. Nondestructive Examination

VI. Recommended Rules for Care and Operation of Heating Boilers

VII. Recommended Guidelines for the Care of Power Boilers

VIII. Rules for Construction of Pressure Vessels

- Division 1
- Division 2: Alternative Rules
- Division 3: Alternative Rules for Construction of High-Pressure Vessels

IX. Welding, Brazing, and Fusing Qualifications

X. Fiber-Reinforced Plastic Pressure Vessels

XI. Rules for Inservice Inspection of Nuclear Power Plant Components

XII. Rules for Construction and Continued Service of Transport Tanks

ASME Boiler and Pressure Vessel Committee

The Boiler and Pressure Vessel Committee was established by the American Society of Mechanical Engineers in 1911. The Committee was established to formulate standard rules for the construction of steam boilers and other pressure vessels. In 2009, the Boiler and Pressure Vessel Committee was superseded by the following committees:

a. Committee on Power Boilers (I)

b. Committee on Materials (II)

c. Committee on Construction of Nuclear Facility Components (III)

d. Committee on Heating Boilers (IV)

e. Committee on Nondestructive Examination (V)

f. Committee on Pressure Vessels (VIII)

g. Committee on Welding, Brazing, and Fusing (IX)

h. Committee on Fiber-Reinforced Plastic Pressure Vessels (X)

i. Committee on Nuclear Inservice Inspection (XI)

j. Committee on Transport Tanks (XII)

k. Technical Oversight Management Committee (TOMC)

The increase in the size of the Code reflects the progress of the industry. To keep up with this spontaneous growth, constant revisions have been required. The ASME Code has been kept up to date by the above Committees (currently consisting of hundreds of volunteer engineers and other technical professionals) which consider the needs of the users, manufacturers, and inspectors of boilers and pressure vessels. In the formulation of its rules for the establishment of design and operating pressures, the Committee considers materials, construction, the methods of fabrication, inspection, certification, and safety devices. The ASME works closely with the American National Standards Institute (ANSI) to assure that the resulting documents meet the latter's criteria for publication as American National Standards.

The members of the Committee do not represent particular organizations or companies but have recognized background and experience by which they are placed in categories which include manufacturers, users of the products for which the codes are written, insurance inspection, regulatory, and general. The Committee meets on a regular basis to consider requests for interpretations, revisions, and additions to Code rules as dictated by advancements in technology. Approved revisions are published at regular intervals in the Editions of the Code.

To illustrate, boilers in 1914 were operating at a maximum pressure of 275 psi and temperature of 600°F. Today, boilers are designed for pressures as high as 5000 psi and temperatures of 1100°F, and pressure vessels for pressures of 3000 psi and over and for temperatures ranging from –350°F to more than 1000°F.

Each new material, design, fabrication method, and protective device brought new problems to the Boiler Code Committee, requiring the expert technical advice of many subcommittees in order to expedite proper additions to and revisions of the Code. As a result of the splendid work done by these committees, the *ASME Boiler and Pressure Vessel Code* has been developed. It is a set of standards that assures every state of the safe design and construction of all boiler and pressure vessels used within its borders and is used around the world as a basis for enhancing public health, safety, and welfare. Many foreign manufacturers are accredited under the provisions of the *ASME Boiler and Pressure Vessel Code.*

Procedure for Obtaining the Code Symbol and Certificate

The users of pressure vessels prefer to order ASME Code vessels because they know that such vessels will be designed, fabricated, and inspected to an approved quality control system in compliance with a safe standard. Pressure vessel manufacturers want the Code symbol and the Certificate of Authorization so that they will be able to bid for Code work, thereby broadening their business opportunities. They also believe that authorization to build Code vessels will enhance the reputation of their shop.

Those companies interested in building Code vessels according to the ASME Section VIII, Division 1, Pressure Vessels Code should access the ASME website, www.asme.org. Under the heading of Certification & Accreditation below the ASME Certification Process, Certification Process is a set of general guidelines to be used for initiating the certification process. The website guidelines identify steps toward *preparation*, *application*, *assessment*, and *certification*. These steps include the following:

- Enter into an agreement with an Authorized Inspection Agency (AIA). This third party in the manufacturer's plant, by virtue of being authorized by the state to do Code inspection, is the legal representation which permits the manufacturer to fabricate under state laws (the ASME Code).
- Prepare a Quality Manual—a written description of the quality and documentation program which is followed by the plant (see Code Appendix 10, Quality Control Systems).
- Along with the preparation of the Quality Manual, indoctrinate and train personnel in the execution and implementation of the quality program.
- Prepare and submit a signed Accreditation and Certification Agreement Form.

A complete list of the steps is included in the referenced website.

The National Board of Boiler and Pressure Vessel Inspectors

The National Board of Boiler and Pressure Vessel Inspectors is an independent, nonprofit organization whose members are the jurisdictional officials responsible for enforcing and administrating the *ASME Boiler and Pressure Vessel Code*. The National Board was first organized in 1919. Since then, it has served for the uniform administration and enforcement of the rules of the *ASME Boiler and Pressure*

Vessel Code. In drafting the first Code, the Boiler Code Committee realized that it had no authority to write the rules to govern administration and that compliance could be made mandatory only by the legislative bodies of states and cities.

Although some authorities thought that uniformity could be achieved by a code of uniform rules, they proved to be mistaken. Various states and cities adopted the ASME Code with the provision that boilers be inspected during construction by an inspector qualified under their own regulations; the boiler was then to be stamped with the individual local stamping to indicate its conformity with these regulations. This requirement created an unwieldy situation, for boilers constructed in strict accordance with the ASME Code still had to be stamped with the local stamping, thereby causing needless delay and expense in the delivery of the vessel.

It was evident that some arrangement had to be made to overcome such difficulties. Therefore, boiler manufacturers met with the chief inspectors of the states and cities that had adopted the ASME Code and formed the National Board of Boiler and Pressure Vessel Inspectors for the purpose of presenting the ASME Code to governing bodies of all states and cities. Their aim was not only to promote safety and uniformity in the construction, installation, and inspection of boilers and pressure vessels but also to establish reciprocity between the political subdivisions of the United States. Such ideals could best be carried out by a central organization under whose auspices chief inspectors, or other officials charged with the enforcement of inspection regulations, could meet and discuss their problems. The efforts of this first group of administrators succeeded in extending the National Board membership to all Canadian provinces and most of the states and cities of the United States. It is now possible for an authorized shop to build a boiler or pressure vessel that will be accepted anywhere in the United States or Canada after it has been inspected by an Authorized Inspector holding a National Board Commission.

The ASME Code requires that Inspectors must meet certain minimum requirements of education and experience and must pass a written examination before they can be commissioned to perform Code inspections. One of the many functions of the National Board is the commissioning of Authorized Inspectors.

Shop Reviews by the National Board

Before the issuance of an ASME Certificate of Authorization to build Code vessels, the manufacturer must have and demonstrate a quality control system. This system has to include a written description explaining in detail the quality-controlled manufacturing process.

Before the issuance of renewal of a Certificate of Authorization, the manufacturer's facilities and organization are subject to a joint review by the inspection agency and the legal jurisdiction concerned. For those areas where there is no jurisdiction, or where a jurisdiction does not review a manufacturer's facility, and/or the jurisdiction is the inspection agency, the function may be carried out by a representative of the National Board of Boiler and Pressure Vessel Inspectors. When the review of the manufacturer's facilities and organization has been made jointly by the inspection agency, and/or the legal jurisdiction, or a representative of the National Board of Boiler and Pressure Vessel Inspectors, a written report will be sent to the American Society of Mechanical Engineers. The National Board of Boiler and Pressure Vessel Inspectors also participates in all nuclear surveys.

The National Board also issues Certificates of Authorization for the use of the National Board R stamp, to be applied to ASME Code and National Board–stamped boilers and/or pressure vessels on which repairs are to be made (see the section Welded Repair or Alteration Procedure in this chapter).

Another important service the National Board performs is the certification of safety valves and safety relief valves. The *ASME Boiler and Pressure Vessel Code* contains specific requirements governing the design and capacity certification of safety valves installed on Code-stamped vessels. The certification tests are conducted at a testing laboratory approved by the ASME Boiler and Pressure Vessel Committee.

In addition to the safety valve certification, the National Board has its own testing laboratory for its experimental work and for certifying ASME safety and relief valves. The facilities of the laboratory are also available to manufacturers and other organizations for research, development, or other test work.

National Board Requirements

All Canadian provinces and most states require boilers and pressure vessels to be inspected during fabrication by an Inspector holding a National Board Commission and then to be stamped with a National Board standard number. Qualified and authorized boiler and pressure vessel manufacturers must be registered with the National Board of Boiler and Pressure Vessel Inspectors, 1055 Crupper Ave., Columbus, OH 43229. (Additional contact information is available at https://www.nationalboard.org.) Code paragraph UG-120 details circumstances wherein Manufacturer's Design Reports are registered and filed with the National Board.

The National Board of Boiler and Pressure Vessel Inspectors is constantly working to assure greater safety of life and property by promoting and securing uniform enforcement of boiler and pressure vessel laws and uniform approval of designs and structural details of these vessels, including the accessories that affect their safe operation; by furthering the establishment of a uniform Code; by espousing one standard of qualifications and examinations for inspectors who are to enforce the requirements of this Code; and by seeing that all relevant data of the Code are made available to the members.

In prior years, the Uniform Boiler and Pressure Vessel Laws Society issued a *Synopsis of Boiler and Pressure Vessel Laws, Rules, and Regulations*, the publishing of which has been assumed by the National Board. NB-370, the National Board Synopsis of Boiler and Pressure Vessel Laws, Rules, and Regulations, is a compilation of jurisdiction laws, rules, and regulations. It features the prevailing requirements, detailed contact information, and regulatory history for each jurisdiction. It gives a summary of the applicable laws of states, provinces, cities, and counties. The Synopsis has proved valuable to users and manufacturers of boilers and pressure vessels. Figure 1.3 shows a map of the political subdivisions of the United States and Canada indicating the requirements for boiler and pressure vessel laws and National Board registration. Figure 1.4 shows boiler and pressure vessel stamping requirements throughout the United States and Canada. And Figure 1.5 shows the requirements for each of the ASME Codes within the United States and Canada.

Code Case Interpretations

As the Code does not cover all the details of design, construction, and materials, pressure vessel manufacturers sometimes have difficulty in interpreting the Code when trying to meet specific customer requirements. In such cases the Authorized Inspector should be consulted. If the Inspector is unable to give an acceptable answer or has any doubts or questions about the proper interpretation of the intent of the Code, the question can be referred to the Inspector Supervisor's office. If the Inspector and his or her supervisor are not able to provide a ruling, the manufacturer may then request the assistance of the Boiler and Pressure Vessel Committee, which meets regularly to consider the inquiries of this nature. In referring questions to the Boiler and Pressure Vessel Committee, it is necessary to submit complete details, the sketches of the construction involved, and references to the applicable paragraphs of the Code; it is also useful to include the opinions expressed by others. See Submittal of Technical Inquiries to the Boiler and Pressure Vessel Standards Committees for guidance in preparing inquiries.

Requests for Code Interpretation should preferably be submitted through the online Interpretation Submittal Form. The form is accessible at http://go.asme.org/InterpretationRequest. If the Inquirer is unable to use the online form, the Inquirer may mail the request to the following address: Secretary, ASME Boiler and Pressure Vessel Committee, Two Park Avenue, New York, NY 10016-5990.

Interpretations are issued in real time in ASME's Interpretations Database at http://go.asme.org/Interpretations. Historical BPVC interpretations may also be found in the Database.

FIGURE 1.3 Political map of boiler and pressure vessel law requirements and National Board registration requirements in the United States and Canada. (*Reprinted with permission of the Na*tional Board of Boiler and Pressure Vessel Inspectors, www.nationalboard.org.) The National Board of Boiler and Pressure Vessel Inspectors ("National Board") does not guarantee that this information or forms are up to date. Always consult the National Board's website for the most current version of applicable forms.

The committees meet regularly to consider proposed additions and revisions to the Code and to formulate Cases to clarify the intent of existing requirements or provide, when the need is urgent, rules for materials or constructions not covered by existing Code rules. Those Cases that have been adopted will appear in the appropriate Code Cases book: *Boilers and Pressure Vessels* or *Nuclear Components*. Supplements will be sent or made available automatically to the purchasers of the Code Cases books up to the publication of the next Code.

UNITED STATES

Jurisdiction	National Board	ASME
Alabama	All	All
Alaska	R, VR	All
Arizona	NB, R, VR	A, E, H, HLW, HV, M, PP, S, U, UM, UV, UD, U2, U3, UV3, V
Arkansas	NR, R, VR	All
California	None***	All
Colorado	NR, R, VR	All
Connecticut	R, VR	A, E, H, HLW, HV, M, N, NA, NPT, NV, PP, S, V
Delaware	NR, R, VR	All
Florida	R, VR	A, E, H, HLW, HV, M, PP, S, V
Georgia	NR, R, VR	A, E, H, HLW, HV, M, N, NA, NPT, NV, PP, RP, S, U, U2, U3, UM, UV, V
Hawaii	R, VR	A, E, H, HLW, HV, M, PP, S, U, UM, UV, V
Idaho	None	None
Illinois	R, VR	A, E, H, HLW, HV, M, N,* NA,* NPT,* NV,* PP, RP, S, U, U2, UM, UV, V
Indiana	NR, R	A, E, H, HLW, HV, M, N, NA, NPT, PP, S, U, U2, U3, UM, UV, V
Iowa	R, VR	A, E, H, HLW, HV, M, N, NA, NPT, PP, S, U, U2, U3, UM, UV, V
Kansas	R, VR	A, E, H, HLW, HV, M, PP, S, U, UM, V
Kentucky	R, VR	A, E, H, HLW, HV, PP, S, U, UV, V
Louisiana	R, VR	A, E, H, HLW, HV, M, S
Maine	R, VR	A, E, H, HLW, HV, M, PP, RP, S, U, U2, U3, UM, UV, V
Maryland	R	All
Massachusetts	NR, R, VR	H, M, N, NA, NPT, NV, PP, S, U, UM
Michigan	R**	A, E, H, HLW, HV, M, N, NA, NPT, NV, PP, S, U, V
Minnesota	NR, R, VR	All
Mississippi	VR	A, E, H, HLW, HV, M, S, U, UV, V
Missouri	R, VR	All
Montana	R, VR	A, E, H, HV, S
Nebraska	R, VR	A, E, H, HLW, HV, M, PP, S, U, U2, U3, V
Nevada	NR, R, VR	A, E, H, HLW, HV, M, NV, PP, S, U, U2, UM, UV, V
New Hampshire	NR, R, VR	A, H, N, NA, NPT, PP, S, U, UV
New Jersey	R	All
New Mexico	None	A, E, H, HLW, HV, PP, S, V
New York	R, VR	A, E, H, M, PP, S, U, V
North Carolina	R, VR	All
North Dakota	NB, R, VR	A, E, H, HLW, HV, M, PP, RP, S, U, U2, U3, UD, UV, UV3, V
Ohio	NR, R, VR	All
Oklahoma	R, VR	A, E, H, HLW, HV, M, PP, S, U, U2, U3, UM, UV, V
Oregon	R, VR	A, H, HLW, PP, S, U, U2, U3, UM, V
Pennsylvania	NR, R, VR	A, E, H, HLW, HV, M, N, NA, NPT, NV, PP, S, U, U2, U3, UM, UV, V
Rhode Island	NR, R, VR	All
South Carolina	R	H, HLW, S, U (fired)
South Dakota	None	A, E, H, HLW, HV, M, PP, S, V
Tennessee	NR, R, VR	All

FIGURE 1.4 Tabulation of boiler and pressure vessel stamping requirements in the United States and Canada. (*Reprinted with permission of the National Board of Boiler and Pressure Vessel Inspectors, www.nationalboard.org.*) The National Board of Boiler and Pressure Vessel Inspectors ("National Board") does not guarantee that this information or forms are up to date. Always consult the National Board's website for the most current version of applicable forms.

Canadian Pressure Vessel Requirements

The United States manufacturers of ASME Code vessels often receive orders for vessels to be installed in Canada. All provinces of Canada have adopted the ASME Code and require vessels to be shop-inspected by an Inspector holding a National Board Commission and

Jurisdiction	National Board	ASME
Texas	NR, R, VR	A, E, H, HLW, HV, M, N, NA, NPT, NV, PP, S, U, UD, U2, UV, V
Utah	NR, R, VR	A, E, H, HLW, HV, M, N, NA, NPT, NV, PP, RP, S, U, U2, U3, UM, UV, V
Vermont	R, VR	A, E, H, HLW, HV, M, PP, S, U, U2, UM, UV, V
Virginia	R	All
Washington	NB, NR, R, VR	All
West Virginia	NB, R, VR	A, E, M, PP, S, V
Wisconsin	NR, R, VR	All
Wyoming	None	U, U2, U3, UV

*Comes under the jurisdiction of the Illinois Emergency Management Agency.
**High-pressure repairs and/or alterations to any boiler.
***Except for refinery and oilfield locations require an R stamp.

CANADA

Jurisdiction	National Board	ASME
Alberta	R*	All when manufactured outside Canada, and all Section VIII, Division 2 and 3 pressure vessels
British Columbia	R, VR	As per adopted codes
Manitoba	R*, VR	Symbol stamp is mandatory when manufactured outside Canada
New Brunswick	NR, R*, VR	H, HLW, N, NV, PP, S, U
Newfoundland and Labrador	R*	All when manufactured outside Canada
Northwest Territories	R*	A, E, H, HLW, HV, M, S, U, U2, UM, UV, V
Nova Scotia	R*, VR	A, H, HLW, HV, M, PP, S, U, U2, U3, UM, UV, V
Nunavut Territory	R*	A, E, H, HLW, HV, M, S, U, U2, UM, UV, V
Ontario	R*	All
Prince Edward Island	NR, R*, VR	H, HLW, HV, M, N, NA, NPT, NV, S, U, U2, UV, V
Québec	R*	All when manufactured in the United States
Saskatchewan	R*	All
Yukon Territory	R*	All

*Accepted

CITIES, COUNTIES, AND TERRITORIES OF THE UNITED STATES

Jurisdiction	National Board	ASME
Albuquerque, New Mexico	R, VR	A, E, H, HLW, HV, M, S, U, UM, UV, V
Buffalo, New York	R, VR	H, S, U
Chicago, Illinois	All	All
Detroit, Michigan	None	All
Los Angeles, California	NR, R, VR	A, H, HLW, HV, M, PP, S, U, UM, UV, V
Miami, Florida	None	None
Miami-Dade County, Florida	None	None
Milwaukee, Wisconsin	NR, R, VR	All
New Orleans, Louisiana	NR, R, VR	All
New York, New York	R, VR	A, H, HLW, M, PP, S, UV, V
Omaha, Nebraska	R, VR	A, E, H, HLW, HV, M, PP, S, U, U2, U3, V
Puerto Rico	NR, R, VR	All
Seattle, Washington	All	All
Spokane, Washington	R, VR	A, E, H, HLW, M, PP, S, U, U2, UV, V
St. Louis, Missouri	R	H, S, U, U2
Washington, D. C.	VR	A, E, H, PP, U

FIGURE 1.4 *(Continued)*

to be stamped with a provincial registration number in addition to the ASME symbol and National Board stamping.

The CRN (Canadian Registration Number) system is the Canadian method of reviewing and registering the design of pressure vessels. All Canadian provinces and territories use the CRN system.

KEY:

I – Power Boilers
II – Materials
III(1) – Nuclear Facility Components
III(2) – Nuclear Facility (Code for Concrete Containments)
IV – Heating Boilers
V – Nondestructive Examination
VI – Care and Operating of Heating Boilers
VII – Care of Power Boilers
VIII(1) – Pressure Vessels
VIII(2) – Pressure Vessels (Alternate Rules)
VIII(3) – Pressure Vessels (Alternate Rules for High Pressure)
IX – Welding and Brazing Qualifications
X – Fiber-Reinforced Plastic Pressure Vessels
XI – Inservice Inspection, Nuclear
XII – Transport Tanks

Y – Required by Law N – Not Required by Law

UNITED STATES

Jurisdiction	I	II	III		IV	V	VI	VII	VIII			IX	X	XI	XII	CODE CASES
			(1)	(2)					(1)	(2)	(3)					
Alabama	Y	Y	N	N	Y	Y	Y	Y	Y	Y	Y	Y	Y	N	N	N
Alaska	Y	Y	N	N	Y	Y	Y	Y	Y	Y	Y	Y	Y	N	–	Y*
Arizona	Y**	Y	N	N	Y	Y	N	N	Y	Y	Y	Y	N	N	–	Y*
Arkansas	Y	Y	Y	Y	Y	Y	Y	Y	Y	Y	Y	Y	Y	Y	–	Y
California	Y	Y	Y	Y	Y	Y	Y	Y	Y	Y	N	Y	Y	Y	–	Y*
Colorado	Y	N	N	N	Y	Y	N	N	Y	Y	Y	Y	Y	N	–	N
Connecticut	Y	Y	Y	N	Y	Y	Y	Y	N	N	N	Y	N	N	–	Y
Delaware	Y	Y	Y	Y	Y	Y	N	N	Y	Y	Y	Y	Y	Y	–	Y
Florida	Y	Y	N	N	Y	Y	Y	Y	N	N	N	Y	N	N	–	N
Georgia	Y	N	Y	Y	Y	N	N	N	Y	Y	Y	N	Y	Y	–	Y
Hawaii	Y	Y	Y	Y	Y	Y	Y	Y	Y	Y	Y	Y	Y	Y	–	Y
Idaho																
Illinois	Y	Y	N	N	Y	Y	Y	Y	Y	Y	Y	Y	Y	N	–	N
Indiana	Y	Y	Y	Y	Y	Y	N	N	Y	Y	Y	Y	Y	Y	–	N
Iowa	Y	Y	Y	Y	Y	Y	Y	Y	Y	Y	Y	Y	Y	N	–	Y*
Kansas	Y	Y	N	N	Y	Y	Y	Y	Y	Y	Y	Y	Y	N	–	N
Kentucky	Y	Y	Y	Y	Y	Y	N	N	Y	Y	N	Y	N	N	–	Y
Louisiana	Y	N	N	N	Y	N	Y	Y	N	N	N	Y	N	N	–	Y
Maine	Y	Y	N	N	Y	Y	Y	Y	Y	Y	Y	Y	Y	N	Y	N
Maryland	Y	Y	Y	Y	Y	Y	Y	Y	Y	Y	Y	Y	Y	Y	–	Y
Massachusetts	Y	Y	Y	N	Y	Y	N	N	Y	N	N	Y	Y	Y	–	N
Michigan	Y	Y	Y	Y	Y	Y	N	N	Y	N	N	Y	N	Y	–	Y
Minnesota	Y	Y	Y	Y	Y	Y	Y	Y	Y	Y	Y	Y	Y	Y	–	Y
Mississippi	Y	N	N	N	Y	N	N	N	Y	N	N	N	N	N	–	Y
Missouri	Y	Y	N	N	Y	Y	Y	Y	Y	Y	N	Y	Y	N	–	Y
Montana	Y	Y	N	N	Y	Y	Y	Y	N	N	N	Y	N	N	–	N
Nebraska	Y	Y	N	N	Y	Y	Y	Y	Y	Y	Y	Y	Y	N	–	Y***
Nevada	Y	Y	N	N	Y	Y	Y	Y	Y	Y	Y	Y	Y	N	–	N
New Hampshire	Y	Y	Y	Y	Y	Y	Y	Y	Y	Y	N	Y	Y	Y	–	N
New Jersey	Y	Y	Y	Y	Y	Y	Y	Y	Y	Y	N	Y	Y	Y	–	Y

Figure 1.5 Tabulation of ASME Boiler and Pressure Vessel Codes and their requirements for usage in the United States and Canada. (*Reprinted with permission of the National Board of Boiler and Pressure Vessel Inspectors, www.nationalboard.org.*) The National Board of Boiler and Pressure Vessel Inspectors ("National Board") does not guarantee that this information or forms are up to date. Always consult the National Board's website for the most current version of applicable forms.

Registration is required for each jurisdiction where the equipment is used. Upon registration, fabrication and inspection are handled the same as with National Board.

Once a design has been approved and registered, any number of vessels of that design can be built and used in the province where it is approved. If a vessel of the same design is to be sent to another

UNITED STATES *continued . . .*

Jurisdiction	I	II	III		IV	V	VI	VII	VIII			IX	X	XI	XII	CODE CASES
			(1)	(2)					(1)	(2)	(3)					
New Mexico	Y	N	N	N	Y	N	N	N	N	N	N	Y	N	N	–	N
New York	Y	Y	N	N	Y	Y	N	N	Y	N	N	N	N	N	–	Y
North Carolina	Y	Y	Y	Y	Y	Y	Y	Y	Y	Y	Y	Y	Y	Y	Y	Y*
North Dakota	Y	Y	N	N	Y	Y	N	N	Y	Y	Y	Y	Y	N	–	–
Ohio	Y	Y	Y	Y	Y	Y	Y	Y	Y	Y	N	Y	Y	Y	–	Y
Oklahoma	Y	Y	N	N	Y	Y	Y	Y	Y	Y	Y	Y	N	N	–	N
Oregon	Y	Y	N	N	Y	Y	N	N	Y	Y	Y	Y	Y	N	–	N
Pennsylvania	Y	Y	Y	Y	Y	Y	N	N	Y	Y	Y	Y	N	Y	–	N
Rhode Island	Y	Y	Y	Y	Y	Y	Y	Y	Y	Y	N	Y	Y	Y	–	Y
South Carolina	Y	N	N	N	Y	N	N	N	Y*	Y*	Y*	Y*	Y*	N	–	N
South Dakota	Y	N	N	N	Y	Y	N	N	N	N	N	Y	N	N	–	N
Tennessee	Y	Y	Y	Y	Y	Y	Y	Y	Y	Y	Y	Y	Y	Y	Y	Y
Texas	Y	N	Y	Y	Y	N	N	N	Y	Y	N	N	N	Y	Y	Y
Utah	Y	Y	Y	Y	Y	Y	Y	Y	Y	Y	Y	Y	Y	Y	–	N
Vermont	Y*	Y*	Y*	Y*	Y*	Y*	Y*	Y*	Y*	Y*	Y*	Y*	Y*	Y*	–	Y*
Virginia	Y	Y	Y	Y	Y	Y	Y	Y	Y	Y	N	Y	Y	Y	–	Y
Washington	Y	N	Y	Y	Y	N	N	N	Y	Y	Y	N	Y	N	Y	Y*
West Virginia	Y	Y	N	N	N	Y	N	N	N	N	N	Y	N	N	–	Y
Wisconsin	Y	Y	N	N	Y	Y	N	N	Y	Y	N	Y	Y	N	–	Y
Wyoming	N	N	N	N	N	N	N	N	Y	Y	Y	N	N	N	–	N

CANADA

Jurisdiction	I	II	III		IV	V	VI	VII	VIII			IX	X	XI	XII	CODE CASES
			(1)	(2)					(1)	(2)	(3)					
Alberta	Y	Y	Y	Y	Y	Y	Y	Y	Y	Y*	Y	Y	Y	Y	Y	N
British Columbia	Y	N	N	N	Y	N	N	N	Y	Y	Y	Y	Y	N	Y	N
Manitoba	Y	Y	Y	Y	Y	Y	Y	Y	Y	Y	N	Y	Y	Y	–	Y
New Brunswick	Y	Y	Y	Y	Y	Y	Y	Y	Y	Y	N	Y	Y	Y	–	Y
Newfoundland and Labrador	Y	Y	N	N	Y	Y	Y	Y	Y	Y	Y	Y	Y	N	–	N
Northwest Territories	Y	Y	N	N	Y	Y	Y	Y	Y	Y	N	Y	N	N	–	N
Nova Scotia	Y	Y	N	N	Y	Y	Y	Y	Y	Y	Y	Y	Y	N	–	Y
Nunavut Terri tory	Y	Y	Y	Y	Y	Y	Y	Y	Y	Y	N	Y	Y	Y	–	N
Ontario	Y	Y	Y	Y	Y	Y	N	N	Y	Y	Y	Y	Y	N	Y	Y
Prince Edward Island	Y	Y	Y	Y	Y	Y	Y	Y	Y	Y	N	Y	Y	Y	–	N
Québec	N	N	N	N	N	N	N	N	N	N	N	Y	N	N	–	N
Saskatchewan	Y	Y	N	N	Y	Y	N	N	Y	Y	Y	Y	Y	N	–	Y*
Yukon Territory	Y	Y	Y	Y	Y	Y	Y	Y	Y	Y	N	Y	Y	N	–	N

* On a conditional or case-by-case basis or as appropriate.
**2017 Edition, Section I, Part PA not permitted.
***Accepted upon publication. The chief boiler inspector will make public notification of any Code Cases that are not approved for use in Nebraska.

FIGURE 1.5 *(Continued)*

Canadian province, the manufacturer's letter to the chief inspector of that province should state the registration number of the original approval. The second province will then use the same registration number with the addition of its own province number. For example, a design first registered in Ontario might have been given the

CITIES, COUNTIES, AND TERRITORIES OF THE UNITED STATES

			III						VIII							
Jurisdiction	I	II	(1)	(2)	IV	V	VI	VII	(1)	(2)	(3)	IX	X	XI	XII	CODE CASES
Albuquerque, New Mexico	Y	N	N	N	Y	N	N	N	N	N	N	Y	N	N	–	Y
Buffalo, New York	Y	N	Y	Y	Y	N	N	N	Y	N	N	N	N	N	–	Y
Chicago, Illinois	Y	Y	Y	Y	Y	Y	Y	Y	Y	Y	N	Y	Y	Y	–	Y
Detroit, Michigan	Y	Y	Y	Y	Y	Y	Y	Y	Y	Y	N	Y	Y	N	–	N
Los Angeles, California	Y	Y	Y	Y	Y	Y	N	N	Y	Y	N	Y	N	N	–	Y
Miami, Florida	Y	N	N	N	Y	N	N	N	Y	Y	N	N	N	N	–	Y
Miami-Dade County, Florida	Y	N	N	N	Y	N	N	N	Y	Y	N	N	N	N	–	Y
Milwaukee, Wisconsin	Y	Y	N	N	Y	Y	N	N	Y	Y	N	Y	Y	N	–	Y
New Orleans, Louisiana	Y	Y	Y	Y	Y	Y	Y	Y	Y	Y	N	Y	Y	Y	–	N
New York, New York	Y	N	N	N	Y	Y	N	N	Y	N	N	Y	N	N	–	Y
Omaha, Nebraska	Y	N	N	N	Y	Y	N	N	Y	Y	Y	Y	Y	Y	–	Y
Puerto Rico	Y	Y	Y	Y	Y	Y	Y	Y	Y	Y	N	Y	Y	Y	–	Y
Seattle, Washington	Y	N	Y	Y	Y	N	N	N	Y	Y	N	N	Y	N	N	Y
Spokane, Washington	Y	Y	N	N	Y	Y	Y	Y	Y	Y	Y	Y	Y	N	Y	Y
St. Louis, Missouri	Y	Y	N	N	Y	Y	Y	Y	Y	Y	Y	Y	Y	N	–	N
Washington, D.C.	Y	Y	Y	Y	Y	Y	Y	Y	Y	Y	N	Y	Y	Y	–	Y

FIGURE 1.5 *(Continued)*

number 764. This number would be followed by a decimal point and then the number 5, which signifies the Province of Ontario. The registration would thus be 764.5. If this same design were used in the Province of Manitoba, it would be given the registration number 764.54, with the figure 4 denoting Manitoba. If the letter C follows the designation of the province of the first registration, it means that the design is subsequently registered across Canada, so the registration in the above example would be 764.5C.

Department of Transportation Regulations

Chapter 11 in this book addresses the recently introduced Section XII, Rules for Construction and Continued Service of Transport Tanks. In 2004, the ASME published the initial edition of Section XII. Prior to that, transport tanks were governed by DOT rules, with an intermingling of ASME Section VIII, Division 1 regulations for some aspects of the tanks. This new Section mandates and requires that all over-the-road specification cargo tank trucks be manufactured by a holder of an ASME Code symbol T stamp. Additionally, and most importantly, all specification cargo tanks will now be required to follow Section XII regulations for in-service inspection, repair, and modifications.

Welded Repair or Alteration Procedure

Repairs or alterations of ASME Code vessels may be made in any shop that manufactures Code vessels or in the field by any welding contractor qualified to make repairs on such vessels. Part 3 of the National Board Inspection Code (NBIC), Repairs and Alterations, has the information and guidance to perform, verify, and document acceptable repairs or alterations to pressure-retaining items regardless of the code of construction. Alternative methods for examination, testing, heat treatment, etc. are provided when the original code of construction requirements cannot be met. Specific acceptable and proven repair methods are also provided.

The National Board also has six Certificates of Authorization for use of the National Board R stamp. Any repair organization may obtain the Certificates of Authorization for the use of this stamp to be applied to the ASME Code and National Board–stamped boilers and/or pressure vessels on which repairs have been made.

Before an R stamp is issued, the firm making repairs must submit a review of its repair methods. It must also have an acceptable written quality control system covering repair design, materials to be used, welding procedure specifications, welders' qualifications, repair methods, and examination. The review will be made by a National Board member jurisdiction where the repair company is located. If the jurisdiction has no National Board member, or upon request by the jurisdiction, the review will be made by a representative of the National Board.

Manufacturers and assemblers who hold ASME Certificates of Authorization and Code stamps—with the exception of H (cast iron), V, UV, and UM stamps—may obtain the National Board R stamp without review provided their quality control system covers repairs.

Whenever the R stamp is to be applied to the National Board–stamped boilers and pressure vessels, all repairs are subject to acceptance by an Inspector who holds a valid National Board Commission and a valid Certificate of Competency issued by a state or jurisdiction of the United States or by a Canadian province.

The Authorized Inspector will be guided by the NBIC as well as by any existing local rules governing the repair of vessels. When the work is to be done on location, repairs to a vessel typically require greater skill than was required when it was first shop-constructed. Utmost consideration must be given to field repairs, which may necessitate cutting, shaping, and welding with portable equipment. The practicability of moving the vessel from the site to a properly equipped repair shop should be considered. Before repairs are made, the method of repair must be approved by an Authorized Inspector. The Inspector will examine the vessel, identify the material to be welded, and compare it with the material to be used in repair. A check is then made to ensure that the welding contractor or shop has a qualified welding procedure for the material being welded and that the welder who does the work is properly qualified to weld that material. If the repair or alteration requires design calculations, the Inspector will review the calculations to assure that the original design requirements are met. For the repair of Section VIII, Division 2 pressure vessels, the design calculations must be reviewed and certified by a Registered Professional Engineer experienced in pressure vessel design.

When the repairs or alterations and the necessary tests have been completed, one or more of the following reports are filled out (see Fig. 1.6): Form R-1, Report of Welded Repair; Form R-2, Report of Alteration; and Form R-3, Report of Parts Fabricated by Welding. (Form R-4, Report Supplementary Sheet, should be used for any information for which there is not enough room on one of the other three forms.) Some states and insurance companies have their own forms which must be signed by the contractor or manufacturer making the repair and by the Authorized Inspector.

When an alteration is made to a vessel, it shall comply with the section of the ASME Code to which the original vessel was constructed to conform. The required inspection shall be made by an Authorized Inspector holding a National Board Commission. No alteration shall be initiated without the approval of the Authorized Inspector.

When alterations are completed, Form R-2 shall be signed and certified, indicating that the alteration has met the applicable requirements. This R-2 report shall record the required data and shall clearly indicate what changes or alterations have been made to the original construction. On the R-2 report, under Remarks, the name of the original manufacturer, manufacturer's serial number, and the National Board number assigned to the vessel by the original manufacturer shall be stated. When the vessel has been altered, an original and one copy of the certified repair or alteration report shall be filed with the National Board of Boiler and Pressure Vessel Inspectors. A copy shall go to the jurisdictional authority, inspection agency, and the customer for either repairs or alterations when so requested by the applicable persons.

When the stamping plate is made for an altered vessel, it shall contain all the information required by the National Board Inspection Code (see Fig. 1.7). Note the difference in the stamping of the nameplate of an altered vessel from the stamping of the nameplate of a repaired vessel (see Fig. 1.8).

The words *Altered* or *Rerated* in letters 5⁄16 in. high shall be stamped on the plate in lieu of the Code symbol, National Board R symbol, and National Board number. The date shown on the nameplate shall be the date of alteration. This nameplate shall be permanently attached to the vessel adjacent to the original manufacturer's nameplate or stamping, which shall remain on the vessel.

If the alteration requires the removal of the original manufacturer's nameplate or stamping, it shall be retained and reattached to the

FORM R-1 REPORT OF REPAIR
in accordance with provisions of the *National Board Inspection Code*

(1) ______ (Authorized Rep. initials)
(2) ______ (Inspectors initials)
(3) ______ (Form "**R**" Registration no.)
(4) ______ (P.O. no., job no., etc.)

1. WORK PERFORMED BY: (5) ______ (name of repair organization)
   ______ (address)
2. OWNER: (6) ______ (name)
   ______ (address)
3. LOCATION OF INSTALLATION: (7) ______ (name)
   ______ (address)
4. ITEM IDENTIFICATION: (8) ______ (boiler, pressure vessel, or piping) NAME OF ORIGINAL MANUFACTURER: (9) ______
5. IDENTIFYING NOS: (10) ______ (mfg. serial no.) (11) ______ (National Board no.) (12) ______ (jurisdiction no.) (13) ______ (other) (14) ______ (year built)
6. NBIC EDITION/ADDENDA: (15) ______ (edition) ______ (addenda)
   Original Code of Construction for Item: (16) ______ (name / section / division) ______ (edition / addenda)
   Construction Code Used for Repair Performed: (17) ______ (name / section / division) ______ (edition / addenda)
7. REPAIR TYPE: (18) ☐ welded ☐ graphite pressure equipment ☐ FRP pressure equipment ☐ DOT
8. DESCRIPTION OF WORK: ☐ Form R-4, Report Supplementary Sheet is attached ☐ FFSA Form (NB-403) is attached
   (use Form R-4, if necessary)
   (19) ______

   (20) ______ (Liquid, Pneumatic, Vacuum, Leak) Pressure Test, if applied (21) ______ psi MAWP (22) ______ psi
9. REPLACEMENT PARTS: (Attached are Manufacturer's Partial Data Reports or Form R-3's properly completed for the following items of this report):
   ______ (name of part, item number, data report type or Certificate of Compliance, mfg's. name and identifying stamp)
   (23) ______
10. REMARKS: (24) ______

FIGURE 1.6 Form R-1, Report of Welded Repair; Form R-2, Report of Alteration; Form R-3, Report of Parts Fabricated by Welding; Form R-4, Report Supplementary Sheet. (*Reprinted with permission of the National Board of Boiler and Pressure Vessel Inspectors, www.nationalboard.org.*) The National Board of Boiler and Pressure Vessel Inspectors ("National Board") does not guarantee that this information or forms are up to date. Always consult the National Board's website for the most current version of applicable forms. Use the adjacent barcode to access the latest edition of the forms on the National Board website.

(3) ______________
(Form "R" Registration no.)
(4) ______________
(P.O. no., job no., etc.)

CERTIFICATE OF COMPLIANCE

I, (25) ______________, certify that to the best of my knowledge and belief the statements made in this report are correct and that all material, construction, and workmanship on this Repair conforms to the *National Board Inspection Code*. National Board "**R**" *Certificate of Authorization* No. (26) ______________ Expiration date: (27) ______________

Repair Organization: (28) ______________

Signed: (29) ______________
(authorized representative)

Date: (30) ______________

CERTIFICATE OF INSPECTION

I, (31) ______________, holding a valid commission issued by The National Board of Boiler and Pressure Vessel Inspectors and certificate of competency, where required, issued by the jurisdiction of (32) ______________ and employed by (33) ______________ of (34) ______________ have inspected the work described in this report on (35) ______________, ______________ and state that to the best of my knowledge and belief, this work complies with the applicable requirements of the *National Board Inspection Code*. By signing this certificate, neither the undersigned nor my employer makes any warranty, expressed or implied, concerning the work described in this report. Furthermore, neither the undersigned nor my employer shall be liable in any manner for any personal injury, property damage, or loss of any kind arising from or connected with this inspection.

Commissions: (36) ______________
(National Board and Jurisdiction no. including endorsement)

Signed: (37) ______________
(Inspector)

Date: (38) ______________

FIGURE 1.6 *(Continued)*

altered vessel. Such relocation shall be noted on the R-1 report subject to the jurisdictional requirements.

Managerial Factors in Handling Code Work

The producer of welded pressure vessels is faced today with many serious problems, not the least of which is managing to stay in business and earn a fair profit.[1] Severe competition exists throughout the plate-fabricating industry, not only because of the growth in the number of fabricating shops but also because of the tremendous technological advances made in recent years. Many of the older, more heavily capitalized plants have been hit hard. Some of these have successfully turned to specialized fields; others produce specialty products that return a good profit when in demand. When demand slackens, these shops often turn to general fabrication, thereby increasing the competitive situation for the general plate fabricator.

For this type of fabricator, who essentially runs a job shop, the profit margin is restricted from the outset because jobs are usually awarded to the lowest bidder. Although there is nothing unusual in this method of competing, it does increase the need to think ahead

[1] The material in this section, Managerial Factors in Handling Code Work, was contributed by C. H. Willer, former editor of *Welding Journal*.

This guide is to be used when completing the National Board Form R-1, Report of Repair. When computer generated, the format of the form shall replicate the type and relative location of the information depicted on the Form R-1, Report of Repair.

When additional space is needed, a Form R-4 shall be used and attached. Information determined to be of a proprietary nature need not be included, but shall be stated on the form.

1) Initials of the National Board "R" *Certificate of Authorization* authorized representative who signed the Certificate of Compliance (See #29)

2) Initials of the Inspector who signed the Certificate of Inspection (See #37).

3) When registering a Form "R" Report with the National Board, this line is solely designated for a unique sequential number assigned by the National Board "R" *Certificate of Authorization* Holder. When the "R" Form is not to be registered, indicate so by "N/A". As described in NBIC Part 3, Paragraph 5.6, a log shall be maintained identifying unique and sequentially numbered Form "R" reports that are registered with the National Board.

4) If applicable, document the unique purchase order, job, or tracking number, assigned by the organization performing work.

5) The name and address of the National Board "R" *Certificate of Authorization* Holder performing the work as it appears on the "*Certificate of Authorization* ".

6) Name and address of the owner of the pressure-retaining item.

7) Name and address of plant or facility where the pressure-retaining item is installed.

8) Description of the pressure-retaining item, such as boiler or pressure vessel, or piping. Include the applicable unit identification.

9) Name of the original manufacturer of the pressure-retaining item. If the original manufacturer is unknown, state "unknown."

10) Document the serial number of the pressure-retaining item if assigned by the original manufacturer. If there is no serial number assigned or it is unknown, state "unknown."

11) When the pressure-retaining item is registered with the National Board, document the applicable registration number. If the pressure-retaining item is installed in Canada, indicate the Canadian design, registration number (CRN), and list the drawing number under "other." If the item is not registered, indicate, "none."

12) Record the Jurisdiction number, if available.

13) Document any other unique number assigned to the pressure-retaining item such as a unique number assigned by the owner or user. If unknown, state "unknown."

14) Identify the year in which fabrication/construction of the item was completed.

15) Indicate edition and addenda (if applicable) of the NBIC under which this work is being performed, as applicable.

16) Indicate the name, section, division, edition, and addenda (if applicable) of the original Code of Construction for the pressure-retaining item.

17) Indicate the name, section, division, edition, and addenda (if applicable) of the Code of Construction used for the work being performed. If code cases are used, they shall be identified in the "Remarks" section.

18) Indicate the type of repair, e.g., welded, graphite pressure equipment, fiber-reinforced plastic pressure equipment or DOT.

19) Provide a detailed summary of the scope of work that was completed. The information to be considered when describing the scope of work should include such items as, the nature of the repair (i.e. welding, bonding, cementing), the specific location of the work performed of the PRI, the steps taken to remove a defect or as allowed by NBIC Part 3, Paragraph 3.3.4.8 to remain in place, the method of repair or alteration described as listed in the examples of NBIC Part 3, Section 3 or applicable supplement, and the acceptance testing and/or examination method used in accordance with the NBIC.

Figure 1.6 (*Continued*)

clearly and to avoid unnecessary costs. For the job shop handling ASME Code vessels as well as non-Code work, the importance of careful planning is greatly increased because of the relatively high costs of Code material and also the variety in techniques and procedures demanded.

A second important factor in the award of a contract is delivery time, where again planning assumes utmost importance. The buyer of a Code vessel is not necessarily concerned if a producer incurs extra expense through carelessness but is very often acutely aware of the delivery time. Good customer relations and even future business depend upon on-time deliveries. Obviously, cost and delivery time are related.

20) Indicate type of pressure test applied (Liquid, pneumatic, vacuum, leak)

21) Indicate test pressure applied.

22) Indicate the maximum allowable working pressure of the pressure-retaining item.

23) As applicable, identify which parts manufactured by welding or bonding were introduced as needed to complete the scope of work. Indicate part, item number, manufacturer's name, stamped identification, and data report type of Certificate of Compliance.

24) Indicate any additional information pertaining to the work involved (e.g., routine repairs, code cases, interpretations used).

25) Type or print the name of the National Board "R" *Certificate of Authorization* authorized representative.

26) Indicate National Board "R" *Certificate of Authorization* number.

27) Indicate month, day, and year the National Board "R" *Certificate of Authorization* expires.

28) Record name of the National Board "R" Certificate of Authorization Holder who performed the described work, using full name as shown on the *Certificate of Authorization* or an abbreviation acceptable to the National Board.

29) Signature of National Board "R" *Certificate of Authorization* authorized representative.

30) Indicate the date the repair was certified.

31) Type or print the name of the Inspector.

32) Indicate Inspector's Jurisdiction.

33) Indicate Inspector's employer.

34) Indicate address of Inspector's employer (city and state or province).

35) Indicate month, day and year of inspection by Inspector. In case of routine repairs this shall be the month, day, and year the Inspector reviews the completed routine repair package.

36) Inspectors National Board Commission number and endorsement that qualifies the Inspector to sign this report, and when required by the Jurisdiction, the applicable state or provincial numbers.

37) Signature of Inspector.

38) Indicate the date the Inspector signs the completed "R" Form.

FIGURE 1.6 *(Continued)*

The job shop, then, is faced with three fundamental problems in the production of Code work:

1. The ability to fabricate in accordance with the Code
2. The necessity for keeping costs at a minimum
3. The importance of making deliveries on time

The question naturally arises, "What concrete steps can be taken to assure the best fulfillment of these requirements?" A few general suggestions bear mention. First, there is no substitute for reading and understanding the ASME Code. Many protests are heard on this score—even from inspectors and engineers—but a knowledge of those sections of the Code that pertain to the type of work in which one is engaged is a necessity. But even this is not enough. Some vessel requirements are extremely simple while others are not. The vast industrial complex of today has created such a wide variety of pressure vessels that they constitute a perpetual challenge to the ingenuity of the fabricator. For the job shop, close cooperation and understanding are required from start to finish. Sales, estimating, design, and purchasing must all be backed up by sound shop experience and a reasonable time schedule.

FORM R-2 REPORT OF ALTERATION
in accordance with provisions of the *National Board Inspection Code*

(1) ______ (Authorized Rep. initials)
(2) ______ (Inspectors initials)
(3) ______ (Form "R" Registration no.)
______ (P.O. no., job no., etc.)

1a. DESIGN PERFORMED BY: (5) ______ (name of "R" organization responsible for design)

______ (address)

1b. CONSTRUCTION PERFORMED BY: (6) ______ (name of "R" organization responsible for construction)

______ (address)

2. OWNER OF PRESSURE RETAINING ITEM: (7) ______ (name)

______ (address)

3. LOCATION OF INSTALLATION: (8) ______ (name)

______ (address)

4. ITEM IDENTIFICATION: (9) ______ (boiler, pressure vessel, or piping) NAME OF ORIGINAL MANUFACTURER: (10) ______

5. IDENTIFYING NOS: (11) ______ (mfg. serial no.) (12) ______ (National Board no.) (13) ______ (jurisdiction no.) (14) ______ (other) (15) ______ (year built)

6. NBIC EDITION/ADDENDA: (16) ______ (edition) ______ (addenda) (11) ______

Original Code of Construction for Item: (17) ______ (name / section / division) ______ (edition / addenda)

Construction Code Used for Alteration Performed: (18) ______ (name / section / division) ______ (edition / addenda)

7a. DESCRIPTION OF DESIGN SCOPE: ☐ Form R-4, Report Supplementary Sheet is attached

(19) ______

7b. DESCRIPTION OF CONSTRUCTION SCOPE: ☐ Form R-4, Report Supplementary Sheet is attached

(20) ______

(21) ______ Pressure Test, if applied (22) ______ psi MAWP (23) ______ psi

Figure 1.6 (*Continued*)

Particular methods for handling an order vary from one company to another. But within the limits set by customer specification and the rules of the Code, the job shop can seek a competitive advantage and avoid unnecessary costs and time losses by determining the following:

1. An overall production schedule
2. Acceptable alternatives in plate material and components

(3) ____________
(Form "R" Registration no.)
(4) ____________
(P.O. no., job no., etc.)

8. REPLACEMENT PARTS: (Attached are Manufacturer's Partial Data Reports or Form R-3's properly completed for the following items of this report):

(name of part, item number, data report type or Certificate of Compliance, mfg's. name and identifying stamp)
(24)

9. REMARKS: (25)

Design Certification

I, (26) ____________, certify that to the best of my knowledge and belief the statements in this report are correct and that the Design Change described in this report conforms to the *National Board Inspection Code.* National Board "**R**" Certificate of Authorization No. (27) ____________ expires on (28) ____________

Date (29) ____, (30) ____________ Signed (31) ____________
(name of design organization) (authorized representative)

certificate of design change review

I, (32) ____________, holding a valid Commission issued by The National Board of Boiler and Pressure Vessel Inspector and certificate of competency, where required, issued by the jurisdiction of (33) ____________ and employed by (34) ____________ (34) of ____________
have reviewed the design change as described in this report and state that to the best of my knowledge and belief such change complies with the applicable requirements of the *National Board Inspection Code*.
By signing this certificate, neither the undersigned nor my employer makes any warranty, expressed or implied, concerning the work described in this report. Furthermore, neither the undersigned nor my employer shall be liable in any manner for any personal injury, property damage or loss of any kind arising from or connected with this inspection.

Date (35) ____________ Signed (36) ____________ Commissions (37) ____________
(inspector) (National Board and jurisdiction no. including endorsement)

CONSTRUCTION CERTIFICATION

I, (38) ____________, certify that to the best of my knowledge and belief the statements in this report are correct and that all material, construction, and workmanship on this Alteration conforms to the National Board Inspection Code. National Board "**R**" Certificate of Authorization No. (39) ____________ expires on (40) ____________

Date (41) ____, (42) ____________ Signed (43) ____________
(name of alteration organization) (authorized representative)

CERTIFICATE OF INSPECTION

I, (44) ____________, holding a valid commission issued by the National Board of Boiler and Pressure Vessel Inspectors and certificate of competency, where required, issued by the Jurisdiction of (45) ____________ and employed by (44) ____________ of (44) ____________
have inspected the work described in this report on (46) ____________, (47) ____________ and state that to the best of my knowledge and belief, this work complies with the applicable requirements of the National Board Inspection Code. By signing this certificate, neither the undersigned nor my employer makes any warranty, expressed or implied, concerning the work described in this report. Furthermore, neither the undersigned nor my employer shall be liable in any manner for any personal injury, property damage, or loss of any kind arising from or connected with this inspection.

Date (48) ____________ Signed (49) ____________ (50) ____________
(inspector) (National Board and jurisdiction no. including endorsement)

Figure 1.6 *(Continued)*

3. Acceptable alternatives in design geometry
4. Acceptable alternatives in fabricating methods
5. Acceptable sources of supply

Some freedom of choice is usually left to the fabricator, and finding the best means of satisfying the requirements at the least cost is the essence of this phase of the manufacturing process.

This guide is to be used when completing the National Board Form R-2, Report of Alteration. When computer generated, the format of the form shall replicate the type and relative location of the information depicted on the Form R-2, Report of Alteration.

When additional space is needed, a Form R-4 shall be used and attached. Information determined to be of a proprietary nature need not be included, but shall be stated on the form.

1) Initials of the National Board "R" Certificate of Authorization authorized representative who registers the Form "R".

2) Initials of the Inspector who Certificate the completed Form "R" for registration.

3) When registering a Form "R" Report with the National Board, this line is solely designated for a unique sequential number assigned by the "R"Certificate Holder. When the "R" Form is not to be registered, indicate so by "N/A". As described in NBIC Part 3, Paragraph 5.6, a log shall be maintained identifying unique and sequentially numbered Form "R" reports that are registered with the National Board. For re-rating only, the Design Organization registers the Form R-2.

4) If applicable, document the unique purchase order, job, or tracking number assigned by organization performing work.

5) The name and address of the National Board "R" Certificate of Authorization holder performing the design as it appears on the "Certificate of Authorization".

6) The name and address of the National Board "R" Certificate of Authorization holder performing the construction activity as it appears on the "Certificate of Authorization".

7) Name and address of the owner of the pressure-retaining item.

8) Name and address of the plant or facility where the pressure-retaining item is installed.

9) Description of the pressure-retaining item, such as boiler, pressure vessel, or piping.

10) Name of the original manufacturer of the pressure-retaining item. If the original manufacturer is unknown, state "unknown."

11) Document the serial number of the pressure-retaining item if assigned by the original manufacturer. If there is no serial number assigned or it is unknown, state "unknown."

12) When the pressure-retaining item is registered with the National Board, document the applicable registration number. If the pressure-retaining item is installed in Canada, indicate the Canadian design, registration number (CRN), and list the drawing number under "other." If the item is not registered, indicate, "none."

13) Record the Jurisdiction Number, if available.

14) Document any other unique number assigned to the pressure retaining item such as a unique number assigned by the owner or user. If unknown, state "unknown."

15) Identify the year in which fabrication/construction of the item was completed.

16) Indicate edition and addenda, if applicable, of the NBIC under which this work is being performed, as applicable.

17) Indicate the name, section, division, edition, and addenda (if applicable) of the original code of construction for the pressure-retaining item.

18) Indicate the name, section, division, edition, and addenda (if applicable) of the construction code used for the work being performed. If code cases are used, they shall be identified in the "Remarks" section.

19) Provide a detailed summary of the scope of design that was performed.

20) The information to be considered when describing the scope of work should include such items as, the nature of the alteration (i.e. welding, bonding, cementing), the specific location of the work performed to the pressure-retaining item, the steps taken to remove a defect or as allowed by NBIC Part 3, Paragraph 3.4.8 to remain in place, and the method of alteration described as listed in the examples of NBIC Part 3, Section 3 or applicable supplement.

Figure 1.6 *(Continued)*

Work-Flow Schedules

By the time an order for a Code job is accepted, it has been estimated, bid, and sold. For many reasons, a job-shop estimator does not completely and finally specify all the details of material design and fabrication at the time of bidding. After an order is received, a work-flow schedule must be drawn up that allots time for shop fabrication, procurement of materials and components, design drafting, and takeoff. The shop fabrication schedule is based on an estimate of the total quantity of work-hours required, the necessary work-hour sequence (allowing for nonadditive, parallel operations), and finally the stipulated delivery date and the total shop load for the calendar

21) Indicate the type of Pressure test performed (Liquid, Pneumatic, Vacuum)

22) Indicate test pressure applied.

23) Indicate the maximum allowable working pressure of the part. (As altered)

24) As applicable, identify which parts manufactured by welding or bonding were introduced as needed to complete the scope of work. Indicatepart, item number, manufacturer's name, stamped identification, and data report type or Certificate of Compliance.

25) Indicate any additional information pertaining to the work involved (e.g., routine repairs, code cases, interpretations used).

26) Type or print name of the National Board "R" Certificate of Authorization authorized representative responsible for design certification.

27) Indicate National Board "R" Certificate or Authorization number.

28) Indicate month, day, and year the National Board "R" Certificate of Authorization expires.

29) Indicate the date the alteration was certified.

30) Record the name of National Board "R" Certificate of Authorization holder who performed the described work, using full name as shown on the "Certificate of Authorization" or an abbreviation acceptable to the National Board.

31) Signature of National Board "R" Certificate of Authorization authorized representative.

32) Type or print the name of Inspector certifying the design review.

33) Indicate the Inspector's Jurisdiction.

34) Indicate the Inspector's employer; also indicate the address of the Inspector's employer (city and state or province).

35) Indicate the month, day and year of the design certification by the Inspector

36) Signature of the Inspector certifying the design review.

37) Inspectors National Board commission number and endorsement that qualifies the Inspector to sign this report, and when required by the Jurisdiction, the applicable State or Provincial numbers.

38) Type or print name of the National Board "R" Certificate of Authorization authorized representative responsible for any construction.

39) Indicate the National Board "R" Certificate or Authorization number.

40) Indicate month, day, and year the National Board "R" Certificate of Authorization expires.

41) Indicate the date the alteration was certified.

42) Record the name of National Board "R" Certificate of Authorization holder who performed the described work, using full name as shown on the Certificate of Authorization or an abbreviation acceptable to the National Board.

43) Signature of National Board "R" Certificate of Authorization authorized representative

44) Type or print the name of Inspector certifying the construction inspection.

45) Indicate the Inspector's Jurisdiction.

46) Indicate the Inspector's employer; also indicate the address of the Inspector's employer (city and state or province).

47) Indicate the month, day and year of the inspection by the Inspector.

48) Indicate the month, day and year the completed Form "R" was signed by the Inspector.

49) Signature of the Inspector certifying the construction inspection.

50) Inspectors National Board commission number and endorsement that qualifies the Inspector to sign this report, and when required by the Jurisdiction, the applicable State or Provincial numbers.

FIGURE 1.6 *(Continued)*

period. This information determines the required delivery date for plate material, purchased components, and shop subassemblies. The lead time for these items sets the date for purchasing and, thus, for the completion of the design, drafting, and takeoff details initially required. An estimate of the time needed for the latter operations indicates the provisional starting date.

Although this simple computation can be made in many different ways, it is most important that it be made in a specific way, for it forms the basis on which all scheduling and expediting are founded.

FORM R-3 REPORT OF PARTS FABRICATED BY WELDING
in accordance with provisions of the *National Board Inspection Code*

(1) ________ (Authorized Rep. initials)
(2) ________ (Inspectors initials)
(3) ________ (Form "R-3" Registration no.)
(5) ________ (P.O. no., job no., etc.)

1. MANUFACTURED BY: (4) ________________ (name of "R" certificate holder)
________________ (address)

2. MANUFACTURED FOR: (6) ________________ (name)
________________ (address)

3. DESIGN CONDITION SPECIFIED BY: (7) ________________ CODE DESIGN BY: (8) ________________

4. DESIGN CODE: (9) ________ ________ (10) ________ (11) ________ (12) ________

5. REPAIR/ALTERATION/MODIFICATION ACTIVITIES

Name of Part	Qty.	Line No.	Manufacturer's Identifying No.	Manufacturer's Drawing No.	MAWP	Shop Hydro PSI	Year Built
(13)	(14)	(15)	(16)	(17)	(18)	(19)	(20)

6. DESCRIPTION OF PARTS

	(a) Connections other than tubes			Heads or Ends			(b) Tubes		
Line No.	Size and Shape	Material Spec. No.	Thickness (in.)	Shape	Thickness (in.)	Material Spec. No.	Diameter (in.)	Thickness (in.)	Material Spec. No.
(15)	(21)	(22)	(23)	(24)	(25)	(26)	(27)	(28)	(29)

7. REMARKS: (30) ________________

FIGURE 1.6 (*Continued*)

If the provisional starting date indicates on-time delivery, there is no problem. The sequence of determined dates becomes the production schedule, and expediting is managed accordingly. If the estimated delivery falls considerably ahead of time, the entire schedule can be appropriately delayed or more time can be allotted to any or all of the production steps. In this situation, the cost of maintaining the inventory must be balanced against the future availability of material in a

(3) ____________
(Form "R-3" Registration no.)
(5) ____________
(P.O. no., job no., etc.)

CERTIFICATE OF COMPLIANCE

I, (31) ____________, certify that to the best of my knowledge and belief the statements made in this report are correct and that all material, fabrication, construction, and workmanship of the described parts conforms to the *National Board Inspection Code* and the standards of construction cited.

National Board "**R**" *Certificate of Authorization* No. (32) ____________ expires on: (33) ________, ________
Date (34) ________, ________ (35) ____________ Signed (36) ____________
(name of "R" Certificate holder) (Authorized Representative)

CERTIFICATE OF INSPECTION

I, (37) ____________, holding a valid commission issued by the National Board of Boiler and Pressure Vessel Inspectors and certificate of competency, where required, issued by the Jurisdiction of (38) ____________ and employed by (39) ____________ of (40) ____________ have inspected the part described in this report on (41) ________, ________ and state that to the best of my knowledge and belief the parts comply with the applicable requirements of the *National Board Inspection Code*.

By signing this certificate, neither the undersigned nor my employer makes any warranty, expressed or implied, concerning the work described in this report. Furthermore, neither the undersigned nor my employer shall be liable in any manner for any personal injury, property damage, or loss of any kind arising from or connected with this inspection.
Date (42) ________, ________ Signed (43) ____________ Commissions (44) ____________
(inspector) (National Board and jurisdiction No. including endorsement)

FIGURE 1.6 *(Continued)*

changing market. It must be remembered that although filler jobs are desirable, a reduction in work rate is not. If the estimated delivery falls after the required date, on the other hand, the operations must be reviewed to determine where and how time can be saved. Sometimes a simple solution is quickly found; more often, pressure has to be applied on all concerned, including the ASME Code Inspector; however, this pressure must be applied in a rational way.

It should be realized that the two hard necessities governing job-shop work—cost saving and time saving—are not always compatible. Earlier deliveries of purchased material usually cost more; heavier pressure on a workforce usually results in overtime. Code restrictions on design, fabrication, and materials narrow the field of action. Nevertheless, certain routine procedures exist that will result in considerable savings.

Mill Orders

From a cost and delivery angle, the design of a pressure vessel and the purchasing of the plate and heads are often related. Thus, the steel mill system of extras, based on the weight of items ordered, varies inversely with the weight—a 10,000 lb item having no extra and lesser quantities taking gradually increased extras according to the published table. The maximum cost difference could amount to

This guide is to be used when completing the National Board Form R-3, Report of Parts Fabricated by Welding. When computer generated, the format of the form shall replicate the type and relative location of the information depicted on the Form R-3, Report of Parts Fabricated by Welding.

1) Initials of the National Board "R" Certificate of Authorization authorized representative who registers the Form "R".

2) Initials of the Inspector who certified the completed Form "R" for registration.

3) When registering a Form "R" Report with the National Board, this line is solely designated for a unique sequential number assigned by the "R" Certificate Holder. When the "R" Form is not to be registered, indicated so by "N/A". As described in NBIC Part 3, Paragraph 5.6, a log shall be maintained identifying sequentially, any Form "R" registered with the National Board.

4) The name and address of the Certificate Holder performing the work as it appears on the *"Certificate of Authorization"*.

5) If applicable, document the unique purchase order, job, or tracking number assigned by organization performing work.

6) Document name and address of organization that purchased the parts for incorporation into the repair or alteration. If the part's origin is unknown or the part was built for stock, so state.

7) Document name of organization responsible for specifying the code design conditions, if known. If origin of design conditions are un known, state "unknown."

8) Document name of organization responsible for performing the code design, if known. If code design organization is unknown, state "unknown."

9) Name, section, and division of the design code, if known. If the design is unknown, state "unknown."

10) Indicate code edition year used for fabrication.

11) Indicate code addenda date used for fabrication, if applicable.

12) Indicate the code paragraph reference for formula used to establish the MAWP, if known. If the code reference of the formula is unknown, state "unknown."

13) If available, identify component by part's original name, function, or use the original equipment manufacturer's "mark or item number."

14) Indicate quantity of named parts.

15) Match line number of part references for Identification of Parts in item 5 and Description of Parts in item 6.

16) Indicate manufacturer's serial number for the named part.

17) Indicate drawing number for the named part.

18) Indicate maximum allowable working pressure for the part, if known.

19) Indicate test pressure applied.

20) Identify the year in which fabrication/construction of the item was completed.

21) Use inside diameter for size: indicate shape as square, round, etc.

22) Indicate the complete material specification number and grade.

23) Indicate nominal thickness of plate and minimum thickness after forming.

24) Indicate share as flat, dished, ellipsoidal, or hemispherical.

25) Indicate minimum thickness after forming.

FIGURE 1.6 *(Continued)*

many dollars a ton. As an item consists of any number of plates of one width and one thickness of a given quality, it pays to design a vessel with equal-width plates (see Appendix A). Various mills also publish extras based on width, and these tables must be consulted. A buyer might be able to save considerably by arbitrarily increasing a plate width a fraction of an inch.

Tonnage mounts up fast in steel plate. The question is often asked, "Is it less expensive to design a given vessel for a higher test steel with less thickness even though the quality extra incurred by the decreased thickness may be higher?" It sometimes works out this way, by using, say, an SA-516 grade 70 steel instead of an SA-285 grade C steel. Fabrication cost differences must also be considered.

26) Indicate the complete material specification number and grade for the head or end.

27) Indicate outside diameter.

28) Indicate minimum thickness of tubes.

29) Indicate the complete material specification number and grade for tubes.

30) Indicate any additional information pertaining to the work involved (e.g. code cases). The part manufacturer is to indicate the extent he has performed any or all of the design function. If only a portion of the design, state which portion.

31) Type or print name of authorized representative of the "R" Certificate Holder attesting to accuracy of the work described.

32) Indicate National Board "R" Certificate of Authorization number.

33) Indicate month, day, and year that the "R" Certificate of Authorization expires.

34) Indicate the date the repair was certified.

35) Record name of "R" Certificate Holder who performed the described work, using full name as shown on the Certificate of Authorization or an abbreviation acceptable to the National Board.

36) Signature of National Board "R" Certificate of Authorization authorized representative.

37) Type or print name of Inspector.

38) Indicate Inspector's Jurisdiction.

39) Indicate Inspector's employer.

40) Indicate address of Inspector's employer (city and state or province).

41) Indicate month, day, and year of inspection by Inspector.

42) Indicate the month, day and year the completed Form "R" was signed by the Inspector.

43) Signature of Inspector.

44) Inspectors National Board commission number and endorsement that qualifies the Inspector to sign this report, and when required by the Jurisdiction, the applicable State or Provincial numbers.

FIGURE 1.6 *(Continued)*

Sometimes the buyer rather than the designer will note that a plate useful for stock inventory can be added to a special item to bring the total weight into a lower class of extras, thus saving money. This is also true of heads. At some mills, discounts for quantity begin with four to six heads, and sizable discounts are offered for larger quantities. It is also highly desirable to attempt to keep head diameters, radii, and thicknesses to the standards carried in mill stocks.

Mill conditions often affect the length of time required to fill an order. As mills have different rolling practices, schedules, minimum-quantity stipulations, base prices, size limits, and the like, all these details should be discussed with a sales representative before attempting to place an order, especially if the buyer is not experienced. The lead time for most mills in normal periods runs about 45 days, but it is not safe to guess.

Warehouse Orders

If delivery requirements make it impossible to obtain steel from a mill, Code plate can be obtained in various sizes from warehouses.

FORM R-4 REPORT SUPPLEMENT SHEET

in accordance with provisions of the *National Board Inspection Code*

(1) ______ (form "R" referenced)

(2) ______ (P.O. no., job no., etc.)

1. WORK PERFORMED BY: (3) ______ (name)

______ (address)

2. OWNER: (4) ______ (name)

______ (address)

3. LOCATION OF INSTALLATION: (5) ______ (name)

______ (address)

REFERENCE LINE NO. (7) CONTINUED FROM FORM R– (6) (8)

Date (9) ______, ______ Signed (10) ______ (authorized representative) Name (11) ______ (Name of "R" certificate holder)

Date (12) ______, ______ Signed (13) ______ (inspector) Commissions (14) ______ (National Board and jurisdiction no. including endorsement)

FIGURE 1.6 (*Continued*)

Quick delivery can be obtained of certain size plates, usually the lighter gages up to ½ in. thick, 96 in. wide, and 10 to 20 ft long. In addition to the higher base price, the buyer must consider the cost of obtaining the plate size desired from the sizes offered. Heads of certain sizes and gages are carried by some warehouses and are also stocked for immediate delivery by some mills. Standard thicknesses of 3⁄16, 1⁄4, 5⁄16, 3⁄8, and 1⁄2 in. are typical of flanged and dished heads with diameters up to 96 in. Special heads are also listed in published catalogs.

This guide is to be used when completing the National Board Form R-4, Report Supplementary Sheet. When computer generated, the format of the form shall replicate the type and relative location of the information depicted on the Form R-4, Report Supplementary Sheet.

1) When registering a Form "R" Report with the National Board, this line is solely designated for a unique sequential number assigned by the "R" Certificate Holder. When the "R" Form is not to be registered, indicate so by "N/A". As described in NBIC Part 3, Paragraph 5.6, a log shall be maintained identifying unique and sequentially numbered Form "R" reports that are registered with the National Board. For re-rating only, the Design Organization registers the Form R-2. Complete information identical to that shown on the Form "R" to which this sheet is supplementary.
2) If applicable, document the unique purchase order, job, or tracking number, assigned by the organization performing work.
3) The name and address of the Certificate Holder performing the work as it appears on the "*Certificate of Authorization*".
4) Name and address of the owner of the pressure-retaining item.
5) Name and address of plant or facility where the pressure-retaining item is installed.
6) Indicate the Form "R" type to which this report is supplementary. Example: Form R-1, Form R-2, Form R-3.
7) Indicate the reference line number from the Form "R" to which this report is supplementary.
8) Complete information for which there was insufficient spare on the reference Form "R".
9) Indicate the date certified.
10) Signature of the repair organizations authorized representative.
11) Record name of "R" Certificate Holder who performed the described work, using full name as shown on the Certificate of Authorization or an abbreviation acceptable to the National Board.
12) Indicate the date the form was completed by the Inspector.
13) Signature of the Inspector.
14) Inspectors National Board commission number and endorsement that qualifies the Inspector to sign this report, and when required by the Jurisdiction, the applicable State or Province numbers.

FIGURE 1.6 (*Continued*)

Shop Organization

By far the biggest opportunities for saving time and money lie in the handling and organizing of Code work in the shop. Unless the shop procedure is properly set up and the work is scheduled in advance for inspection, extra cost and delay are likely.

Each visit by a Code Inspector represents a direct cash outlay both for the time used by the Inspector and for the time taken by personnel to show and review the work. Many shops adopt the practice of arranging for periodic visits so that as much work as possible can be grouped for one visit and the availability of the Inspector can be relied upon. A lack of planning often results in a last-minute rush to get ready with possible overtime expense or in an incomplete presentation that necessitates an additional inspection for some overlooked detail. If a job is not carefully watched, it may progress beyond the point at which an Inspector can make a satisfactory investigation of a completed operation.

Similarly, the vessel may be found to contain components for which the Code quality standard cannot be established. Either situation may result in costly rework or even complete loss. Sometimes it seems that in the complexity of a busy job shop, it is almost impossible to plan very far ahead. A lack of planning, however, only means that more plans must be made, discarded, revised, and started over again.

Before work is started, the complete specifications should be reviewed, the sequence of work operations determined, and a rough

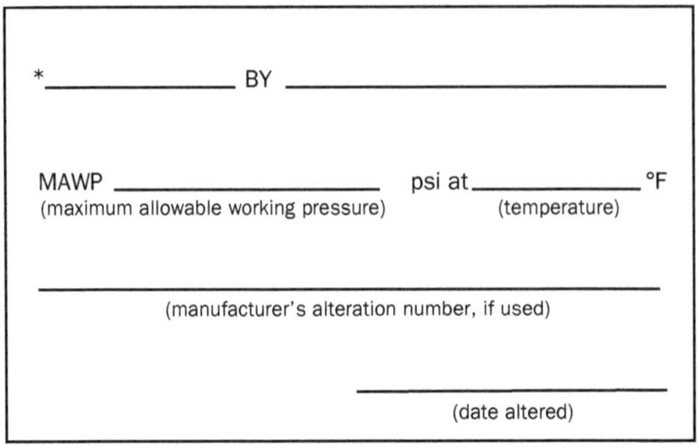

Figure 1.7 Stamping or nameplate of an altered boiler or pressure vessel. (*Reprinted with permission from the National Board Inspection Code, 1992 ed., p. 49. Copyright 1992 National Board of Boiler and Pressure Vessel Inspectors.*) The National Board of Boiler and Pressure Vessel Inspectors ("National Board") does not guarantee that this information or forms are up to date. Always consult the National Board's website for the most current version of applicable forms.

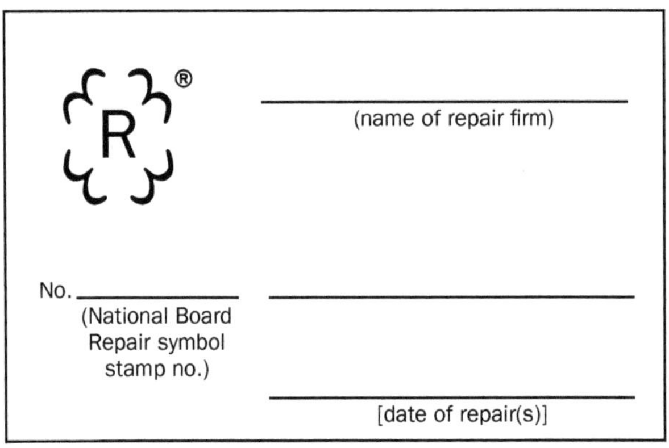

Figure 1.8 Stamping or nameplate of a boiler or pressure vessel repaired by welding. (*Reprinted with permission from the National Board Inspection Code, 1992 ed., p. 49. Copyright 1992 National Board of Boiler and Pressure Vessel Inspectors.*) The National Board of Boiler and Pressure Vessel Inspectors ("National Board") does not guarantee that this information or forms are up to date. Always consult the National Board's website for the most current version of applicable forms.

shop schedule, including the expected dates of inspection formulated. Without this schedule a final delivery date cannot be safely determined. The job must follow this schedule closely under responsible supervision.

When Code material is received, it should be carefully marked and checked. At least by the time of layout, the reference numbers should be checked against material certification and the required physical and chemical properties proved. At this time also, material should be checked for thickness and surface defects. This simple routine should help to prevent unnecessary and costly rework and loss of time.

CHAPTER 2

Descriptive Guide to the ASME Code Section VIII, Division 1, Pressure Vessels

An attempt is made in this chapter to assist those wishing to find specific information quickly without having to wade through the large fund of specifications and recommendations encompassed by the Code. The figures and tables outline the designs, materials, and performance requirements; indicate the conditions for which precautions are necessary; and list the Code paragraphs to be read for proper information. Experience has shown that many of these items are often overlooked. Also included is a handy reference chart (see Fig. 2.1) that graphically illustrates the various parts of a pressure vessel and the Code paragraphs that apply to each. Table 2.1 lists the various classes of materials from Section VIII, Division 1; the covering Code section part; applicable references for the stress value tables found in Section II, Part D; and additional helpful hints. It should be remembered, however, that there is no substitute for reading the Code itself.

Requirements for Establishing Design Thickness Based on the Degree of Radiography (Code Pars. UW-11 and UW-12)

The requirements based on the degree of radiographic examination of butt welds in pressure vessels constructed to the requirements of Section VIII, Division 1 of the *ASME Boiler and Pressure Vessel Code* have been the subject of numerous inquiries for interpretation from the manufacturers, designers, Authorized Inspectors, and users of

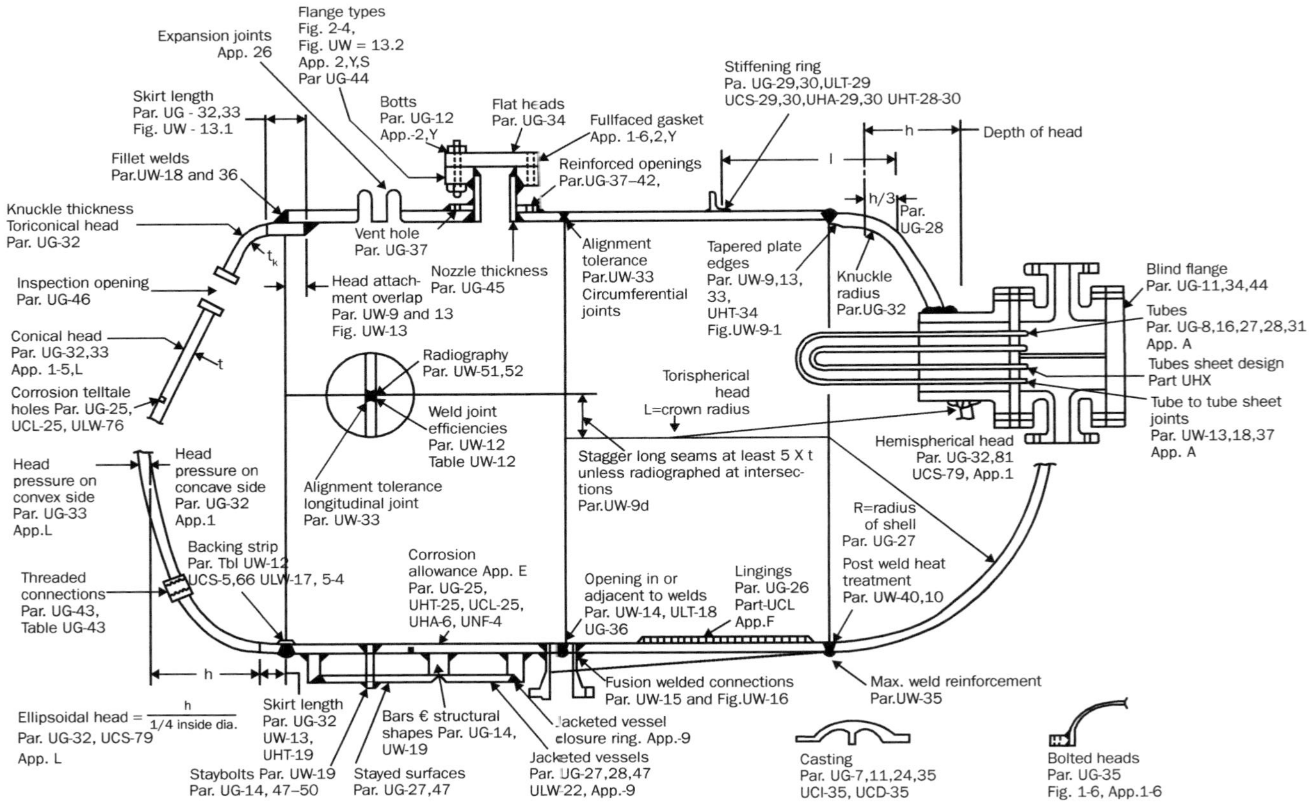

FIGURE 2.1 Pressure vessel reference

Material	Covering Code part	Applicable Code stress value tables	Remarks
Carbon and low-alloy steels	UCS	Code Section II, Part D, Table 1A	Basis for establishing stress values—Code Appendix P, UG-23 Low-temperature service requires use of notch-tough materials—Code Pars. UCS-65, UCS-66, UCS-67, UCS-68, UG-84, Code Figs. UCS-66, UCS-66M, UCS-66.1, UCS-66.1M, UCS-66.2, Code Table UCS-66 Creep strength is essential in high-temperature operation Design temperature—Code Par. UG-20 Design pressure—Code Par. UG-21 Temperature above 800°F may cause the carbide phase of carbon steel to convert to graphite Pipe and tubes—Code Pars. UG-8, UG-10, UG-16, UG-31, UCS-9, UCS-27 Creep and rupture properties—Code Par. UCS-A-2
Nonferrous metals	UNF	Code Section II, Part D, Table 1B	Basis for establishing values—Code Appendix P, UG-23 Metal characteristics—Code Part UNF, Appendix UNF-A Low-temperature operation—Code Par. UNF-65 Nonferrous castings—Code Par. UNF-8

TABLE 2.1 Classes of Materials

Material	Covering Code part	Applicable Code stress value tables	Remarks
High-alloy steels	UHA	Code Section II, Part D, Table 1A	Selection and treatment of high-alloy steels—Code Part UHA, Appendix UHA-A Inspection and tests—Code Pars. UHA-33, UHA-34, UHA-50, UHA-51, UHA-52 Liquid penetrant examination required if shell thickness exceeds 3⁄4 in.—in all austenitic chromium–nickel alloy steel and austenitic/ferritic duplex steel—Code Par. UHA-34 Low-temperature service—Code Pars. UHA-51, UG-84 High-alloy castings—Code Pars. UHA-8, UG-7, UG-11, UG-24, Code Appendix 7
Cast iron	UCI	UCI-23	Service Restrictions—Code Par. UCI-2 Selection of materials—Code Pars. UCI-1, UCI-3, UCI-5, UCI-12, UG-10, UG-11 Inspection and tests—Code Pars. UCI-3, UCI-90, UCI-99, UCI-101 Repairs in cast-iron materials—Code Par. UCI-78
Integrally clad plate, weld metal overlay, or applied linings	UCL	(See Code Pars. UCL-11, UCL 23.)	Suggest careful study of Part UCL

Table 2.1 *(Contd)*

Material	Covering Code part	Applicable Code stress value tables	Remarks
			Selection of materials—Code Pars. UCL-1, UCL-3, UCL-10, UCS-5, UF-5, ULW-5, UCL-11, UCL-12, UG-10 Qualification of welding procedure—Code Par. UCL-40 Postweld heat treatment—Code Pars. UCL-34, UCS-56 Inspection and test—Code Pars. UCL-50, UCL-51, UCL-52 Use of linings—Code Par. UG-26 and Code Appendix F
Welded and seamless pipes and tubes (carbon and low-alloy steels)	UCS	Code Section II, Part D, Table 1A	Thickness under internal pressure—Code Par. UG-27 Thickness under external pressure—Code Par. UG-28 Provide additional thickness when tubes are threaded and when corrosion, erosion, or wear caused by cleaning is expected—Code Par. UG-31 For calculating the required thickness, use minimum pipe wall thickness rather than nominal— Code Par. UG-16

Table 2.1 *(Contd)*

Material	Covering Code part	Applicable Code stress value tables	Remarks
			30 in. maximum on welded pipe made by open-hearth, basic oxygen, or electric-furnace process—Code Par. UCS-27
Welded and seamless pipe (high-alloy steels)	UHA	Code Section II, Part D, Table 1A	
Forgings	UF	Code Section II, Part D, Table 1A	Materials—Code Pars. UG-6, UG-11, UF-6, UCS-7 Welding—Code Par. UF-32
Low-temperature materials	ULT	ULT-23	An operation at very low temperatures requires the use of notch-tough materials
Layered construction	ULW		Vessels having a shell and/or heads made up of two or more separate layers—Code Par. ULW-2
Ferritic steels with tensile properties enhanced by heat treatment	UHT	Code Section II, Part D, Table 1A	Scope—Code Par. UHT-1 For marking on plate or stamping, use *low-stress* stamps—Code Par. UHT-86

Table 2.1 (*Contd*)

these vessels. Most of the questions have been concerned with the examination requirements for circumferential butt welds that attach seamless heads to shell sections.

The Code has always been clear regarding the requirements for butt welds attaching shell sections which have fully radiographed, longitudinal butt welds and when a joint efficiency of 1.00 is used in the calculation of shell thickness. A revision given in the Summer 1969 Addenda provided, for the first time, a requirement for the radiographic examination of the butt weld attaching a seamless head to a seamless section in order to be consistent with the rules for butt-welded sections. However, the unwillingness to accept this new requirement generated further inquiries. Consequently, the Committee reopened the subject for possible further revision.

In an effort to achieve uniformity, *The Transactions of the ASME Series J Journal of Pressure Vessel Technology,* in the February 1975 issue, and in subsequent follow-up article in February 1977, included examples by G. M. Eisenberg, Senior Pressure Vessel Engineering Administrator, clarifying the design factors that were used as a guide. The National Board of Boiler and Pressure Vessel Inspectors also published these examples in their April 1975 *National Board Bulletin.*

The degrees of radiographic examination shown in Code Par. UW-11 include full, spot, and no radiography requirements. The factors for allowable stress and joint efficiency to be used in calculations for shell sections and head thicknesses are set forth in Code Par. UW-12 and Code Table UW-12. The use of spot radiography, in lieu of partial radiography on Category B or C butt welds, provided a relaxation of the previous requirements. [Refer to Code Par. UW-11(a)(5)(b); partial radiography is no longer addressed by Section VIII, Division 1.]

The Code details the joint efficiency requirements and the percentage of the tabulated allowable stress which shall be used in head and shell calculations for both circumferential (hoop) and longitudinal (axial) stresses.

In the 1986 Addenda to ASME Section VIII, Division 1, extensive revisions to the past rules that affect the minimum design thickness requirements used in the design of Code pressure parts were made. These new revisions have continued to evolve slowly to the rules which are in effect today. Additionally, these new requirements adversely affect how the maximum allowable working pressure (MAWP) or design thicknesses of materials used in Code construction are determined. The new rules are used to determine the method for the selection of joint efficiencies and/or stress multipliers and how to apply them in the design of Section VIII, Division 1 pressure vessels.

Under the current requirements, the designer must consider each ASME Code Section VIII, Division 1, Table UW-12 joint separately for the value of E which will be utilized in the formulas of Code Pars. UG-27, UG-32, UG-34, and UG-37, and Code Appendix 1, Appendix 9-6, and Appendix 13. The symbol E is formally known to mean *joint efficiency* or *efficiency.* Today these terms may also be known as a *stress multiplier* which can be either in the form of a joint efficiency or stress reduction factor.

Under the revised rules, the designer would be allowed to employ spot radiography of the circumferential joints in a vessel. This could possibly be the case even if the designer had elected to perform full radiography of the intersecting longitudinal butt weld. For example,

Type of joint and radiography[1, 2]	Efficiency, percent	Code references
Double-welded butt joints (Type 1)		Par. UW-11, UW-12, UW-35, Table UW-12
Fully radiographed	100	Par. UW-51
Spot radiographed	85	Par. UW-52
No radiograph	70	
Single-welded butt joints (backing strip left in place) (Type 2)		Par. UW-11, UW-12, UW-35, Table UW-12
Fully radiographed	90	Par. UW-51
Spot radiographed	80	Par. UW-52
No radiograph	65	
Single-welded butt joints, no backing strip (Type 3), limited to circumferential joints only, not over 5⁄8 in. thick, and not over 24 in. outside diameter	60	Table UW-12
Fillet weld lap joints and single-welded butt circumferential joints		Table UW-12
Seamless vessel sections or heads (spot-radiographed)	100	Par. UW-12(d)
Seamless vessel sections or heads (no radiography)	85	Par. UW-12(d)

[1]See Table UW-12 for *type number* description
[2]See UW-3 and Fig. UW-3 for *weld joint category* description

TABLE 2.2 Welded Joint Efficiency Values

one approach may be to fully radiograph the longitudinal butt joints of a vessel utilizing $E = 1.00$ in the formula. The designer may then choose to apply a lesser degree of radiographic examination (e.g., spot radiography) for the circumferential butt joints using $E = 0.85$.

It should be noted with caution that the Code only allows one level of radiographic examination difference between longitudinal butt joints (or seamless equivalents) and the intersecting circumferential Table UW-12 butt joints in vessels, vessel sections, and heads [see Code Par. UW-11(a)(5)(b)].

The Code recognizes that seamless vessel sections (e.g., pipe used for shells of vessels) or seamless heads and their intersecting Category B and C butt joints are considered equivalent to the welded parts of the same geometry. This rule applies only when the longitudinal joint is welded from both the inside and the outside of the weld joint preparation or when some other process is used which will ensure that the butt joint has complete penetration at the joint.

Remarks	Code references
Full radiography	
Full radiograph for all joints over 1½ in. or for lesser thickness as defined in adjacent references	Pars. UW-11(a)2, UCS-57, UNF-57, UHA-33, UCL-35, UCL-36, UHT-57, ULW-51,ULW-52(d), ULW-54, ULT-57
Circumferential butt joints of nozzles and communicating chambers with diameter not exceeding NPS 10 or with thickness not exceeding 1⅛ in. do not require radiography	Par. UW-11(a)4
Vessels containing lethal substances must be fully radiographed	Pars. UW-11(a)1, UW-2
Efficiency of fully radiographed joints	Par. UW-12, Table UW-12
Radiographic acceptance standards	Par. UW-51
Surface weld buildup	Par. UW-42
Carbon and low-alloy steel materials	Pars. UCS-19, UCS-57
High-alloy steel materials	Par. UHA-33
Clad-plate materials	Par. UCL-35
Nonferrous materials	Par. UNF-57
Ferritic steels whose tensile properties have been enhanced by heat treatment	Par. UHT-57
Layered vessel inner shells and heads	Par. ULW-51
Welded joints in layers of layered vessels	Par. ULW-52(d)
Butt welded joints in layered vessels	Par. ULW-54
Low-temperature vessels	Par. ULT-57
Spot radiography	
General recommendations	Par. UW-11(b)
Spot radiographing required for higher weld joint efficiencies	Pars. UW-12, UW-52, Table UW-12
Minimum number of spots	Par. UW-52(b)
Minimum length of spot radiograph, 6 in.	Par. UW-52(c)
Surface weld buildup	Par. UW-42
If cladding is included in computing required thickness, spot radiograph is a minimum requirement	Par. UCL-23(c)
Vessels having 50 ft or less of main seams require one spot examination; larger vessels require one spot for each 50 ft of welding	Par. UW-52(b)1
Additional spots may be selected to examine the welding of each welder or welding operator	Par. UW-52(b)2

Table 2.3 Radiographic Requirements

Remarks	Code references
Repair welding defects in material; forgings	Pars. UF-37, UCD-78
General requirements for castings	Par. UG-24
Low-temperature vessels	Par. ULT-57
High-alloy steel vessels	Par. UHA-33
Ferritic steel whose tensile properties have been enhanced by heat treatment	Par. UHT-57
Joints welded with austenitic chromium–nickel steel	Pars. UCL-35, UCL-36
Repair welds in layered construction	Pars. ULW-51, ULW-57

TABLE 2.3 *(Contd)*

For calculations involving the circumferential stresses of seamless parts, or for the thicknesses of seamless shells or heads, the value of E may equal 1.00 when the spot radiographic requirements of Code Par. UW-11(a)(5)(b) are met. Additionally, the value of E may equal 0.85 and may be utilized in the formula if the designer elects not to perform any radiography.

When the pressure welding processes are to be utilized in the production of Code pressure vessels, such as flash induction welding, pressure gas welding, and inertia and continuous drive friction welding, the value $E = 0.80$ may continue to be used in Code calculations provided the welding process used is permitted by the rules of Section VIII, Division 1 for the materials to be welded and provided that such welds are tested. Such test specimens shall be representative of the production welding on each vessel. These specimens may be taken from the shell of the vessel or from a separate "runoff" tab attached as a protrusion of the longitudinal joint of the vessel. If there is no longitudinal joint, a separate test plate of the same base material, thickness, and welding materials which will be used on the production piece shall be used. Each test piece shall be tested with one reduced section tension test specimen and two side bend test specimens which shall meet the requirements of the ASME Code Section IX Pars. QW-150 and QW-160.

As a result of the joint efficiency requirement allowing a per-joint factor, many vessels now built will be utilizing the RT-4 when marking the ASME Code data plate. This marking applies only when a part of the vessel satisfies the requirements of UW-11(a) or when RT-1 [the complete vessel satisfies the full radiographic requirements of UW-11(a)], RT-2 [the complete vessel satisfies the radiograph requirements of UW-11(a)(5) and the spot radiographic requirements of UW-11(a)(5)(b)], or RT-3 [the complete vessel satisfies the spot radiographic requirements of UW-11(b)] does not apply.

The Remarks section of the Manufacturer's Data Report must also list the various combinations of joint efficiencies, E. In cases where complex vessels are designed with various sections, a joint efficiency map may be generated and documented on the U-4 form of the ASME Manufacturer's Data Report Supplementary Sheet.

The flow charts depicted in Figs. 2.2 to 2.5 indicate the required joint efficiencies for various components as well as clearly outline where the designer may find the applicable joint efficiencies for various weld categories in the ASME Code Section VIII, Division 1.

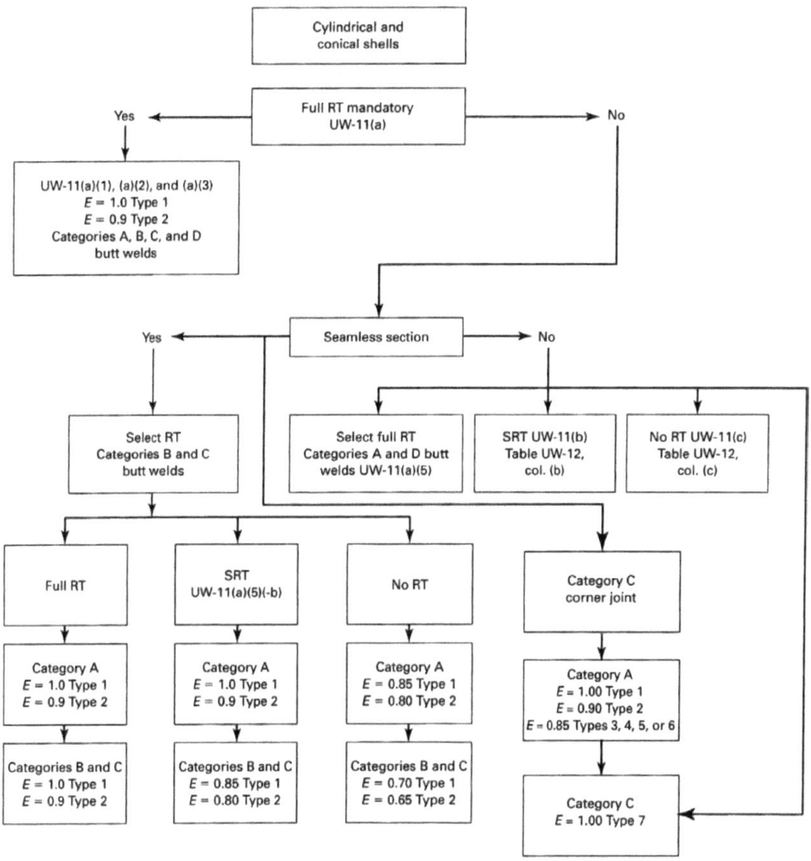

GENERAL NOTES:

(a) Thickness required for longitudinal stress in conical sections is as follows: $t = PD/[4 \cos(\alpha)(SE + 0.4P)]$.

(b) See UW-11(a)(4) for Categories B and C butt welds in nozzles and communicating chambers equal to or less than NPS 10 and thickness $1\frac{1}{8}$ in. (30 mm) or less.

(c) Type 2 joints not allowed for Category A weld joints for UW-2(c) designs.

(d) Type 2 joints allowed for Category A weld joints for UW-2(b) designs for austenitic stainless steel Type 304.

FIGURE 2.2 Joint efficiency and weld joint type—Cylinders and cones (Code Fig. L-1.4-1). *(Courtesy of the American Society of Mechanical Engineers.)*

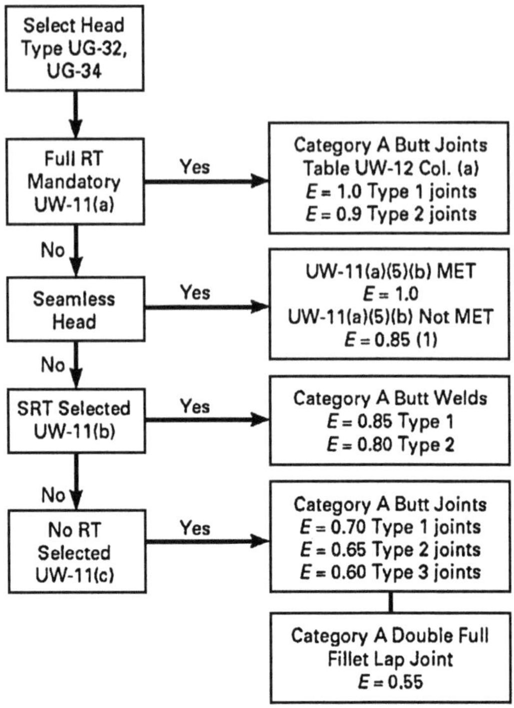

NOTES:
(1) See UW-12(d) when head-to-shell attachment weld is Type No. 3, 4, 5, or 6.

FIGURE 2.3 Joint efficiency and weld joint type—Heads (Code Fig. L-1.4-2). *(Courtesy of the American Society of Mechanical Engineers.)*

Use of Internal-Pressure Cylindrical Shell Thickness Tables

Tables 2.4*a* to 2.4*f* may be applied to seamless shells (joint efficiency, $E = 1.00$). These may also be applied to shells with double-welded butt longitudinal joints that have been fully radiographed ($E = 1.00$), spot radiographed ($E = 0.85$), or not radiographed ($E = 0.70$). They may also be applied to single-welded butt joints with a backing strip that are fully radiographed ($E = 0.90$), spot radiographed ($E = 0.80$), or not radiographed ($E = 0.65$). The required shell thickness may be determined by the following steps:

1. Determine the appropriate table corresponding to the desired joint efficiency. Note that the value of joint efficiency for each table is printed in the upper left-hand corner.
2. Locate the desired diameter. The diameter values are printed across the top of the page for each table.

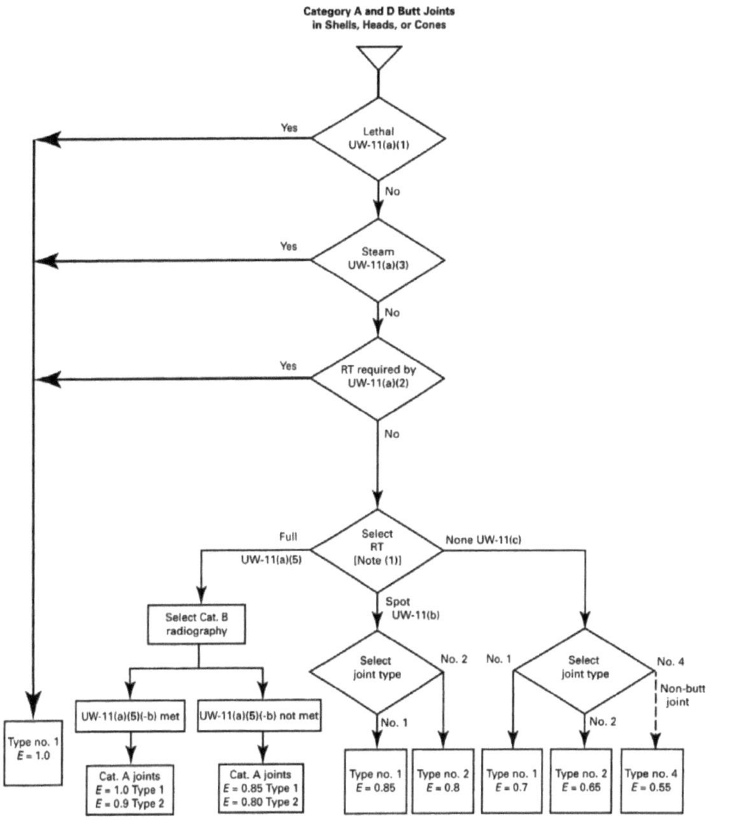

GENERAL NOTE: Use value of *E* obtained from UW-12 in appropriate formulas for circumferential stress or stress on Category A welds, such as
UG-27(c)(1) — cylindrical shells
UG-32(c) — ellipsoidal heads
UG-32(d) — torispherical heads
UG-32(e) — hemispherical heads
UG-32(f) — conical heads and sections
UG-34(c)(2) — flat heads

NOTE:
(1) See UW-11(a)(5)(-b).

Figure 2.4 Joint efficiencies for Categories A and D welded joints in shells, heads, or cones (Code Fig. L-1.4-3). (*Courtesy of the American Society of Mechanical Engineers.*)

3. Locate the desired pressure. The pressure values are located in the left column.
4. Read the thickness corresponding to the diameter and pressure.

Each table is based on an allowable stress value of 13,800 psi. For other stress values, multiply the thickness read from the table by the following *factor*.

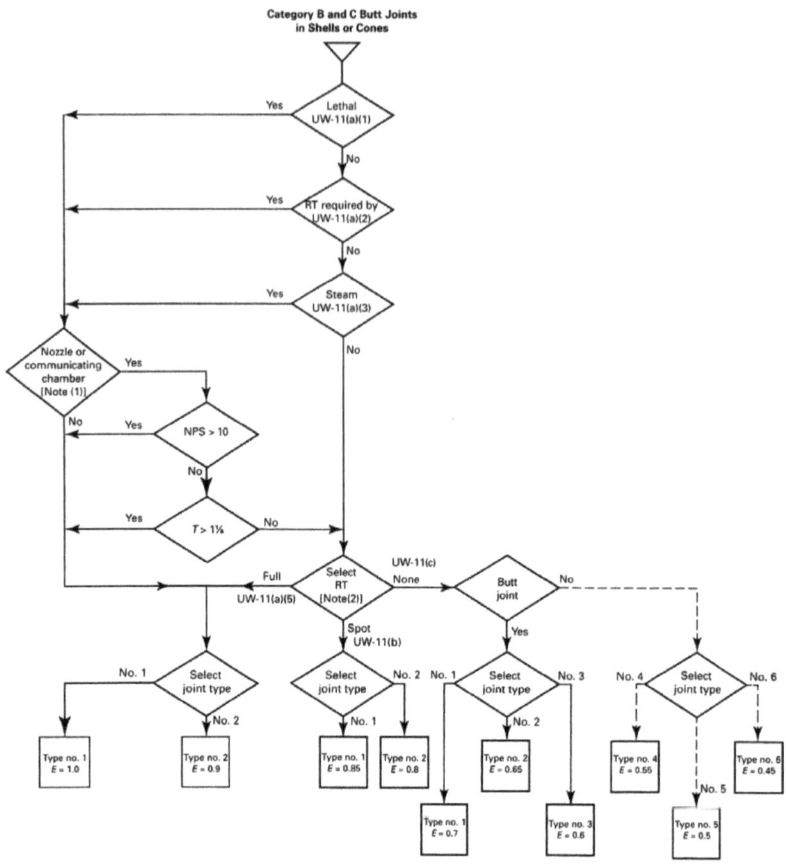

GENERAL NOTE: Use value of E obtained from UW-12 in appropriate formulas for longitudinal stress, such as UG-27(c)(2).

NOTES:
(1) See UHT-57(a).
(2) See UW-11(a)(5)(-b).

FIGURE 2.5 Joint efficiencies for Categories B and C welded joints in shells or cones (Code Fig. L-1.4-4). (*Courtesy of the American Society of Mechanical Engineers.*)

$$\text{Factor} = \frac{14{,}178}{(\text{allowable stress})^{1.003}}$$

The user is cautioned that the thicknesses calculated using the tables with the above factor are more than adequate for the purpose of estimation, but these are not exact. Calculations with appropriate Code equations are required for all formal Code documentation.

Thickness equations for cylindrical shells may be found in Table 2.5.

SHELL THICKNESS TABLES IN INCHES

JOINT EFF.=1.00 ALLOWABLE STRESS VALUE= 13800. PSI

	INSIDE DIAMETER,IN.																				
PSI	12.	18.	24.	30.	36.	42.	48.	54.	60.	66.	72.	78.	84.	90.	96.	102.	108.	114.	120.	126.	132.
15.	0.007	0.010	0.013	0.016	0.020	0.023	0.026	0.029	0.033	0.036	0.039	0.042	0.046	0.049	0.052	0.055	0.059	0.062	0.065	0.069	0.072
20.	0.009	0.013	0.017	0.022	0.026	0.030	0.035	0.039	0.044	0.048	0.052	0.057	0.061	0.065	0.070	0.074	0.078	0.083	0.087	0.091	0.096
25.	0.011	0.016	0.022	0.027	0.033	0.038	0.044	0.049	0.054	0.060	0.065	0.071	0.076	0.082	0.087	0.092	0.098	0.103	0.109	0.114	0.120
30.	0.013	0.020	0.026	0.033	0.039	0.046	0.052	0.059	0.065	0.072	0.078	0.085	0.091	0.098	0.104	0.111	0.118	0.124	0.131	0.137	0.144
35.	0.015	0.023	0.030	0.038	0.046	0.053	0.061	0.069	0.076	0.084	0.091	0.099	0.107	0.114	0.122	0.130	0.137	0.145	0.152	0.160	0.168
40.	0.017	0.026	0.035	0.044	0.052	0.061	0.070	0.078	0.087	0.096	0.105	0.113	0.122	0.131	0.139	0.148	0.157	0.166	0.174	0.183	0.192
45.	0.020	0.029	0.039	0.049	0.059	0.069	0.078	0.088	0.098	0.108	0.118	0.127	0.137	0.147	0.157	0.167	0.176	0.186	0.196	0.206	0.216
50.	0.022	0.033	0.044	0.054	0.065	0.076	0.087	0.098	0.109	0.120	0.131	0.142	0.153	0.163	0.174	0.185	0.196	0.207	0.218	0.229	0.240
55.	0.024	0.036	0.048	0.060	0.072	0.084	0.096	0.108	0.120	0.132	0.144	0.156	0.168	0.180	0.192	0.204	0.216	0.228	0.240	0.252	0.264
60.	0.026	0.039	0.052	0.065	0.078	0.092	0.105	0.118	0.131	0.144	0.157	0.170	0.183	0.196	0.209	0.222	0.235	0.248	0.262	0.275	0.288
65.	0.028	0.043	0.057	0.071	0.085	0.099	0.113	0.128	0.142	0.156	0.170	0.184	0.198	0.213	0.227	0.241	0.255	0.269	0.283	0.298	0.312
70.	0.031	0.046	0.061	0.076	0.092	0.107	0.122	0.137	0.153	0.168	0.183	0.198	0.214	0.229	0.244	0.259	0.275	0.290	0.305	0.321	0.336
75.	0.033	0.049	0.065	0.082	0.098	0.115	0.131	0.147	0.164	0.180	0.196	0.213	0.229	0.245	0.262	0.278	0.294	0.311	0.327	0.344	0.360
80.	0.035	0.052	0.070	0.087	0.105	0.122	0.140	0.157	0.175	0.192	0.209	0.227	0.244	0.262	0.279	0.297	0.314	0.332	0.349	0.366	0.384
85.	0.037	0.056	0.074	0.093	0.111	0.130	0.148	0.167	0.185	0.204	0.223	0.241	0.260	0.278	0.297	0.315	0.334	0.352	0.371	0.389	0.408
90.	0.039	0.059	0.079	0.098	0.118	0.137	0.157	0.177	0.196	0.216	0.236	0.255	0.275	0.295	0.314	0.334	0.354	0.373	0.393	0.412	0.432
95.	0.041	0.062	0.083	0.104	0.124	0.145	0.166	0.187	0.207	0.228	0.249	0.270	0.290	0.311	0.332	0.353	0.373	0.394	0.415	0.435	0.456
100.	0.044	0.066	0.087	0.109	0.131	0.153	0.175	0.197	0.218	0.240	0.262	0.284	0.306	0.328	0.349	0.371	0.393	0.415	0.437	0.459	0.480
105.	0.046	0.069	0.092	0.115	0.138	0.161	0.183	0.206	0.229	0.252	0.275	0.298	0.321	0.344	0.367	0.390	0.413	0.436	0.459	0.482	0.504
110.	0.048	0.072	0.096	0.120	0.144	0.168	0.192	0.216	0.240	0.264	0.288	0.312	0.336	0.360	0.384	0.408	0.433	0.457	0.481	0.505	0.529
115.	0.050	0.075	0.101	0.126	0.151	0.176	0.201	0.226	0.251	0.276	0.302	0.327	0.352	0.377	0.402	0.427	0.452	0.477	0.503	0.528	0.553
120.	0.052	0.079	0.105	0.131	0.157	0.184	0.210	0.236	0.262	0.288	0.315	0.341	0.367	0.393	0.420	0.446	0.472	0.498	0.524	0.551	0.577
125.	0.055	0.082	0.109	0.137	0.164	0.191	0.219	0.246	0.273	0.301	0.328	0.355	0.383	0.410	0.437	0.464	0.492	0.519	0.546	0.574	0.601
130.	0.057	0.085	0.114	0.142	0.171	0.199	0.227	0.256	0.284	0.313	0.341	0.369	0.398	0.426	0.455	0.483	0.512	0.540	0.568	0.597	0.625
135.	0.059	0.089	0.118	0.148	0.177	0.207	0.236	0.266	0.295	0.325	0.354	0.384	0.413	0.443	0.472	0.502	0.531	0.561	0.590	0.620	0.649
140.	0.061	0.092	0.122	0.153	0.184	0.214	0.245	0.276	0.306	0.337	0.367	0.398	0.429	0.459	0.490	0.521	0.551	0.582	0.612	0.643	0.674
145.	0.063	0.095	0.127	0.159	0.190	0.222	0.254	0.285	0.317	0.349	0.381	0.412	0.444	0.476	0.508	0.539	0.571	0.603	0.634	0.666	0.698
150.	0.066	0.098	0.131	0.164	0.197	0.230	0.263	C.295	0.328	0.361	0.394	0.427	0.460	0.492	0.525	0.558	0.591	0.624	0.656	0.689	0.722
160.	0.070	0.105	0.140	0.175	0.210	0.245	0.280	0.315	0.350	0.385	0.420	0.455	0.490	0.525	0.560	0.595	0.630	0.665	0.701	0.736	0.771
170.	0.074	0.112	0.149	0.186	0.223	0.261	0.298	0.335	0.372	0.410	0.447	0.484	0.521	0.558	0.596	0.633	0.670	0.707	0.745	0.782	0.819
180.	0.079	0.118	0.158	0.197	0.237	0.276	0.316	0.355	0.394	0.434	0.473	0.513	0.552	0.592	0.631	0.670	0.710	0.749	0.789	0.828	0.868
190.	0.083	0.125	0.167	0.208	0.250	0.292	0.333	0.375	0.416	0.458	0.500	0.541	0.583	0.625	0.666	0.708	0.750	0.791	0.833	0.875	0.916
200.	0.088	0.132	0.175	0.219	0.263	0.307	0.351	0.395	0.439	0.482	0.526	0.570	0.614	0.658	0.702	0.746	0.789	0.833	0.877	0.921	0.965
210.	0.092	0.138	0.184	0.230	0.276	0.323	0.369	0.415	0.461	0.507	0.553	0.599	0.645	0.691	0.737	0.783	0.829	0.875	0.921	0.968	1.014
220.	0.097	0.145	0.193	0.241	0.290	0.338	0.386	0.435	0.483	0.531	0.579	0.628	0.676	0.724	0.773	0.821	0.869	0.917	0.966	1.014	1.062
230.	0.101	0.152	0.202	0.253	0.303	0.354	0.404	0.455	0.505	0.556	0.606	0.657	0.707	0.758	0.808	0.859	0.909	0.960	1.010	1.061	1.111
240.	0.105	0.158	0.211	0.264	0.316	0.369	0.422	0.475	0.527	0.580	0.633	0.685	0.738	0.791	0.844	0.896	0.949	1.002	1.054	1.107	1.160
250.	0.110	0.165	0.220	0.275	0.330	0.385	0.440	0.495	0.549	0.604	0.659	0.714	0.769	0.824	0.879	0.934	0.989	1.044	1.099	1.154	1.209
260.	0.114	0.172	0.229	0.286	0.343	0.400	0.457	0.515	0.572	0.629	0.686	0.743	0.800	0.858	0.915	0.972	1.029	1.086	1.143	1.201	1.258
270.	0.119	0.178	0.238	0.297	0.356	0.416	0.475	0.535	0.594	0.653	0.713	0.772	0.832	0.891	0.950	1.010	1.069	1.128	1.188	1.247	1.307
280.	0.123	0.185	0.246	0.308	0.370	0.431	0.493	0.555	0.616	0.678	0.739	0.801	0.863	0.924	0.986	1.048	1.109	1.171	1.232	1.294	1.356
290.	0.128	0.192	0.255	0.319	0.383	0.447	0.511	0.575	0.638	0.702	0.766	0.830	0.894	0.958	1.022	1.085	1.149	1.213	1.277	1.341	1.405
300.	0.132	0.198	0.264	0.330	0.396	0.463	0.529	0.595	0.661	0.727	0.793	0.859	0.925	0.991	1.057	1.123	1.189	1.256	1.322	1.388	1.454
310.	0.137	0.205	0.273	0.342	0.410	0.478	0.546	0.615	0.683	0.751	0.820	0.888	0.956	1.025	1.093	1.161	1.230	1.298	1.366	1.435	1.503
320.	0.141	0.212	0.282	0.353	0.423	0.494	0.564	0.635	0.705	0.776	0.847	0.917	0.988	1.058	1.129	1.199	1.270	1.340	1.411	1.481	1.552
330.	0.146	0.218	0.291	0.364	0.437	0.509	0.582	0.655	0.728	0.801	0.873	0.946	1.019	1.092	1.165	1.237	1.310	1.383	1.456	1.528	1.601
340.	0.150	0.225	0.300	0.375	0.450	0.525	0.600	0.675	0.750	0.825	0.900	0.975	1.050	1.125	1.200	1.275	1.350	1.425	1.500	1.575	1.650
350.	0.155	0.232	0.309	0.386	0.464	0.541	0.618	0.695	0.773	0.850	0.927	1.004	1.082	1.159	1.236	1.313	1.391	1.468	1.545	1.623	1.700
360.	0.159	0.239	0.318	0.398	0.477	0.557	0.636	0.716	0.795	0.875	0.954	1.034	1.113	1.193	1.272	1.352	1.431	1.511	1.590	1.670	1.749
370.	0.163	C.245	0.327	0.409	0.490	0.572	0.654	0.736	0.817	0.899	0.981	1.063	1.144	1.226	1.308	1.390	1.471	1.553	1.635	1.717	1.798

TABLE 2.4a Shell Thickness Tables in Inches: Joint Efficiency = 1.0

SHELL THICKNESS TABLES IN INCHES

JOINT EFF.=0.90   ALLOWABLE STRESS VALUE= 13800. PSI

PSI	INSIDE DIAMETER, IN. 12.	18.	24.	30.	36.	42.	48.	54.	60.	66.	72.	78.	84.	90.	96.	102.	108.	114.	120.	126.	132.
15.	0.007	0.011	0.015	0.018	0.022	0.025	0.029	0.033	0.036	0.040	0.044	0.047	0.051	0.054	0.058	0.062	0.065	0.069	0.073	0.076	0.080
20.	0.010	0.015	0.019	0.024	0.029	0.034	0.039	0.044	0.048	0.053	0.058	0.063	0.068	0.073	0.077	0.082	0.087	0.092	0.097	0.102	0.106
25.	0.012	0.018	0.024	0.030	0.036	0.042	0.048	0.054	0.060	0.067	0.073	0.079	0.085	0.091	0.097	0.103	0.109	0.115	0.121	0.127	0.133
30.	0.015	0.022	0.029	0.036	0.044	0.051	0.058	0.065	0.073	0.080	0.087	0.094	0.102	0.109	0.116	0.123	0.131	0.138	0.145	0.152	0.160
35.	0.017	0.025	0.034	0.042	0.051	0.059	0.068	0.076	0.085	0.093	0.102	0.110	0.119	0.127	0.135	0.144	0.152	0.161	0.169	0.178	0.186
40.	0.019	0.029	0.039	0.048	0.058	0.068	0.077	0.087	0.097	0.106	0.116	0.126	0.136	0.145	0.155	0.165	0.174	0.184	0.194	0.203	0.213
45.	0.022	0.033	0.044	0.054	0.065	0.076	0.087	0.098	0.109	0.120	0.131	0.142	0.153	0.163	0.174	0.185	0.196	0.207	0.218	0.229	0.240
50.	0.024	0.036	0.048	0.061	0.073	0.085	0.097	0.109	0.121	0.133	0.145	0.157	0.169	0.182	0.194	0.206	0.218	0.230	0.242	0.254	0.266
55.	0.027	0.040	0.053	0.067	0.080	0.093	0.107	0.120	0.133	0.147	0.160	0.173	0.186	0.200	0.213	0.226	0.240	0.253	0.266	0.280	0.293
60.	0.029	0.044	0.058	0.073	0.087	0.102	0.116	0.131	0.145	0.160	0.174	0.189	0.203	0.218	0.233	0.247	0.262	0.276	0.291	0.305	0.320
65.	0.031	0.047	0.063	0.079	0.094	0.110	0.126	0.142	0.157	0.173	0.189	0.205	0.220	0.236	0.252	0.268	0.283	0.299	0.315	0.331	0.346
70.	0.034	0.051	0.068	0.085	0.102	0.119	0.136	0.153	0.170	0.187	0.204	0.221	0.238	0.254	0.271	0.288	0.305	0.322	0.339	0.356	0.373
75.	0.036	0.055	0.073	0.091	0.109	0.127	0.145	0.164	0.182	0.200	0.218	0.236	0.255	0.273	0.291	0.309	0.327	0.345	0.364	0.382	0.400
80.	0.039	0.058	0.078	0.097	0.116	0.136	0.155	0.175	0.194	0.213	0.233	0.252	0.272	0.291	0.310	0.330	0.349	0.369	0.388	0.407	0.427
85.	0.041	0.062	0.082	0.103	0.124	0.144	0.165	0.186	0.206	0.227	0.247	0.268	0.289	0.309	0.330	0.350	0.371	0.392	0.412	0.433	0.454
90.	0.044	0.066	0.087	0.109	0.131	0.153	0.175	0.197	0.218	0.240	0.262	0.284	0.306	0.328	0.349	0.371	0.393	0.415	0.437	0.459	0.480
95.	0.046	0.069	0.092	0.115	0.138	0.161	0.184	0.207	0.231	0.254	0.277	0.300	0.323	0.346	0.369	0.392	0.415	0.438	0.461	0.484	0.507
100.	0.049	0.073	0.097	0.121	0.146	0.170	0.194	0.218	0.243	0.267	0.291	0.316	0.340	0.364	0.388	0.413	0.437	0.461	0.485	0.510	0.534
105.	0.051	0.076	0.102	0.127	0.153	0.178	0.204	0.229	0.255	0.280	0.306	0.331	0.357	0.382	0.408	0.433	0.459	0.484	0.510	0.535	0.561
110.	0.053	0.080	0.107	0.134	0.160	0.187	0.214	0.240	0.267	0.294	0.321	0.347	0.374	0.401	0.427	0.454	0.481	0.508	0.534	0.561	0.588
115.	0.056	0.084	0.112	0.140	0.168	0.196	0.223	0.251	0.279	0.307	0.335	0.363	0.391	0.419	0.447	0.475	0.503	0.531	0.559	0.587	0.615
120.	0.058	0.087	0.117	0.146	0.175	0.204	0.233	0.262	0.292	0.321	0.350	0.379	0.408	0.437	0.466	0.496	0.525	0.554	0.583	0.612	0.641
125.	0.061	0.091	0.122	0.152	0.182	0.213	0.243	0.273	0.304	0.334	0.365	0.395	0.425	0.456	0.486	0.516	0.547	0.577	0.608	0.638	0.668
130.	0.063	0.095	0.126	0.158	0.190	0.221	0.253	0.284	0.316	0.348	0.379	0.411	0.442	0.474	0.506	0.537	0.569	0.600	0.632	0.664	0.695
135.	0.066	0.098	0.131	0.164	0.197	0.230	0.263	0.295	0.328	0.361	0.394	0.427	0.460	0.492	0.525	0.558	0.591	0.624	0.656	0.689	0.722
140.	0.068	0.102	0.136	0.170	0.204	0.238	0.272	0.306	0.340	0.375	0.409	0.443	0.477	0.511	0.545	0.579	0.613	0.647	0.681	0.715	0.749
145.	0.071	0.106	0.141	0.176	0.212	0.247	0.282	0.317	0.353	0.388	0.423	0.459	0.494	0.529	0.564	0.600	0.635	0.670	0.705	0.741	0.776
150.	0.073	0.109	0.146	0.182	0.219	0.255	0.292	0.328	0.365	0.401	0.438	0.474	0.511	0.547	0.584	0.620	0.657	0.693	0.730	0.766	0.803
160.	0.078	0.117	0.156	0.195	0.234	0.273	0.312	0.351	0.389	0.428	0.467	0.506	0.545	0.584	0.623	0.662	0.701	0.740	0.779	0.818	0.857
170.	0.083	0.124	0.166	0.207	0.248	0.290	0.331	0.373	0.414	0.455	0.497	0.538	0.580	0.621	0.662	0.704	0.745	0.787	0.828	0.869	0.911
180.	0.088	0.132	0.175	0.219	0.263	0.307	0.351	0.395	0.439	0.482	0.526	0.570	0.614	0.658	0.702	0.746	0.789	0.833	0.877	0.921	0.965
190.	0.093	0.139	0.185	0.232	0.278	0.324	0.371	0.417	0.463	0.510	0.556	0.602	0.648	0.695	0.741	0.787	0.834	0.880	0.926	0.973	1.019
200.	0.098	0.146	0.195	0.244	0.293	0.341	0.390	0.439	0.488	0.537	0.585	0.634	0.683	0.732	0.780	0.829	0.878	0.927	0.976	1.024	1.073
210.	0.102	0.154	0.205	0.256	0.307	0.359	0.410	0.461	0.512	0.564	0.615	0.666	0.717	0.769	0.820	0.871	0.922	0.974	1.025	1.076	1.127
220.	0.107	0.161	0.215	0.269	0.322	0.376	0.430	0.483	0.537	0.591	0.645	0.698	0.752	0.806	0.859	0.913	0.967	1.021	1.074	1.128	1.182
230.	0.112	0.169	0.225	0.281	0.337	0.393	0.449	0.506	0.562	0.618	0.674	0.730	0.787	0.843	0.899	0.955	1.011	1.067	1.124	1.180	1.236
240.	0.117	0.176	0.235	0.293	0.352	0.411	0.469	0.528	0.587	0.645	0.704	0.762	0.821	0.880	0.938	0.997	1.056	1.114	1.173	1.232	1.290
250.	0.122	0.183	0.244	0.306	0.367	0.428	0.489	0.550	0.611	0.672	0.733	0.795	0.856	0.917	0.978	1.039	1.100	1.161	1.222	1.284	1.345
260.	0.127	0.191	0.254	0.318	0.382	0.445	0.509	0.572	0.636	0.700	0.763	0.827	0.890	0.954	1.018	1.081	1.145	1.208	1.272	1.336	1.399
270.	0.132	0.198	0.264	0.330	0.396	0.463	0.529	0.595	0.661	0.727	0.793	0.859	0.925	0.991	1.057	1.123	1.189	1.256	1.322	1.388	1.454
280.	0.137	0.206	0.274	0.343	0.411	0.480	0.548	0.617	0.686	0.754	0.823	0.891	0.960	1.028	1.097	1.166	1.234	1.303	1.371	1.440	1.508
290.	0.142	0.213	0.284	0.355	0.426	0.497	0.568	0.639	0.710	0.781	0.853	0.924	0.995	1.066	1.137	1.208	1.279	1.350	1.421	1.492	1.563
300.	0.147	0.221	0.294	0.368	0.441	0.515	0.588	0.662	0.735	0.809	0.882	0.956	1.029	1.103	1.176	1.250	1.324	1.397	1.471	1.544	1.618
310.	0.152	0.228	0.304	0.380	0.456	0.532	0.608	0.684	0.760	0.836	0.912	0.988	1.064	1.140	1.216	1.292	1.368	1.444	1.520	1.596	1.672
320.	0.157	0.236	0.314	0.393	0.471	0.550	0.628	0.707	0.785	0.864	0.942	1.021	1.099	1.178	1.256	1.335	1.413	1.492	1.570	1.649	1.727
330.	0.162	0.243	0.324	0.405	0.486	0.567	0.648	0.729	0.810	0.891	0.972	1.053	1.134	1.215	1.296	1.377	1.458	1.539	1.620	1.701	1.782
340.	0.167	0.250	0.334	0.417	0.501	0.584	0.668	0.751	0.835	0.918	1.002	1.085	1.169	1.252	1.336	1.419	1.503	1.586	1.670	1.753	1.837
350.	0.172	0.258	0.344	0.430	0.516	0.602	0.688	0.774	0.860	0.946	1.032	1.118	1.204	1.290	1.376	1.462	1.548	1.634	1.720	1.806	1.892
360.	0.177	0.265	0.354	0.442	0.531	0.619	0.708	0.796	0.885	0.973	1.062	1.150	1.239	1.327	1.416	1.504	1.593	1.681	1.770	1.858	1.947
370.	0.182	0.273	0.364	0.455	0.546	0.637	0.728	0.819	0.910	1.001	1.092	1.183	1.274	1.365	1.456	1.547	1.638	1.729	1.820	1.911	2.002

TABLE 2.4*b* Shell Thickness Tables in Inches: Joint Efficiency = 0.90

SHELL THICKNESS TABLES IN INCHES

JOINT EFF.=0.85 ALLOWABLE STRESS VALUE= 13800. PSI

PSI	INSIDE DIAMETER, IN. 12.	18.	24.	30.	36.	42.	48.	54.	60.	66.	72.	78.	84.	90.	96.	102.	108.	114.	120.	126.	132.
15.	0.008	0.012	0.015	0.019	0.023	0.027	0.031	0.035	0.038	0.042	0.046	0.050	0.054	0.058	0.061	0.065	0.069	0.073	0.077	0.081	0.084
20.	0.010	0.015	0.020	0.026	0.031	0.036	0.041	0.046	0.051	0.056	0.061	0.067	0.072	0.077	0.082	0.087	0.092	0.097	0.102	0.108	0.113
25.	0.013	0.019	0.026	0.032	0.038	0.045	0.051	0.058	0.064	0.070	0.077	0.083	0.090	0.096	0.102	0.109	0.115	0.122	0.128	0.134	0.141
30.	0.015	0.023	0.031	0.038	0.046	0.054	0.061	0.069	0.077	0.085	0.092	0.100	0.108	0.115	0.123	0.131	0.138	0.146	0.154	0.161	0.169
35.	0.018	0.027	0.036	0.045	0.054	0.063	0.072	0.081	0.090	0.099	0.108	0.117	0.126	0.135	0.143	0.152	0.161	0.170	0.179	0.188	0.197
40.	0.021	0.031	0.041	0.051	0.062	0.072	0.082	0.092	0.103	0.113	0.123	0.133	0.144	0.154	0.164	0.174	0.185	0.195	0.205	0.215	0.226
45.	0.023	0.035	0.046	0.058	0.069	0.081	0.092	0.104	0.115	0.127	0.138	0.150	0.161	0.173	0.185	0.196	0.208	0.219	0.231	0.242	0.254
50.	0.026	0.038	0.051	0.064	0.077	0.090	0.103	0.115	0.128	0.141	0.154	0.167	0.179	0.192	0.205	0.218	0.231	0.244	0.256	0.269	0.282
55.	0.028	0.042	0.056	0.071	0.085	0.099	0.113	0.127	0.141	0.155	0.169	0.183	0.197	0.212	0.226	0.240	0.254	0.268	0.282	0.296	0.310
60.	0.031	0.046	0.062	0.077	0.092	0.108	0.123	0.139	0.154	0.169	0.185	0.200	0.215	0.231	0.246	0.262	0.277	0.292	0.308	0.323	0.339
65.	0.033	0.050	0.067	0.083	0.100	0.117	0.133	0.150	0.167	0.183	0.200	0.217	0.234	0.250	0.267	0.284	0.300	0.317	0.334	0.350	0.367
70.	0.036	0.054	0.072	0.090	0.108	0.126	0.144	0.162	0.180	0.198	0.216	0.234	0.252	0.270	0.287	0.305	0.323	0.341	0.359	0.377	0.395
75.	0.039	0.058	0.077	0.096	0.116	0.135	0.154	0.173	0.193	0.212	0.231	0.250	0.270	0.289	0.308	0.327	0.347	0.366	0.385	0.404	0.424
80.	0.041	0.062	0.082	0.103	0.123	0.144	0.164	0.185	0.205	0.226	0.247	0.267	0.288	0.308	0.329	0.349	0.370	0.390	0.411	0.431	0.452
85.	0.044	0.066	0.087	0.109	0.131	0.153	0.175	0.197	0.218	0.240	0.262	0.284	0.306	0.328	0.349	0.371	0.393	0.415	0.437	0.459	0.480
90.	0.046	0.069	0.092	0.116	0.139	0.162	0.185	0.208	0.231	0.254	0.277	0.301	0.324	0.347	0.370	0.393	0.416	0.439	0.462	0.486	0.509
95.	0.049	0.073	0.098	0.122	0.146	0.171	0.195	0.220	0.244	0.269	0.293	0.317	0.342	0.366	0.391	0.415	0.439	0.464	0.488	0.513	0.537
100.	0.051	0.077	0.103	0.129	0.154	0.180	0.206	0.231	0.257	0.283	0.308	0.334	0.360	0.386	0.411	0.437	0.463	0.488	0.514	0.540	0.566
105.	0.054	0.081	0.108	0.135	0.162	0.189	0.216	0.243	0.270	0.297	0.324	0.351	0.378	0.405	0.432	0.459	0.486	0.513	0.540	0.567	0.594
110.	0.057	0.085	0.113	0.141	0.170	0.198	0.226	0.255	0.283	0.311	0.340	0.368	0.396	0.424	0.453	0.481	0.509	0.538	0.566	0.594	0.622
115.	0.059	0.089	0.118	0.148	0.178	0.207	0.237	0.266	0.296	0.325	0.355	0.385	0.414	0.444	0.473	0.503	0.533	0.562	0.592	0.621	0.651
120.	0.062	0.093	0.124	0.154	0.185	0.216	0.247	0.278	0.309	0.340	0.371	0.401	0.432	0.463	0.494	0.525	0.556	0.587	0.618	0.648	0.679
125.	0.064	0.097	0.129	0.161	0.193	0.225	0.257	0.290	0.322	0.354	0.386	0.418	0.450	0.483	0.515	0.547	0.579	0.611	0.644	0.676	0.708
130.	0.067	0.100	0.134	0.167	0.201	0.234	0.268	0.301	0.335	0.368	0.402	0.435	0.469	0.502	0.536	0.569	0.602	0.636	0.669	0.703	0.736
135.	0.070	0.104	0.139	0.174	0.209	0.243	0.278	0.313	0.348	0.382	0.417	0.452	0.487	0.522	0.556	0.591	0.626	0.661	0.695	0.730	0.765
140.	0.072	0.108	0.144	0.180	0.216	0.252	0.289	0.325	0.361	0.397	0.433	0.469	0.505	0.541	0.577	0.613	0.649	0.685	0.721	0.757	0.793
145.	0.075	0.112	0.149	0.187	0.224	0.262	0.299	0.336	0.374	0.411	0.448	0.486	0.523	0.560	0.598	0.635	0.673	0.710	0.747	0.785	0.822
150.	0.077	0.116	0.155	0.193	0.232	0.271	0.309	0.348	0.387	0.425	0.464	0.503	0.541	0.580	0.619	0.657	0.696	0.735	0.773	0.812	0.851
160.	0.083	0.124	0.165	0.206	0.248	0.289	0.330	0.371	0.413	0.454	0.495	0.536	0.578	0.619	0.660	0.701	0.743	0.784	0.825	0.866	0.908
170.	0.088	0.132	0.175	0.219	0.263	0.307	0.351	0.395	0.439	0.482	0.526	0.570	0.614	0.658	0.702	0.746	0.789	0.833	0.877	0.921	0.965
180.	0.093	0.139	0.186	0.232	0.279	0.325	0.372	0.418	0.465	0.511	0.558	0.604	0.650	0.697	0.743	0.790	0.836	0.883	0.929	0.976	1.022
190.	0.098	0.147	0.196	0.245	0.294	0.343	0.393	0.442	0.491	0.540	0.589	0.638	0.687	0.736	0.785	0.834	0.883	0.932	0.981	1.030	1.080
200.	0.103	0.155	0.207	0.258	0.310	0.362	0.413	0.465	0.517	0.568	0.620	0.672	0.724	0.775	0.827	0.879	0.930	0.982	1.034	1.085	1.137
210.	0.109	0.163	0.217	0.271	0.326	0.380	0.434	0.489	0.543	0.597	0.651	0.706	0.760	0.814	0.869	0.923	0.977	1.032	1.086	1.140	1.194
220.	0.114	0.171	0.228	0.285	0.341	0.398	0.455	0.512	0.569	0.626	0.683	0.740	0.797	0.854	0.911	0.967	1.024	1.081	1.138	1.195	1.252
230.	0.119	0.179	0.238	0.298	0.357	0.417	0.476	0.536	0.595	0.655	0.714	0.774	0.833	0.893	0.952	1.012	1.071	1.131	1.190	1.250	1.310
240.	0.124	0.186	0.249	0.311	0.373	0.435	0.497	0.559	0.621	0.684	0.746	0.808	0.870	0.932	0.994	1.056	1.119	1.181	1.243	1.305	1.367
250.	0.130	0.194	0.259	0.324	0.389	0.453	0.518	0.583	0.648	0.712	0.777	0.842	0.907	0.972	1.036	1.101	1.166	1.231	1.295	1.360	1.425
260.	0.135	0.202	0.270	0.337	0.404	0.472	0.539	0.607	0.674	0.741	0.809	0.876	0.943	1.011	1.078	1.146	1.213	1.280	1.348	1.415	1.483
270.	0.140	0.210	0.280	0.350	0.420	0.490	0.560	0.630	0.700	0.770	0.840	0.910	0.980	1.050	1.120	1.190	1.260	1.330	1.400	1.470	1.540
280.	0.145	0.218	0.291	0.363	0.436	0.509	0.581	0.654	0.727	0.799	0.872	0.944	1.017	1.090	1.162	1.235	1.308	1.380	1.453	1.526	1.598
290.	0.151	0.226	0.301	0.376	0.452	0.527	0.602	0.678	0.753	0.828	0.903	0.979	1.054	1.129	1.205	1.280	1.355	1.430	1.506	1.581	1.656
300.	0.156	0.234	0.312	0.390	0.468	0.545	0.623	0.701	0.779	0.857	0.935	1.013	1.091	1.169	1.247	1.325	1.403	1.481	1.558	1.636	1.714
310.	0.161	0.242	0.322	0.403	0.483	0.564	0.644	0.725	0.806	0.886	0.967	1.047	1.128	1.208	1.289	1.370	1.450	1.531	1.611	1.692	1.772
320.	0.166	0.250	0.333	0.416	0.499	0.582	0.666	0.749	0.832	0.915	0.998	1.082	1.165	1.248	1.331	1.414	1.498	1.581	1.664	1.747	1.830
330.	0.172	0.258	0.343	0.429	0.515	0.601	0.687	0.773	0.858	0.944	1.030	1.116	1.202	1.288	1.374	1.459	1.545	1.631	1.717	1.803	1.889
340.	0.177	0.265	0.354	0.442	0.531	0.619	0.708	0.796	0.885	0.973	1.062	1.150	1.239	1.327	1.416	1.504	1.593	1.681	1.770	1.858	1.947
350.	0.182	0.273	0.365	0.456	0.547	0.638	0.729	0.820	0.911	1.003	1.094	1.185	1.276	1.367	1.458	1.549	1.641	1.732	1.823	1.914	2.005
360.	0.188	0.281	0.375	0.469	0.563	0.657	0.750	0.844	0.938	1.032	1.126	1.219	1.313	1.407	1.501	1.595	1.688	1.782	1.876	1.970	2.064
370.	0.193	0.289	0.386	0.482	0.579	0.675	0.772	0.868	0.965	1.061	1.157	1.254	1.350	1.447	1.543	1.640	1.736	1.833	1.929	2.026	2.122

TABLE 2.4c Shell Thickness Tables in Inches: Joint Efficiency = 0.85

SHELL THICKNESS TABLES IN INCHES

JOINT EFF.=0.80 ALLOWABLE STRESS VALUE= 13800. PSI

PSI	INSIDE DIAMETER,IN. 12.	18.	24.	30.	36.	42.	48.	54.	60.	66.	72.	78.	84.	90.	96.	102.	108.	114.	120.	126.	132.
15.	0.008	0.012	0.016	0.020	0.024	0.029	0.033	0.037	0.041	0.045	0.049	0.053	0.057	0.061	0.065	0.069	0.073	0.078	0.082	0.086	0.090
20.	0.011	0.016	0.022	0.027	0.033	0.038	0.044	0.049	0.054	0.060	0.065	0.071	0.076	0.082	0.087	0.092	0.098	0.103	0.109	0.114	0.120
25.	0.014	0.020	0.027	0.034	0.041	0.048	0.054	0.061	0.068	0.075	0.082	0.088	0.095	0.102	0.109	0.116	0.122	0.129	0.136	0.143	0.150
30.	0.016	0.024	0.033	0.041	0.049	0.057	0.065	0.073	0.082	0.090	0.098	0.106	0.114	0.122	0.131	0.139	0.147	0.155	0.163	0.171	0.180
35.	0.019	0.029	0.038	0.048	0.057	0.067	0.076	0.086	0.095	0.105	0.114	0.124	0.133	0.143	0.152	0.162	0.172	0.181	0.191	0.200	0.210
40.	0.022	0.033	0.044	0.054	0.065	0.076	0.087	0.098	0.109	0.120	0.131	0.142	0.153	0.163	0.174	0.185	0.196	0.207	0.218	0.229	0.240
45.	0.025	0.037	0.049	0.061	0.074	0.086	0.098	0.110	0.123	0.135	0.147	0.159	0.172	0.184	0.196	0.208	0.221	0.233	0.245	0.257	0.270
50.	0.027	0.041	0.054	0.068	0.082	0.095	0.109	0.123	0.136	0.150	0.163	0.177	0.191	0.204	0.218	0.232	0.245	0.259	0.272	0.286	0.300
55.	0.030	0.045	0.060	0.075	0.090	0.105	0.120	0.135	0.150	0.165	0.180	0.195	0.210	0.225	0.240	0.255	0.270	0.285	0.300	0.315	0.330
60.	0.033	0.049	0.065	0.082	0.098	0.115	0.131	0.147	0.164	0.180	0.196	0.213	0.229	0.245	0.262	0.278	0.294	0.311	0.327	0.344	0.360
65.	0.035	0.053	0.071	0.089	0.106	0.124	0.142	0.160	0.177	0.195	0.213	0.230	0.248	0.266	0.284	0.301	0.319	0.337	0.355	0.372	0.390
70.	0.038	0.057	0.076	0.095	0.115	0.134	0.153	0.172	0.191	0.210	0.229	0.248	0.267	0.286	0.306	0.325	0.344	0.363	0.382	0.401	0.420
75.	0.041	0.061	0.082	0.102	0.123	0.143	0.164	0.184	0.205	0.225	0.246	0.266	0.286	0.307	0.327	0.348	0.368	0.389	0.409	0.430	0.450
80.	0.044	0.066	0.087	0.109	0.131	0.153	0.175	0.197	0.218	0.240	0.262	0.284	0.306	0.328	0.349	0.371	0.393	0.415	0.437	0.459	0.480
85.	0.046	0.070	0.093	0.116	0.139	0.162	0.186	0.209	0.232	0.255	0.278	0.302	0.325	0.348	0.371	0.394	0.418	0.441	0.464	0.487	0.511
90.	0.049	0.074	0.098	0.123	0.147	0.172	0.197	0.221	0.246	0.270	0.295	0.319	0.344	0.369	0.393	0.418	0.442	0.467	0.492	0.516	0.541
95.	0.052	0.078	0.104	0.130	0.156	0.182	0.208	0.234	0.259	0.285	0.311	0.337	0.363	0.389	0.415	0.441	0.467	0.493	0.519	0.545	0.571
100.	0.055	0.082	0.109	0.137	0.164	0.191	0.219	0.246	0.273	0.301	0.328	0.355	0.383	0.410	0.437	0.464	0.492	0.519	0.546	0.574	0.601
105.	0.057	0.086	0.115	0.143	0.172	0.201	0.230	0.258	0.287	0.316	0.344	0.373	0.402	0.430	0.459	0.488	0.517	0.545	0.574	0.603	0.631
110.	0.060	0.090	0.120	0.150	0.180	0.210	0.241	0.271	0.301	0.331	0.361	0.391	0.421	0.451	0.481	0.511	0.541	0.571	0.601	0.631	0.662
115.	0.063	0.094	0.126	0.157	0.189	0.220	0.252	0.283	0.314	0.346	0.377	0.409	0.440	0.472	0.503	0.535	0.566	0.597	0.629	0.660	0.692
120.	0.066	0.098	0.131	0.164	0.197	0.230	0.263	0.295	0.328	0.361	0.394	0.427	0.460	0.492	0.525	0.558	0.591	0.624	0.656	0.689	0.722
125.	0.068	0.103	0.137	0.171	0.205	0.239	0.274	0.308	0.342	0.376	0.410	0.445	0.479	0.513	0.547	0.581	0.616	0.650	0.684	0.718	0.752
130.	0.071	0.107	0.142	0.178	0.213	0.249	0.285	0.320	0.356	0.391	0.427	0.463	0.498	0.534	0.569	0.605	0.640	0.676	0.712	0.747	0.783
135.	0.074	0.111	0.148	0.185	0.222	0.259	0.296	0.333	0.370	0.407	0.443	0.480	0.517	0.554	0.591	0.628	0.665	0.702	0.739	0.776	0.813
140.	0.077	0.115	0.153	0.192	0.230	0.268	0.307	0.345	0.383	0.422	0.460	0.498	0.537	0.575	0.613	0.652	0.690	0.728	0.767	0.805	0.843
145.	0.079	0.119	0.159	0.199	0.238	0.278	0.318	0.357	0.397	0.437	0.477	0.516	0.556	0.596	0.635	0.675	0.715	0.755	0.794	0.834	0.874
150.	0.082	0.123	0.164	0.205	0.247	0.288	0.329	0.370	0.411	0.452	0.493	0.534	0.575	0.616	0.658	0.699	0.740	0.781	0.822	0.863	0.904
160.	0.088	0.132	0.175	0.219	0.263	0.307	0.351	0.395	0.439	0.482	0.526	0.570	0.614	0.658	0.702	0.746	0.789	0.833	0.877	0.921	0.965
170.	0.093	0.140	0.187	0.233	0.280	0.326	0.373	0.420	0.466	0.513	0.560	0.606	0.653	0.699	0.746	0.793	0.839	0.886	0.933	0.979	1.026
180.	0.099	0.148	0.198	0.247	0.296	0.346	0.395	0.445	0.494	0.543	0.593	0.642	0.692	0.741	0.790	0.840	0.889	0.939	0.988	1.037	1.087
190.	0.104	0.157	0.209	0.261	0.313	0.365	0.417	0.470	0.522	0.574	0.626	0.678	0.730	0.783	0.835	0.887	0.939	0.991	1.043	1.096	1.148
200.	0.110	0.165	0.220	0.275	0.330	0.385	0.440	0.495	0.549	0.604	0.659	0.714	0.769	0.824	0.879	0.934	0.989	1.044	1.099	1.154	1.209
210.	0.115	0.173	0.231	0.289	0.346	0.404	0.462	0.520	0.577	0.635	0.693	0.750	0.808	0.866	0.924	0.981	1.039	1.097	1.154	1.212	1.270
220.	0.121	0.182	0.242	0.303	0.363	0.424	0.484	0.545	0.605	0.666	0.726	0.787	0.847	0.908	0.968	1.029	1.089	1.150	1.210	1.271	1.331
230.	0.127	0.190	0.253	0.316	0.380	0.443	0.506	0.570	0.633	0.696	0.759	0.823	0.886	0.949	1.013	1.076	1.139	1.203	1.266	1.329	1.392
240.	0.132	0.198	0.264	0.330	0.396	0.463	0.529	0.595	0.661	0.727	0.793	0.859	0.925	0.991	1.057	1.123	1.189	1.256	1.322	1.388	1.454
250.	0.138	0.207	0.275	0.344	0.413	0.482	0.551	0.620	0.689	0.758	0.826	0.895	0.964	1.033	1.102	1.171	1.240	1.309	1.377	1.446	1.515
260.	0.143	0.215	0.287	0.358	0.430	0.502	0.573	0.645	0.717	0.788	0.860	0.932	1.003	1.075	1.147	1.218	1.290	1.362	1.433	1.505	1.577
270.	0.149	0.223	0.298	0.372	0.447	0.521	0.596	0.670	0.745	0.819	0.894	0.968	1.042	1.117	1.191	1.266	1.340	1.415	1.489	1.564	1.638
280.	0.155	0.232	0.309	0.386	0.464	0.541	0.618	0.695	0.773	0.850	0.927	1.004	1.082	1.159	1.236	1.313	1.391	1.468	1.545	1.623	1.700
290.	0.160	0.240	0.320	0.400	0.480	0.560	0.641	0.721	0.801	0.881	0.961	1.041	1.121	1.201	1.281	1.361	1.441	1.521	1.601	1.681	1.761
300.	0.166	0.249	0.331	0.414	0.497	0.580	0.663	0.746	0.829	0.912	0.994	1.077	1.160	1.243	1.326	1.409	1.492	1.575	1.657	1.740	1.823
310.	0.171	0.257	0.343	0.428	0.514	0.600	0.685	0.771	0.857	0.943	1.028	1.114	1.200	1.285	1.371	1.457	1.542	1.628	1.714	1.799	1.885
320.	0.177	0.265	0.354	0.442	0.531	0.619	0.708	0.796	0.885	0.973	1.062	1.150	1.239	1.327	1.416	1.504	1.593	1.681	1.770	1.858	1.947
330.	0.183	0.274	0.365	0.457	0.548	0.639	0.730	0.822	0.913	1.004	1.096	1.187	1.278	1.370	1.461	1.552	1.644	1.735	1.826	1.918	2.009
340.	0.188	0.282	0.377	0.471	0.565	0.659	0.753	0.847	0.941	1.035	1.130	1.224	1.318	1.412	1.506	1.600	1.694	1.788	1.883	1.977	2.071
350.	0.194	0.291	0.388	0.485	0.582	0.679	0.776	0.873	0.970	1.066	1.163	1.260	1.357	1.454	1.551	1.648	1.745	1.842	1.939	2.036	2.133
360.	0.200	0.299	0.399	0.499	0.599	0.698	0.798	0.898	0.998	1.098	1.197	1.297	1.397	1.497	1.596	1.696	1.796	1.896	1.996	2.095	2.195
370.	0.205	0.308	0.410	0.513	0.616	0.718	0.821	0.923	1.026	1.129	1.231	1.334	1.436	1.539	1.642	1.744	1.847	1.950	2.052	2.155	2.257

TABLE 2.4d Shell Thickness Tables in Inches: Joint Efficiency = 0.80

SHELL THICKNESS TABLES IN INCHES

JOINT EFF.=0.70 ALLOWABLE STRESS VALUE= 13800. PSI

	INSIDE DIAMETER,IN.																				
PSI	12.	18.	24.	30.	36.	42.	48.	54.	60.	66.	72.	78.	84.	90.	96.	102.	108.	114.	120.	126.	132.
15.	0.009	0.014	0.019	0.023	0.028	0.033	0.037	0.042	0.047	0.051	0.056	0.061	0.065	0.070	0.075	0.079	0.084	0.089	0.093	0.098	0.103
20.	0.012	0.019	0.025	0.031	0.037	0.044	0.050	0.056	0.062	0.068	0.075	0.081	0.087	0.093	0.100	0.106	0.112	0.118	0.124	0.131	0.137
25.	0.016	0.023	0.031	0.039	0.047	0.054	0.062	0.070	0.078	0.086	0.093	0.101	0.109	0.117	0.124	0.132	0.140	0.148	0.156	0.163	0.171
30.	0.019	0.028	0.037	0.047	0.056	0.065	0.075	0.084	0.093	0.103	0.112	0.121	0.131	0.140	0.149	0.159	0.168	0.177	0.187	0.196	0.205
35.	0.022	0.033	0.044	0.054	0.065	0.076	0.087	0.098	0.109	0.120	0.131	0.142	0.153	0.163	0.174	0.185	0.196	0.207	0.218	0.229	0.240
40.	0.025	0.037	0.050	0.062	0.075	0.087	0.100	0.112	0.125	0.137	0.149	0.162	0.174	0.187	0.199	0.212	0.224	0.237	0.249	0.262	0.274
45.	0.028	0.042	0.056	0.070	0.084	0.098	0.112	0.126	0.140	0.154	0.168	0.182	0.196	0.210	0.224	0.238	0.252	0.266	0.280	0.294	0.308
50.	0.031	0.047	0.062	0.078	0.093	0.109	0.125	0.140	0.156	0.171	0.187	0.202	0.218	0.234	0.249	0.265	0.280	0.296	0.312	0.327	0.343
55.	0.034	0.051	0.069	0.086	0.103	0.120	0.137	0.154	0.171	0.189	0.206	0.223	0.240	0.257	0.274	0.291	0.309	0.326	0.343	0.360	0.377
60.	0.037	0.056	0.075	0.094	0.112	0.131	0.150	0.168	0.187	0.206	0.224	0.243	0.262	0.281	0.299	0.318	0.337	0.355	0.374	0.393	0.411
65.	0.041	0.061	0.081	0.101	0.122	0.142	0.162	0.182	0.203	0.223	0.243	0.263	0.284	0.304	0.324	0.345	0.365	0.385	0.405	0.426	0.446
70.	0.044	0.066	0.087	0.109	0.131	0.153	0.175	0.197	0.218	0.240	0.262	0.284	0.306	0.328	0.349	0.371	0.393	0.415	0.437	0.459	0.480
75.	0.047	0.070	0.094	0.117	0.140	0.164	0.187	0.211	0.234	0.257	0.281	0.304	0.328	0.351	0.374	0.398	0.421	0.445	0.468	0.491	0.515
80.	0.050	0.075	0.100	0.125	0.150	0.175	0.200	0.225	0.250	0.275	0.300	0.325	0.350	0.375	0.400	0.424	0.449	0.474	0.499	0.524	0.549
85.	0.053	0.080	0.106	0.133	0.159	0.186	0.212	0.239	0.265	0.292	0.318	0.345	0.372	0.398	0.425	0.451	0.478	0.504	0.531	0.557	0.584
90.	0.056	0.084	0.112	0.141	0.169	0.197	0.225	0.253	0.281	0.309	0.337	0.365	0.394	0.422	0.450	0.478	0.506	0.534	0.562	0.590	0.618
95.	0.059	0.089	0.119	0.148	0.178	0.208	0.237	0.267	0.297	0.326	0.356	0.386	0.415	0.445	0.475	0.505	0.534	0.564	0.594	0.623	0.653
100.	0.063	0.094	0.125	0.156	0.188	0.219	0.250	0.281	0.313	0.344	0.375	0.406	0.438	0.469	0.500	0.531	0.563	0.594	0.625	0.656	0.688
105.	0.066	0.098	0.131	0.164	0.197	0.230	0.263	0.295	0.328	0.361	0.394	0.427	0.460	0.492	0.525	0.558	0.591	0.624	0.656	0.689	0.722
110.	0.069	0.103	0.138	0.172	0.206	0.241	0.275	0.310	0.344	0.378	0.413	0.447	0.482	0.516	0.550	0.585	0.619	0.654	0.688	0.722	0.757
115.	0.072	0.108	0.144	0.180	0.216	0.252	0.288	0.324	0.360	0.396	0.432	0.468	0.504	0.540	0.576	0.612	0.647	0.683	0.719	0.755	0.791
120.	0.075	0.113	0.150	0.188	0.225	0.263	0.300	0.338	0.375	0.413	0.451	0.488	0.526	0.563	0.601	0.638	0.676	0.713	0.751	0.788	0.826
125.	0.078	0.117	0.156	0.196	0.235	0.274	0.313	0.352	0.391	0.430	0.469	0.509	0.548	0.587	0.626	0.665	0.704	0.743	0.782	0.822	0.861
130.	0.081	0.122	0.163	0.204	0.244	0.285	0.326	0.366	0.407	0.448	0.488	0.529	0.570	0.611	0.651	0.692	0.733	0.773	0.814	0.855	0.895
135.	0.085	0.127	0.169	0.211	0.254	0.296	0.338	0.381	0.423	0.465	0.507	0.550	0.592	0.634	0.676	0.719	0.761	0.803	0.846	0.888	0.930
140.	0.088	0.132	0.175	0.219	0.263	0.307	0.351	0.395	0.439	0.482	0.526	0.570	0.614	0.658	0.702	0.746	0.789	0.833	0.877	0.921	0.965
145.	0.091	0.136	0.182	0.227	0.273	0.318	0.364	0.409	0.454	0.500	0.545	0.591	0.636	0.682	0.727	0.772	0.818	0.863	0.909	0.954	1.000
150.	0.094	0.141	0.188	0.235	0.282	0.329	0.376	0.423	0.470	0.517	0.564	0.611	0.658	0.705	0.752	0.799	0.846	0.893	0.940	0.987	1.034
160.	0.100	0.151	0.201	0.251	0.301	0.351	0.402	0.452	0.502	0.552	0.602	0.652	0.703	0.753	0.803	0.853	0.903	0.954	1.004	1.054	1.104
170.	0.107	0.160	0.213	0.267	0.320	0.374	0.427	0.480	0.534	0.587	0.640	0.694	0.747	0.800	0.854	0.907	0.960	1.014	1.067	1.121	1.174
180.	0.113	0.170	0.226	0.283	0.339	0.396	0.452	0.509	0.565	0.622	0.678	0.735	0.791	0.848	0.905	0.961	1.018	1.074	1.131	1.187	1.244
190.	0.119	0.179	0.239	0.299	0.358	0.418	0.478	0.537	0.597	0.657	0.717	0.776	0.836	0.896	0.955	1.015	1.075	1.135	1.194	1.254	1.314
200.	0.126	0.189	0.252	0.314	0.377	0.440	0.503	0.566	0.629	0.692	0.755	0.818	0.881	0.943	1.006	1.069	1.132	1.195	1.258	1.321	1.384
210.	0.132	0.198	0.264	0.330	0.396	0.463	0.529	0.595	0.661	0.727	0.793	0.859	0.925	0.991	1.057	1.123	1.189	1.256	1.322	1.388	1.454
220.	0.139	0.208	0.277	0.346	0.416	0.485	0.554	0.623	0.693	0.762	0.831	0.901	0.970	1.039	1.108	1.178	1.247	1.316	1.385	1.455	1.524
230.	0.145	0.217	0.290	0.362	0.435	0.507	0.580	0.652	0.725	0.797	0.870	0.942	1.014	1.087	1.159	1.232	1.304	1.377	1.449	1.522	1.594
240.	0.151	0.227	0.303	0.378	0.454	0.530	0.605	0.681	0.757	0.832	0.908	0.984	1.059	1.135	1.211	1.286	1.362	1.438	1.513	1.589	1.665
250.	0.158	0.237	0.315	0.394	0.473	0.552	0.631	0.710	0.789	0.868	0.946	1.025	1.104	1.183	1.262	1.341	1.420	1.498	1.577	1.656	1.735
260.	0.164	0.246	0.328	0.410	0.492	0.574	0.657	0.739	0.821	0.903	0.985	1.067	1.149	1.231	1.313	1.395	1.477	1.559	1.641	1.723	1.806
270.	0.171	0.256	0.341	0.426	0.512	0.597	0.682	0.768	0.853	0.938	1.023	1.109	1.194	1.279	1.364	1.450	1.535	1.620	1.706	1.791	1.876
280.	0.177	0.265	0.354	0.442	0.531	0.619	0.708	0.796	0.885	0.973	1.062	1.150	1.239	1.327	1.416	1.504	1.593	1.681	1.770	1.858	1.947
290.	0.183	0.275	0.367	0.459	0.550	0.642	0.734	0.825	0.917	1.009	1.101	1.192	1.284	1.376	1.467	1.559	1.651	1.743	1.834	1.926	2.018
300.	0.190	0.285	0.380	0.475	0.570	0.665	0.759	0.854	0.949	1.044	1.139	1.234	1.329	1.424	1.519	1.614	1.709	1.804	1.899	1.994	2.089
310.	0.196	0.294	0.393	0.491	0.589	0.687	0.785	0.883	0.982	1.080	1.178	1.276	1.374	1.472	1.571	1.669	1.767	1.865	1.963	2.061	2.160
320.	0.203	0.304	0.406	0.507	0.608	0.710	0.811	0.913	1.014	1.115	1.217	1.318	1.420	1.521	1.622	1.724	1.825	1.926	2.028	2.129	2.231
330.	0.209	0.314	0.419	0.523	0.628	0.732	0.837	0.942	1.046	1.151	1.256	1.360	1.465	1.569	1.674	1.779	1.883	1.988	2.093	2.197	2.302
340.	0.216	0.324	0.431	0.539	0.647	0.755	0.863	0.971	1.079	1.187	1.294	1.402	1.510	1.618	1.726	1.834	1.942	2.049	2.157	2.265	2.373
350.	0.222	0.333	0.444	0.556	0.667	0.778	0.889	1.000	1.111	1.222	1.333	1.444	1.556	1.667	1.778	1.889	2.000	2.111	2.222	2.333	2.444
360.	0.229	0.343	0.457	0.572	0.686	0.801	0.915	1.029	1.144	1.258	1.372	1.487	1.601	1.715	1.830	1.944	2.058	2.173	2.287	2.402	2.516
370.	0.235	0.353	0.470	0.588	0.706	0.823	0.941	1.058	1.176	1.294	1.411	1.529	1.647	1.764	1.882	1.999	2.117	2.235	2.352	2.470	2.587

TABLE 2.4e Shell Thickness Tables in Inches: Joint Efficiency = 0.70

SHELL THICKNESS TABLES IN INCHES

JOINT EFF.=0.65 ALLOWABLE STRESS VALUE= 13800. PSI

PSI	INSIDE DIAMETER, IN. 12.	18.	24.	30.	36.	42.	48.	54.	60.	66.	72.	78.	84.	90.	96.	102.	108.	114.	120.	126.	132.
15.	0.010	0.015	0.020	0.025	0.030	0.035	0.040	0.045	0.050	0.055	0.060	0.065	0.070	0.075	0.080	0.085	0.090	0.095	0.100	0.105	0.110
20.	0.013	0.020	0.027	0.033	0.040	0.047	0.054	0.060	0.067	0.074	0.080	0.087	0.094	0.100	0.107	0.114	0.121	0.127	0.134	0.141	0.147
25.	0.017	0.025	0.034	0.042	0.050	0.059	0.067	0.075	0.084	0.092	0.101	0.109	0.117	0.126	0.134	0.142	0.151	0.159	0.168	0.176	0.184
30.	0.020	0.030	0.040	0.050	0.060	0.070	0.080	0.090	0.101	0.111	0.121	0.131	0.141	0.151	0.161	0.171	0.181	0.191	0.201	0.211	0.221
35.	0.023	0.035	0.047	0.059	0.070	0.082	0.094	0.106	0.117	0.129	0.141	0.153	0.164	0.176	0.188	0.199	0.211	0.223	0.235	0.246	0.258
40.	0.027	0.040	0.054	0.067	0.080	0.094	0.107	0.121	0.134	0.148	0.161	0.174	0.188	0.201	0.215	0.228	0.241	0.255	0.268	0.282	0.295
45.	0.030	0.045	0.060	0.075	0.091	0.106	0.121	0.136	0.151	0.166	0.181	0.196	0.211	0.226	0.242	0.257	0.272	0.287	0.302	0.317	0.332
50.	0.034	0.050	0.067	0.084	0.101	0.117	0.134	0.151	0.168	0.185	0.201	0.218	0.235	0.252	0.268	0.285	0.302	0.319	0.336	0.352	0.369
55.	0.037	0.055	0.074	0.092	0.111	0.129	0.148	0.166	0.185	0.203	0.222	0.240	0.258	0.277	0.295	0.314	0.332	0.351	0.369	0.388	0.406
60.	0.040	0.060	0.081	0.101	0.121	0.141	0.161	0.181	0.201	0.222	0.242	0.262	0.282	0.302	0.322	0.343	0.363	0.383	0.403	0.423	0.443
65.	0.044	0.066	0.087	0.109	0.131	0.153	0.175	0.197	0.218	0.240	0.262	0.284	0.306	0.328	0.349	0.371	0.393	0.415	0.437	0.459	0.480
70.	0.047	0.071	0.094	0.118	0.141	0.165	0.188	0.212	0.235	0.259	0.282	0.306	0.329	0.353	0.376	0.400	0.423	0.447	0.470	0.494	0.517
75.	0.050	0.076	0.101	0.126	0.151	0.176	0.202	0.227	0.252	0.277	0.303	0.328	0.353	0.378	0.403	0.429	0.454	0.479	0.504	0.529	0.555
80.	0.054	0.081	0.108	0.134	0.161	0.188	0.215	0.242	0.269	0.296	0.323	0.350	0.377	0.403	0.430	0.457	0.484	0.511	0.538	0.565	0.592
85.	0.057	0.086	0.114	0.143	0.172	0.200	0.229	0.257	0.286	0.314	0.343	0.372	0.400	0.429	0.457	0.486	0.515	0.543	0.572	0.600	0.629
90.	0.061	0.091	0.121	0.151	0.182	0.212	0.242	0.273	0.303	0.333	0.363	0.394	0.424	0.454	0.485	0.515	0.545	0.575	0.606	0.636	0.666
95.	0.064	0.096	0.128	0.160	0.192	0.224	0.256	0.288	0.320	0.352	0.384	0.416	0.448	0.480	0.512	0.544	0.576	0.608	0.640	0.671	0.703
100.	0.067	0.101	0.135	0.168	0.202	0.236	0.269	0.303	0.337	0.370	0.404	0.438	0.471	0.505	0.539	0.572	0.606	0.640	0.673	0.707	0.741
105.	0.071	0.106	0.141	0.177	0.212	0.248	0.283	0.318	0.354	0.389	0.424	0.460	0.495	0.530	0.566	0.601	0.637	0.672	0.707	0.743	0.778
110.	0.074	0.111	0.148	0.185	0.222	0.259	0.296	0.334	0.371	0.408	0.445	0.482	0.519	0.556	0.593	0.630	0.667	0.704	0.741	0.778	0.815
115.	0.078	0.116	0.155	0.194	0.233	0.271	0.310	0.349	0.388	0.426	0.465	0.504	0.543	0.581	0.620	0.659	0.698	0.736	0.775	0.814	0.853
120.	0.081	0.121	0.162	0.202	0.243	0.283	0.324	0.364	0.405	0.445	0.486	0.526	0.566	0.607	0.647	0.688	0.728	0.769	0.809	0.850	0.890
125.	0.084	0.126	0.169	0.211	0.253	0.295	0.337	0.379	0.422	0.464	0.506	0.548	0.590	0.632	0.675	0.717	0.759	0.801	0.843	0.885	0.927
130.	0.088	0.132	0.175	0.219	0.263	0.307	0.351	0.395	0.439	0.482	0.526	0.570	0.614	0.658	0.702	0.746	0.789	0.833	0.877	0.921	0.965
135.	0.091	0.137	0.182	0.228	0.273	0.319	0.364	0.410	0.456	0.501	0.547	0.592	0.638	0.683	0.729	0.775	0.820	0.866	0.911	0.957	1.002
140.	0.095	0.142	0.189	0.236	0.284	0.331	0.378	0.425	0.473	0.520	0.567	0.614	0.662	0.709	0.756	0.804	0.851	0.898	0.945	0.993	1.040
145.	0.098	0.147	0.196	0.245	0.294	0.343	0.392	0.441	0.490	0.539	0.588	0.637	0.686	0.735	0.784	0.832	0.881	0.930	0.979	1.028	1.077
150.	0.101	0.152	0.203	0.253	0.304	0.355	0.405	0.456	0.507	0.557	0.608	0.659	0.709	0.760	0.811	0.861	0.912	0.963	1.014	1.064	1.115
160.	0.108	0.162	0.216	0.270	0.325	0.379	0.433	0.487	0.541	0.595	0.649	0.703	0.757	0.811	0.865	0.920	0.974	1.028	1.082	1.136	1.190
170.	0.115	0.173	0.230	0.288	0.345	0.403	0.460	0.518	0.575	0.633	0.690	0.748	0.805	0.863	0.920	0.978	1.035	1.093	1.150	1.208	1.265
180.	0.122	0.183	0.244	0.305	0.366	0.427	0.487	0.548	0.609	0.670	0.731	0.792	0.853	0.914	0.975	1.036	1.097	1.158	1.219	1.280	1.341
190.	0.129	0.193	0.257	0.322	0.386	0.451	0.515	0.579	0.644	0.708	0.772	0.837	0.901	0.965	1.030	1.094	1.159	1.223	1.287	1.352	1.416
200.	0.136	0.203	0.271	0.339	0.407	0.475	0.542	0.610	0.678	0.746	0.814	0.881	0.949	1.017	1.085	1.153	1.220	1.288	1.356	1.424	1.492
210.	0.142	0.214	0.285	0.356	0.427	0.499	0.570	0.641	0.712	0.784	0.855	0.926	0.997	1.069	1.140	1.211	1.282	1.353	1.425	1.496	1.567
220.	0.149	0.224	0.299	0.373	0.448	0.523	0.597	0.672	0.747	0.821	0.896	0.971	1.045	1.120	1.195	1.270	1.344	1.419	1.494	1.568	1.643
230.	0.156	0.234	0.313	0.391	0.469	0.547	0.625	0.703	0.781	0.859	0.938	1.016	1.094	1.172	1.250	1.328	1.406	1.484	1.563	1.641	1.719
240.	0.163	0.245	0.326	0.408	0.489	0.571	0.653	0.734	0.816	0.897	0.979	1.061	1.142	1.224	1.305	1.387	1.468	1.550	1.632	1.713	1.795
250.	0.170	0.255	0.340	0.425	0.510	0.595	0.680	0.765	0.850	0.935	1.020	1.105	1.190	1.276	1.361	1.446	1.531	1.616	1.701	1.786	1.871
260.	0.177	0.265	0.354	0.442	0.531	0.619	0.708	0.796	0.885	0.973	1.062	1.150	1.239	1.327	1.416	1.504	1.593	1.681	1.770	1.858	1.947
270.	0.184	0.276	0.368	0.460	0.552	0.644	0.736	0.828	0.920	1.012	1.104	1.196	1.287	1.379	1.471	1.563	1.655	1.747	1.839	1.931	2.023
280.	0.191	0.286	0.382	0.477	0.573	0.668	0.763	0.859	0.954	1.050	1.145	1.241	1.336	1.431	1.527	1.622	1.718	1.813	1.909	2.004	2.100
290.	0.198	0.297	0.396	0.495	0.593	0.692	0.791	0.890	0.989	1.088	1.187	1.286	1.385	1.484	1.583	1.681	1.780	1.879	1.978	2.077	2.176
300.	0.205	0.307	0.410	0.512	0.614	0.717	0.819	0.922	1.024	1.126	1.229	1.331	1.433	1.536	1.638	1.741	1.843	1.945	2.048	2.150	2.253
310.	0.212	0.318	0.423	0.529	0.635	0.741	0.847	0.953	1.059	1.165	1.270	1.376	1.482	1.588	1.694	1.800	1.906	2.012	2.117	2.223	2.329
320.	0.219	0.328	0.437	0.547	0.656	0.766	0.875	0.984	1.094	1.203	1.312	1.422	1.531	1.640	1.750	1.859	1.969	2.078	2.187	2.297	2.406
330.	0.226	0.339	0.451	0.564	0.677	0.790	0.903	1.016	1.129	1.241	1.354	1.467	1.580	1.693	1.806	1.919	2.031	2.144	2.257	2.370	2.483
340.	0.233	0.349	0.465	0.582	0.698	0.815	0.931	1.047	1.164	1.280	1.396	1.513	1.629	1.745	1.862	1.978	2.094	2.211	2.327	2.444	2.560
350.	0.240	0.360	0.479	0.599	0.719	0.839	0.959	1.079	1.199	1.318	1.438	1.558	1.678	1.798	1.918	2.038	2.158	2.277	2.397	2.517	2.637
360.	0.247	0.370	0.493	0.617	0.740	0.864	0.987	1.110	1.234	1.357	1.480	1.604	1.727	1.851	1.974	2.097	2.221	2.344	2.467	2.591	2.714
370.	0.254	0.381	0.508	0.634	0.761	0.888	1.015	1.142	1.269	1.396	1.523	1.650	1.776	1.903	2.030	2.157	2.284	2.411	2.538	2.665	2.791

TABLE 2.4f Shell Thickness Tables in Inches: Joint Efficiency = 0.65

Equations	Remarks	Code References
Cylindrical shells under internal pressure		
$t = \frac{PR}{SE - 0.6P}$	Equation is used when t is less than ½R or P is less than 0.365SE	Par. UG-27
or $P = \frac{SEt}{R + 0.6t}$	Circumferential joint formula in Code Par. UG-27 applies only if the joint efficiency is less than one-half the longitudinal joint efficiency or if the loadings described in Code Par. UG-22 cause bending or tension	Par. UG-27, UG-22
in which t = minimum thickness, in	Thick cylindrical shells	Appendix 1-2
P = allowable pressure, psi; S = allowable stress, psi	Formulas in terms of the outside radius may be used in place of those in Par. UG-27	Appendix 1-1 to 1-3
	Out-of-roundness measure	Par. UG-80
E = joint efficiency, percent	Welded joint efficiencies	Par. UW-12 Table UW-12
R = inside radius, in	Stress values of materials (use appropriate table)	Code Section II, Part D, Tables UC-23, UCD-23, ULT-23
	Corrosion allowance for carbon-steel vessels	Par. UG-25
Spherical shells under internal pressure		
$t = \frac{PR}{2SE - 0.2P}$	Equation used when t is less than 0.356R or P is less than 0.665SE (unknowns same as above)	Par. UG-27
or $P = \frac{2SEt}{R + 0.2t}$	Thick spherical shells	Appendix 1-2
	Stress values of materials	Code Section II, Part D, Tables UC-23, UCD-23, ULT-23

Table 2.5 Calculations for Cylindrical and Spherical Shells under Internal Pressure

Use of Simplified External-Pressure Cylindrical Shell Thickness Charts

The charts in Figs. 2.6 and 2.7 give the shell thickness required for carbon steel and type 304 stainless steel vessels under external pressure at a glance, thereby eliminating the many steps required by Code Par. UG-28; Code Section II, Part D, Subpart 3 for carbon or low-alloy steel with a minimum yield strength of 30,000 psi and above and for

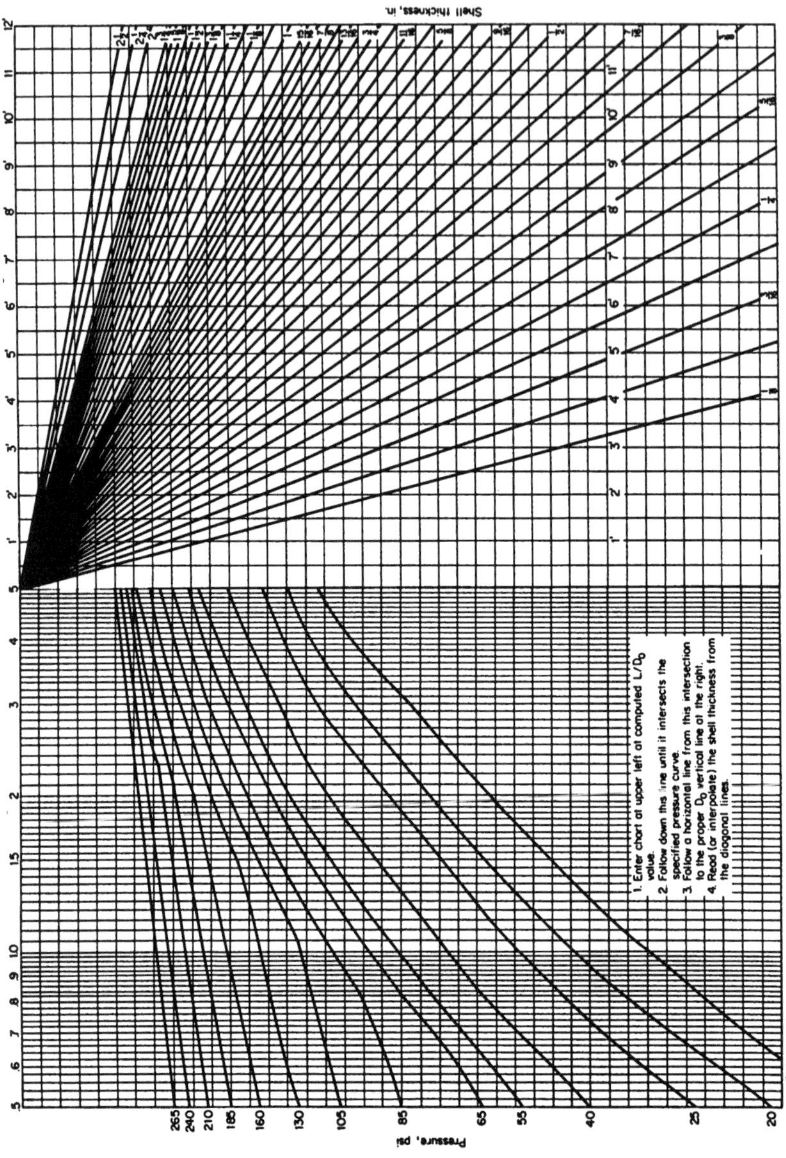

FIGURE 2.6 Simplified external-pressure shell thickness chart: Carbon steel yield strength is 30,000 to 38,000 psi; temperature is 300°F.

type 304, 405, and 410 stainless steels. The temperature limits of Code Par. UG-20(c) shall be taken into consideration. Table 2.6 sets forth the required calculations for cylindrical and spherical shells under external pressure.

To obtain the required thickness from Figs. 2.6 and 2.7: (1) place a rule horizontally at the intersection of the applicable L/D_0 and

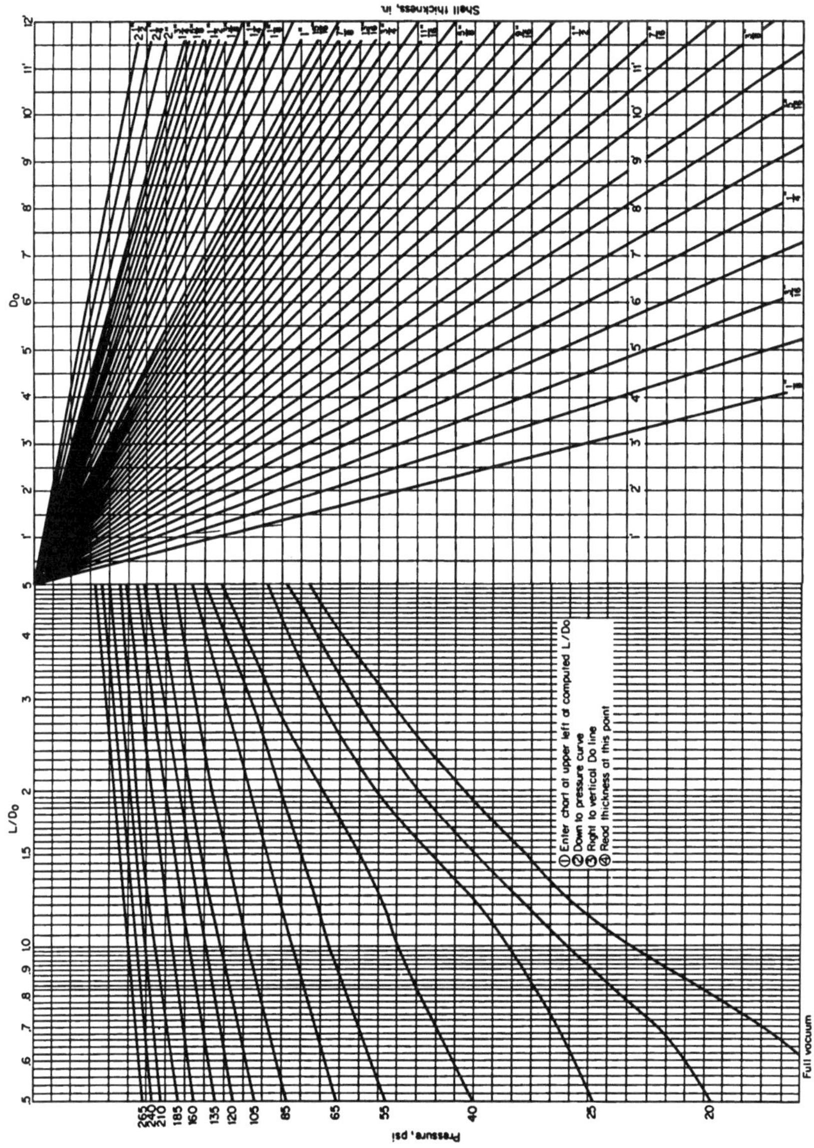

FIGURE 2.7 Simplified external-pressure shell thickness chart: Stainless steel (18 Cr–8 Ni) type 304; temperature, 400°F.

pressure lines on the left-hand side of the chart and (2) read the thickness value where the rule crosses the vertical D_0 line on the right-hand side of the chart. Other common problems may also be easily solved. The allowable external pressure of an existing (or proposed) vessel may be determined by: (1) entering the chart on the right-hand

Equations	Remarks	Code references
Cylindrical and spherical shells under external pressure		
Vessels intended for service under external working pressure of 15 psi or less		Par. UG-28(f)
Use equations in Code Pars. UG-28(c) and (d) and thickness charts for the applicable material (figure numbers of charts lists under Code references)	Carbon-steel shells	Pars. UG-28, UCS-28 Figs. UG-28, UG-29.1, UG-29.2 Code Section II, Part D, subpart 3
For stiffening rings, use equations in Code Pars. UG-29, UG-30	Nonferrous metal shells	Pars. UG-28, UNF-28 Code Section II, Part D, subpart 3
	High-alloy steel shells	Pars. UG-28, UHA-28 Code Section II, Part D, subpart 3
	Cast-iron shells	Pars. UG-28, UCI-28
	Clad-steel shells	Pars. UG-28, UCL-28
	Corrugated shells	Pars. UCS-28(b)
Stiffening rings for cylindrical shells under external pressure		
		Pars. UG-28, UG-29 Figs. UG-29.1, UG-29.2
	Avoid stress	Appendix G
	Stiffening rings may be placed outside or inside a vessel	Pars. UG-30 Figs. UG-29.1, UG-30 UG-29.2
	External material charts and tabular values	Code Section II, Part D, Subpart 2

*Rules in Par. PFT-19 of Power Boiler Code may also be used

Table 2.6 Calculations for Cylindrical and Spherical Shells under External Pressure

side at the proper thickness and diameter and (2) moving horizontally to the left and reading the pressure at the vertical L/D_0 line. A similar procedure may be used to find the maximum allowable effective length L for the spacing of stiffening rings at a specified pressure (see Figs. 2.6 and 2.7).

Equations	Remarks	Code references
Stiffening rings for cylindrical shells under external pressure (*contd*)		
	Stiffening rings may be attached to shell by continuous or intermittent welding	Par. UG-30
	Total length of intermittent welding on each side of stiffening ring shall be:	
	(a) Rings on outside not less than one-half outside circumference of the vessel	
	(b) Rings on inside not less than one-third of the circumference of the vessel	Fig. UG-30
	Maximum spacing of intermittent weld 8x thickness of shell for external rings 12x thickness of shell for internal rings	
	Drainage—Opening at bottom for drainage	Fig. UG-29.1
	Vent—Opening at top for vent	Fig. UG-29.1
	Maximum are of shell left unsupported	Fig. UG-29.1, UG-29.2

Table 2.6 (*Contd*)

Use of Internal-Pressure Thickness Tables for Ellipsoidal and Torispherical Heads

The head thickness tables may be applied to ellipsoidal 2:1 heads or torispherical heads (see Tables 2.7*a* to 2.7*l*). These may also be applied to seamless heads (E = 1.0) and to double-welded butt heads that have been fully radiographed (E = 1.0), spot radiographed (E = 0.85), or not radiographed (E = 0.70). Additionally, these may be applied to single-welded butt joints with a backing strip that are fully radiographed (E = 0.90), spot radiographed (E = 0.80), or not radiographed (E = 0.65). The required head thickness may be determined by the following steps:

1. Determine the appropriate table corresponding to the desired joint efficiency. Note that the value of joint efficiency for each table is printed in the upper left-hand corner.

TORISPHERICAL HEAD THICKNESS TABLES IN INCHES

JOINT EFF.=1.00 ALLOWABLE STRESS VALUE =13800. PSI

	INSIDE CROWN RADIUS,INCHES																				
PSI	12.	18.	24.	30.	36.	42.	48.	54.	60.	66.	72.	78.	84.	90.	96.	102.	108.	114.	120.	126.	132.
15.	0.012	0.017	0.023	0.029	0.035	0.040	0.046	0.052	0.058	0.063	0.069	0.075	0.081	0.087	0.092	0.098	0.104	0.110	0.115	0.121	0.127
20.	0.015	0.023	0.031	0.038	0.046	0.054	0.062	0.069	0.077	0.085	0.092	0.100	0.108	0.115	0.123	0.131	0.139	0.146	0.154	0.162	0.169
25.	0.019	0.029	0.038	0.048	0.058	0.067	0.077	0.087	0.096	0.106	0.115	0.125	0.135	0.144	0.154	0.164	0.173	0.183	0.192	0.202	0.212
30.	0.023	0.035	0.046	0.058	0.069	0.081	0.092	0.104	0.115	0.127	0.139	0.150	0.162	0.173	0.185	0.196	0.208	0.219	0.231	0.242	0.254
35.	0.027	0.040	0.054	0.067	0.081	0.094	0.108	0.121	0.135	0.148	0.162	0.175	0.189	0.202	0.216	0.229	0.242	0.256	0.269	0.283	0.296
40.	0.031	0.046	0.062	0.077	0.092	0.108	0.123	0.139	0.154	0.169	0.185	0.200	0.216	0.231	0.246	0.262	0.277	0.293	0.308	0.323	0.339
45.	0.035	0.052	0.069	0.087	0.104	0.121	0.139	0.156	0.173	0.191	0.208	0.225	0.242	0.260	0.277	0.294	0.312	0.329	0.346	0.364	0.381
50.	0.038	0.058	0.077	0.096	0.115	0.135	0.154	0.173	0.192	0.212	0.231	0.250	0.269	0.289	0.308	0.327	0.346	0.366	0.385	0.404	0.423
55.	0.042	0.064	0.085	0.106	0.127	0.148	0.169	0.191	0.212	0.233	0.254	0.275	0.296	0.318	0.339	0.360	0.381	0.402	0.423	0.445	0.466
60.	0.046	0.069	0.092	0.115	0.139	0.162	0.185	0.208	0.231	0.254	0.277	0.300	0.323	0.346	0.370	0.393	0.416	0.439	0.462	0.485	0.508
65.	0.050	0.075	0.100	0.125	0.150	0.175	0.200	0.225	0.250	0.275	0.300	0.325	0.350	0.375	0.400	0.425	0.450	0.475	0.500	0.525	0.550
70.	0.054	0.081	0.108	0.135	0.162	0.189	0.216	0.243	0.269	0.296	0.323	0.350	0.377	0.404	0.431	0.458	0.485	0.512	0.539	0.566	0.593
75.	0.058	0.087	0.115	0.144	0.173	0.202	0.231	0.260	0.289	0.318	0.346	0.375	0.404	0.433	0.462	0.491	0.520	0.549	0.577	0.606	0.635
80.	0.062	0.092	0.123	0.154	0.185	0.216	0.246	0.277	0.308	0.339	0.370	0.400	0.431	0.462	0.493	0.524	0.554	0.585	0.616	0.647	0.678
85.	0.065	0.098	0.131	0.164	0.196	0.229	0.262	0.295	0.327	0.360	0.393	0.425	0.458	0.491	0.524	0.556	0.589	0.622	0.655	0.687	0.720
90.	0.069	0.104	0.139	0.173	0.208	0.243	0.277	0.312	0.347	0.381	0.416	0.450	0.485	0.520	0.554	0.589	0.624	0.658	0.693	0.728	0.762
95.	0.073	0.110	0.146	0.183	0.219	0.256	0.293	0.329	0.366	0.402	0.439	0.476	0.512	0.549	0.585	0.622	0.658	0.695	0.732	0.768	0.805
100.	0.077	0.116	0.154	0.193	0.231	0.270	0.308	0.347	0.385	0.424	0.462	0.501	0.539	0.578	0.616	0.655	0.693	0.732	0.770	0.809	0.847
105.	0.081	0.121	0.162	0.202	0.243	0.283	0.323	0.364	0.404	0.445	0.485	0.526	0.566	0.606	0.647	0.687	0.728	0.768	0.809	0.849	0.890
110.	0.085	0.127	0.169	0.212	0.254	0.297	0.339	0.381	0.424	0.466	0.508	0.551	0.593	0.635	0.678	0.720	0.762	0.805	0.847	0.890	0.932
115.	0.089	0.133	0.177	0.221	0.266	0.310	0.354	0.399	0.443	0.487	0.531	0.576	0.620	0.664	0.709	0.753	0.797	0.841	0.886	0.930	0.974
120.	0.092	0.139	0.185	0.231	0.277	0.323	0.370	0.416	0.462	0.508	0.555	0.601	0.647	0.693	0.739	0.786	0.832	0.878	0.924	0.970	1.017
125.	0.096	0.144	0.193	0.241	0.289	0.337	0.385	0.433	0.481	0.530	0.578	0.626	0.674	0.722	0.770	0.818	0.867	0.915	0.963	1.011	1.059
130.	0.100	0.150	0.200	0.250	0.300	0.350	0.401	0.451	0.501	0.551	0.601	0.651	0.701	0.751	0.801	0.851	0.901	0.951	1.001	1.051	1.102
135.	0.104	0.156	0.208	0.260	0.312	0.364	0.416	0.468	0.520	0.572	0.624	0.676	0.728	0.780	0.832	0.884	0.936	0.988	1.040	1.092	1.144
140.	0.108	0.162	0.216	0.270	0.324	0.377	0.431	0.485	0.539	0.593	0.647	0.701	0.755	0.809	0.863	0.917	0.971	1.025	1.078	1.132	1.186
145.	0.112	0.168	0.223	0.279	0.335	0.391	0.447	0.503	0.559	0.614	0.670	0.726	0.782	0.838	0.894	0.949	1.005	1.061	1.117	1.173	1.229
150.	0.116	0.173	0.231	0.289	0.347	0.404	0.462	0.520	0.578	0.636	0.693	0.751	0.809	0.867	0.924	0.982	1.040	1.098	1.156	1.213	1.271
160.	0.123	0.185	0.247	0.308	0.370	0.431	0.493	0.555	0.616	0.678	0.740	0.801	0.863	0.925	0.986	1.048	1.109	1.171	1.233	1.294	1.356
170.	0.131	0.196	0.262	0.327	0.393	0.458	0.524	0.589	0.655	0.720	0.786	0.851	0.917	0.982	1.048	1.113	1.179	1.244	1.310	1.375	1.441
180.	0.139	0.208	0.277	0.347	0.416	0.485	0.555	0.624	0.694	0.763	0.832	0.902	0.971	1.040	1.110	1.179	1.248	1.318	1.387	1.456	1.526
190.	0.146	0.220	0.293	0.366	0.439	0.512	0.586	0.659	0.732	0.805	0.879	0.952	1.025	1.098	1.171	1.245	1.318	1.391	1.464	1.537	1.611
200.	0.154	0.231	0.308	0.385	0.462	0.539	0.617	0.694	0.771	0.848	0.925	1.002	1.079	1.156	1.233	1.310	1.387	1.464	1.541	1.618	1.696
210.	0.162	0.243	0.324	0.405	0.486	0.566	0.647	0.728	0.809	0.890	0.971	1.052	1.133	1.214	1.295	1.376	1.457	1.538	1.619	1.699	1.780
220.	0.170	0.254	0.339	0.424	0.509	0.594	0.678	0.763	0.848	0.933	1.017	1.102	1.187	1.272	1.357	1.441	1.526	1.611	1.696	1.781	1.865
230.	0.177	0.266	0.355	0.443	0.532	0.621	0.709	0.798	0.886	0.975	1.064	1.152	1.241	1.330	1.418	1.507	1.596	1.684	1.773	1.862	1.950
240.	0.185	0.278	0.370	0.463	0.555	0.648	0.740	0.833	0.925	1.018	1.110	1.203	1.295	1.388	1.480	1.573	1.665	1.758	1.850	1.943	2.035
250.	0.193	0.289	0.385	0.482	0.578	0.675	0.771	0.867	0.964	1.060	1.156	1.253	1.349	1.446	1.542	1.638	1.735	1.831	1.927	2.024	2.120
260.	0.200	0.301	0.401	0.501	0.601	0.702	0.802	0.902	1.002	1.103	1.203	1.303	1.403	1.503	1.604	1.704	1.804	1.904	2.005	2.105	2.205
270.	0.208	0.312	0.416	0.520	0.625	0.729	0.833	0.937	1.041	1.145	1.249	1.353	1.457	1.561	1.666	1.770	1.874	1.978	2.082	2.186	2.290
280.	0.216	0.324	0.432	0.540	0.648	0.756	0.864	0.972	1.080	1.188	1.295	1.403	1.511	1.619	1.727	1.835	1.943	2.051	2.159	2.267	2.375
290.	0.224	0.335	0.447	0.559	0.671	0.783	0.895	1.006	1.118	1.230	1.342	1.454	1.566	1.677	1.789	1.901	2.013	2.125	2.236	2.348	2.460
300.	0.231	0.347	0.463	0.578	0.694	0.810	0.925	1.041	1.157	1.273	1.388	1.504	1.620	1.735	1.851	1.967	2.082	2.198	2.314	2.429	2.545
310.	0.239	0.359	0.478	0.598	0.717	0.837	0.956	1.076	1.196	1.315	1.435	1.554	1.674	1.793	1.913	2.032	2.152	2.271	2.391	2.511	2.630
320.	0.247	0.370	0.494	0.617	0.740	0.864	0.987	1.111	1.234	1.358	1.481	1.604	1.728	1.851	1.975	2.098	2.221	2.345	2.468	2.592	2.715
330.	0.255	0.382	0.509	0.636	0.764	0.891	1.018	1.146	1.273	1.400	1.527	1.655	1.782	1.909	2.037	2.164	2.291	2.418	2.546	2.673	2.800
340.	0.262	0.393	0.525	0.656	0.787	0.918	1.049	1.180	1.311	1.443	1.574	1.705	1.836	1.967	2.098	2.230	2.361	2.492	2.623	2.754	2.885
350.	0.270	0.405	0.540	0.675	0.810	0.945	1.080	1.215	1.350	1.485	1.620	1.755	1.890	2.025	2.160	2.295	2.430	2.565	2.700	2.835	2.970
360.	0.278	0.417	0.556	0.694	0.833	0.972	1.111	1.250	1.389	1.528	1.667	1.805	1.944	2.083	2.222	2.361	2.500	2.639	2.778	2.917	3.055
370.	0.286	0.428	0.571	0.714	0.857	0.999	1.142	1.285	1.428	1.570	1.713	1.856	1.999	2.141	2.284	2.427	2.570	2.712	2.855	2.998	3.141

Table 2.7a Torispherical Head Thickness Tables in Inches: Joint Efficiency = 1.0

TORISPHERICAL HEAD THICKNESS TABLES IN INCHES

JOINT EFF.=0.90 ALLOWABLE STRESS VALUE =13800. PSI

	INSIDE CROWN RADIUS,INCHES																				
PSI	12.	18.	24.	30.	36.	42.	48.	54.	60.	66.	72.	78.	84.	90.	96.	102.	108.	114.	120.	126.	132.
15.	0.013	0.019	0.026	0.032	0.038	0.045	0.051	0.058	0.064	0.071	0.077	0.083	0.090	0.096	0.103	0.109	0.115	0.122	0.128	0.135	0.141
20.	0.017	0.026	0.034	0.043	0.051	0.060	0.068	0.077	0.086	0.094	0.103	0.111	0.120	0.128	0.137	0.145	0.154	0.162	0.171	0.180	0.188
25.	0.021	0.032	0.043	0.053	0.064	0.075	0.086	0.096	0.107	0.118	0.128	0.139	0.150	0.160	0.171	0.182	0.192	0.203	0.214	0.225	0.235
30.	0.026	0.038	0.051	0.064	0.077	0.090	0.103	0.115	0.128	0.141	0.154	0.167	0.180	0.192	0.205	0.218	0.231	0.244	0.257	0.269	0.282
35.	0.030	0.045	0.060	0.075	0.090	0.105	0.120	0.135	0.150	0.165	0.180	0.195	0.210	0.225	0.239	0.254	0.269	0.284	0.299	0.314	0.329
40.	0.034	0.051	0.068	0.086	0.103	0.120	0.137	0.154	0.171	0.188	0.205	0.222	0.239	0.257	0.274	0.291	0.308	0.325	0.342	0.359	0.376
45.	0.038	0.058	0.077	0.096	0.115	0.135	0.154	0.173	0.192	0.212	0.231	0.250	0.269	0.289	0.308	0.327	0.346	0.366	0.385	0.404	0.423
50.	0.043	0.064	0.086	0.107	0.128	0.150	0.171	0.192	0.214	0.235	0.257	0.278	0.299	0.321	0.342	0.364	0.385	0.406	0.428	0.449	0.470
55.	0.047	0.071	0.094	0.118	0.141	0.165	0.188	0.212	0.235	0.259	0.282	0.306	0.329	0.353	0.376	0.400	0.423	0.447	0.470	0.494	0.518
60.	0.051	0.077	0.103	0.128	0.154	0.180	0.205	0.231	0.257	0.282	0.308	0.334	0.359	0.385	0.411	0.436	0.462	0.488	0.513	0.539	0.565
65.	0.056	0.083	0.111	0.139	0.167	0.195	0.222	0.250	0.278	0.306	0.334	0.361	0.389	0.417	0.445	0.473	0.500	0.528	0.556	0.584	0.612
70.	0.060	0.090	0.120	0.150	0.180	0.210	0.240	0.269	0.299	0.329	0.359	0.389	0.419	0.449	0.479	0.509	0.539	0.569	0.599	0.629	0.659
75.	0.064	0.096	0.128	0.160	0.193	0.225	0.257	0.289	0.321	0.353	0.385	0.417	0.449	0.481	0.513	0.545	0.578	0.610	0.642	0.674	0.706
80.	0.068	0.103	0.137	0.171	0.205	0.240	0.274	0.308	0.342	0.376	0.411	0.445	0.479	0.513	0.548	0.582	0.616	0.650	0.684	0.719	0.753
85.	0.073	0.109	0.145	0.182	0.218	0.255	0.291	0.327	0.364	0.400	0.436	0.473	0.509	0.545	0.582	0.618	0.655	0.691	0.727	0.764	0.800
90.	0.077	0.116	0.154	0.193	0.231	0.270	0.308	0.347	0.385	0.424	0.462	0.501	0.539	0.578	0.616	0.655	0.693	0.732	0.770	0.809	0.847
95.	0.081	0.122	0.163	0.203	0.244	0.285	0.325	0.366	0.406	0.447	0.488	0.528	0.569	0.610	0.650	0.691	0.732	0.772	0.813	0.854	0.894
100.	0.086	0.128	0.171	0.214	0.257	0.300	0.342	0.385	0.428	0.471	0.513	0.556	0.599	0.642	0.685	0.727	0.770	0.813	0.856	0.899	0.941
105.	0.090	0.135	0.180	0.225	0.270	0.315	0.359	0.404	0.449	0.494	0.539	0.584	0.629	0.674	0.719	0.764	0.809	0.854	0.899	0.944	0.988
110.	0.094	0.141	0.188	0.235	0.282	0.329	0.377	0.424	0.471	0.518	0.565	0.612	0.659	0.706	0.753	0.800	0.847	0.894	0.941	0.988	1.036
115.	0.098	0.148	0.197	0.246	0.295	0.344	0.394	0.443	0.492	0.541	0.591	0.640	0.689	0.738	0.787	0.837	0.886	0.935	0.984	1.033	1.083
120.	0.103	0.154	0.205	0.257	0.308	0.359	0.411	0.462	0.514	0.565	0.616	0.668	0.719	0.770	0.822	0.873	0.924	0.976	1.027	1.078	1.130
125.	0.107	0.160	0.214	0.267	0.321	0.374	0.428	0.481	0.535	0.588	0.642	0.695	0.749	0.802	0.856	0.909	0.963	1.016	1.070	1.123	1.177
130.	0.111	0.167	0.223	0.278	0.334	0.389	0.445	0.501	0.556	0.612	0.668	0.723	0.779	0.835	0.890	0.946	1.001	1.057	1.113	1.168	1.224
135.	0.116	0.173	0.231	0.289	0.347	0.404	0.462	0.520	0.578	0.636	0.693	0.751	0.809	0.867	0.924	0.982	1.040	1.098	1.156	1.213	1.271
140.	0.120	0.180	0.240	0.300	0.360	0.419	0.479	0.539	0.599	0.659	0.719	0.779	0.839	0.899	0.959	1.019	1.079	1.139	1.198	1.258	1.318
145.	0.124	0.186	0.248	0.310	0.372	0.434	0.497	0.559	0.621	0.683	0.745	0.807	0.869	0.931	0.993	1.055	1.117	1.179	1.241	1.303	1.365
150.	0.128	0.193	0.257	0.321	0.385	0.449	0.514	0.578	0.642	0.706	0.770	0.835	0.899	0.963	1.027	1.092	1.156	1.220	1.284	1.348	1.413
160.	0.137	0.205	0.274	0.342	0.411	0.479	0.548	0.616	0.685	0.753	0.822	0.890	0.959	1.027	1.096	1.164	1.233	1.301	1.370	1.438	1.507
170.	0.146	0.218	0.291	0.364	0.437	0.509	0.582	0.655	0.728	0.801	0.873	0.946	1.019	1.092	1.164	1.237	1.310	1.383	1.456	1.528	1.601
180.	0.154	0.231	0.308	0.385	0.462	0.539	0.617	0.694	0.771	0.848	0.925	1.002	1.079	1.156	1.233	1.310	1.387	1.464	1.541	1.618	1.696
190.	0.163	0.244	0.325	0.407	0.488	0.569	0.651	0.732	0.814	0.895	0.976	1.058	1.139	1.220	1.302	1.383	1.464	1.546	1.627	1.708	1.790
200.	0.171	0.257	0.343	0.428	0.514	0.600	0.685	0.771	0.856	0.942	1.028	1.113	1.199	1.285	1.370	1.456	1.542	1.627	1.713	1.799	1.884
210.	0.180	0.270	0.360	0.450	0.540	0.630	0.719	0.809	0.899	0.989	1.079	1.169	1.259	1.349	1.439	1.529	1.619	1.709	1.799	1.889	1.979
220.	0.188	0.283	0.377	0.471	0.565	0.660	0.754	0.848	0.942	1.036	1.131	1.225	1.319	1.413	1.508	1.602	1.696	1.790	1.884	1.979	2.073
230.	0.197	0.296	0.394	0.493	0.591	0.690	0.788	0.887	0.985	1.084	1.182	1.281	1.379	1.478	1.576	1.675	1.773	1.872	1.970	2.069	2.167
240.	0.206	0.308	0.411	0.514	0.617	0.720	0.822	0.925	1.028	1.131	1.234	1.336	1.439	1.542	1.645	1.748	1.851	1.953	2.056	2.159	2.262
250.	0.214	0.321	0.428	0.535	0.643	0.750	0.857	0.964	1.071	1.178	1.285	1.392	1.499	1.606	1.714	1.821	1.928	2.035	2.142	2.249	2.356
260.	0.223	0.334	0.446	0.557	0.668	0.780	0.891	1.003	1.114	1.225	1.337	1.448	1.559	1.671	1.782	1.894	2.005	2.116	2.228	2.339	2.451
270.	0.231	0.347	0.463	0.578	0.694	0.810	0.925	1.041	1.157	1.273	1.388	1.504	1.620	1.735	1.851	1.967	2.082	2.198	2.314	2.429	2.545
280.	0.240	0.360	0.480	0.600	0.720	0.840	0.960	1.080	1.200	1.320	1.440	1.560	1.680	1.800	1.920	2.040	2.160	2.280	2.400	2.520	2.640
290.	0.249	0.373	0.497	0.621	0.746	0.870	0.994	1.118	1.243	1.367	1.491	1.616	1.740	1.864	1.988	2.113	2.237	2.361	2.486	2.610	2.734
300.	0.257	0.386	0.514	0.643	0.771	0.900	1.029	1.157	1.286	1.414	1.543	1.671	1.800	1.929	2.057	2.186	2.314	2.443	2.571	2.700	2.829
310.	0.266	0.399	0.531	0.664	0.797	0.930	1.063	1.196	1.329	1.462	1.594	1.727	1.860	1.993	2.126	2.259	2.392	2.524	2.657	2.790	2.923
320.	0.274	0.411	0.549	0.686	0.823	0.960	1.097	1.234	1.372	1.509	1.646	1.783	1.920	2.057	2.195	2.332	2.469	2.606	2.743	2.880	3.018
330.	0.283	0.424	0.566	0.707	0.849	0.990	1.132	1.273	1.415	1.556	1.698	1.839	1.980	2.122	2.263	2.405	2.546	2.688	2.829	2.971	3.112
340.	0.292	0.437	0.583	0.729	0.875	1.020	1.166	1.312	1.458	1.603	1.749	1.895	2.041	2.186	2.332	2.478	2.624	2.769	2.915	3.061	3.207
350.	0.300	0.450	0.600	0.750	0.900	1.050	1.200	1.351	1.501	1.651	1.801	1.951	2.101	2.251	2.401	2.551	2.701	2.851	3.001	3.151	3.301
360.	0.309	0.463	0.617	0.772	0.926	1.081	1.235	1.389	1.544	1.698	1.852	2.007	2.161	2.315	2.470	2.624	2.778	2.933	3.087	3.242	3.396
370.	0.317	0.476	0.635	0.793	0.952	1.111	1.269	1.428	1.587	1.745	1.904	2.063	2.221	2.380	2.539	2.697	2.856	3.015	3.173	3.332	3.491

TABLE 2.7*b* Torispherical Head Thickness Tables in Inches: Joint Efficiency = 0.90

TORISPHERICAL HEAD THICKNESS TABLES IN INCHES

JOINT EFF.=0.85 ALLOWABLE STRESS VALUE =13800. PSI

	INSIDE CROWN RADIUS,INCHES																				
PSI	12.	18.	24.	30.	36.	42.	48.	54.	60.	66.	72.	78.	84.	90.	96.	102.	108.	114.	120.	126.	132.
15.	0.014	0.020	0.027	0.034	0.041	0.048	0.054	0.061	0.068	0.075	0.081	0.088	0.095	0.102	0.109	0.115	0.122	0.129	0.136	0.143	0.149
20.	0.018	0.027	0.036	0.045	0.054	0.063	0.072	0.081	0.091	0.100	0.109	0.118	0.127	0.136	0.145	0.154	0.163	0.172	0.181	0.190	0.199
25.	0.023	0.034	0.045	0.057	0.068	0.079	0.091	0.102	0.113	0.125	0.136	0.147	0.158	0.170	0.181	0.192	0.204	0.215	0.226	0.238	0.249
30.	0.027	0.041	0.054	0.068	0.082	0.095	0.109	0.122	0.136	0.149	0.163	0.177	0.190	0.204	0.217	0.231	0.245	0.258	0.272	0.285	0.299
35.	0.032	0.048	0.063	0.079	0.095	0.111	0.127	0.143	0.158	0.174	0.190	0.206	0.222	0.238	0.254	0.269	0.285	0.301	0.317	0.333	0.349
40.	0.036	0.054	0.072	0.091	0.109	0.127	0.145	0.163	0.181	0.199	0.217	0.235	0.254	0.272	0.290	0.308	0.326	0.344	0.362	0.380	0.398
45.	0.041	0.061	0.082	0.102	0.122	0.143	0.163	0.183	0.204	0.224	0.245	0.265	0.285	0.306	0.326	0.346	0.367	0.387	0.408	0.428	0.448
50.	0.045	0.068	0.091	0.113	0.136	0.159	0.181	0.204	0.226	0.249	0.272	0.294	0.317	0.340	0.362	0.385	0.408	0.430	0.453	0.476	0.498
55.	0.050	0.075	0.100	0.125	0.149	0.174	0.199	0.224	0.249	0.274	0.299	0.324	0.349	0.374	0.399	0.423	0.448	0.473	0.498	0.523	0.548
60.	0.054	0.082	0.109	0.136	0.163	0.190	0.217	0.245	0.272	0.299	0.326	0.353	0.380	0.408	0.435	0.462	0.489	0.516	0.544	0.571	0.598
65.	0.059	0.088	0.118	0.147	0.177	0.206	0.236	0.265	0.294	0.324	0.353	0.383	0.412	0.442	0.471	0.500	0.530	0.559	0.589	0.618	0.648
70.	0.063	0.095	0.127	0.159	0.190	0.222	0.254	0.285	0.317	0.349	0.380	0.412	0.444	0.476	0.507	0.539	0.571	0.602	0.634	0.666	0.698
75.	0.068	0.102	0.136	0.17C	0.204	0.238	0.272	0.306	0.340	0.374	0.408	0.442	0.476	0.510	0.544	0.578	0.612	0.645	0.679	0.713	0.747
80.	0.072	0.109	0.145	0.181	0.217	0.254	0.290	0.326	0.362	0.399	0.435	0.471	0.507	0.544	0.580	0.616	0.652	0.689	0.725	0.761	0.797
85.	0.077	0.116	0.154	0.193	0.231	0.270	0.308	0.347	0.385	0.424	0.462	0.501	0.539	0.578	0.616	0.655	0.693	0.732	0.770	0.809	0.847
90.	0.082	0.122	0.163	0.204	0.245	0.285	0.326	0.367	0.408	0.449	0.489	0.530	0.571	0.612	0.652	0.693	0.734	0.775	0.815	0.856	0.897
95.	0.086	0.129	0.172	0.215	0.258	0.301	0.344	0.387	0.430	0.473	0.516	0.560	0.603	0.646	0.689	0.732	0.775	0.818	0.861	0.904	0.947
100.	0.091	0.136	0.181	0.227	0.272	0.317	0.362	0.408	0.453	0.498	0.544	0.589	0.634	0.680	0.725	0.770	0.816	0.861	0.906	0.951	0.997
105.	0.095	0.143	0.190	0.238	0.285	0.333	0.381	0.428	0.476	0.523	0.571	0.618	0.666	0.714	0.761	0.809	0.856	0.904	0.951	0.999	1.047
110.	0.100	0.150	0.199	0.249	0.299	0.349	0.399	0.449	0.498	0.548	0.598	0.648	0.698	0.748	0.797	0.847	0.897	0.947	0.997	1.047	1.097
115.	0.104	0.156	0.208	0.261	0.313	0.365	0.417	0.469	0.521	0.573	0.625	0.677	0.730	0.782	0.834	0.886	0.938	0.990	1.042	1.094	1.146
120.	0.109	0.163	0.218	0.272	0.326	0.381	0.435	0.489	0.544	0.598	0.653	0.707	0.761	0.816	0.870	0.924	0.979	1.033	1.088	1.142	1.196
125.	0.113	0.170	0.227	0.283	0.340	0.397	0.453	0.510	0.566	0.623	0.680	0.736	0.793	0.850	0.906	0.963	1.020	1.076	1.133	1.190	1.246
130.	0.118	0.177	0.236	0.295	0.353	0.412	0.471	0.530	0.589	0.648	0.707	0.766	0.825	0.884	0.943	1.002	1.060	1.119	1.178	1.237	1.296
135.	0.122	0.184	0.245	0.306	0.367	0.428	0.489	0.551	0.612	0.673	0.734	0.795	0.857	0.918	0.979	1.040	1.101	1.162	1.224	1.285	1.346
140.	0.127	0.190	0.254	0.317	0.381	0.444	0.508	0.571	0.635	0.698	0.761	0.825	0.888	0.952	1.015	1.079	1.142	1.206	1.269	1.332	1.396
145.	0.131	0.197	0.263	0.329	0.394	0.460	0.526	0.591	0.657	0.723	0.789	0.854	0.920	0.986	1.052	1.117	1.183	1.249	1.314	1.380	1.446
150.	0.136	0.204	0.272	0.340	0.408	0.476	0.544	0.612	0.680	0.748	0.816	0.884	0.952	1.020	1.088	1.156	1.224	1.292	1.360	1.428	1.496
160.	0.145	0.218	0.290	0.363	0.435	0.508	0.580	0.653	0.725	0.798	0.870	0.943	1.015	1.088	1.160	1.233	1.306	1.378	1.451	1.523	1.596
170.	0.154	0.231	0.308	0.385	0.462	0.539	0.617	0.694	0.771	0.848	0.925	1.002	1.079	1.156	1.233	1.310	1.387	1.464	1.541	1.618	1.696
180.	0.163	0.245	0.326	0.408	0.490	0.571	0.653	0.734	0.816	0.898	0.979	1.061	1.143	1.224	1.306	1.387	1.469	1.551	1.632	1.714	1.795
190.	0.172	0.258	0.345	0.431	0.517	0.603	0.689	0.775	0.861	0.948	1.034	1.120	1.206	1.292	1.378	1.465	1.551	1.637	1.723	1.809	1.895
200.	0.181	0.272	0.363	0.453	0.544	0.635	0.726	0.816	0.907	0.998	1.088	1.179	1.270	1.360	1.451	1.542	1.632	1.723	1.814	1.905	1.995
210.	0.190	0.286	0.381	0.476	0.571	0.667	0.762	0.857	0.952	1.048	1.143	1.238	1.333	1.429	1.524	1.619	1.714	1.809	1.905	2.000	2.095
220.	0.200	0.299	0.399	0.499	0.599	0.698	0.798	0.898	0.998	1.098	1.197	1.297	1.397	1.497	1.596	1.696	1.796	1.896	1.996	2.095	2.195
230.	0.209	0.313	0.417	0.522	0.626	0.730	0.835	0.939	1.043	1.148	1.252	1.356	1.461	1.565	1.669	1.773	1.878	1.982	2.086	2.191	2.295
240.	0.218	0.327	0.435	0.544	0.653	0.762	0.871	0.980	1.089	1.198	1.306	1.415	1.524	1.633	1.742	1.851	1.960	2.068	2.177	2.286	2.395
250.	0.227	0.340	0.454	0.567	0.680	0.794	0.907	1.021	1.134	1.248	1.361	1.474	1.588	1.701	1.815	1.928	2.041	2.155	2.268	2.382	2.495
260.	0.236	0.354	0.472	0.590	0.708	0.826	0.944	1.062	1.180	1.298	1.416	1.533	1.651	1.769	1.887	2.005	2.123	2.241	2.359	2.477	2.595
270.	0.245	0.368	0.490	0.613	0.735	0.858	0.980	1.103	1.225	1.348	1.470	1.593	1.715	1.838	1.960	2.083	2.205	2.328	2.450	2.573	2.695
280.	0.254	0.381	0.508	0.635	0.762	0.889	1.016	1.143	1.271	1.398	1.525	1.652	1.779	1.906	2.033	2.160	2.287	2.414	2.541	2.668	2.795
290.	0.263	0.395	0.526	0.658	0.790	0.921	1.053	1.184	1.316	1.448	1.579	1.711	1.842	1.974	2.106	2.237	2.369	2.500	2.632	2.764	2.895
300.	0.272	0.408	0.545	0.681	0.817	0.953	1.089	1.225	1.362	1.498	1.634	1.770	1.906	2.042	2.178	2.315	2.451	2.587	2.723	2.859	2.995
310.	0.281	0.422	0.563	0.704	0.844	0.985	1.126	1.266	1.407	1.548	1.688	1.829	1.970	2.111	2.251	2.392	2.533	2.673	2.814	2.955	3.095
320.	0.291	0.436	0.581	0.726	0.872	1.017	1.162	1.307	1.453	1.598	1.743	1.888	2.034	2.179	2.324	2.469	2.615	2.760	2.905	3.050	3.196
330.	0.300	0.449	0.599	0.749	0.899	1.049	1.198	1.348	1.498	1.648	1.798	1.947	2.097	2.247	2.397	2.547	2.697	2.846	2.996	3.146	3.296
340.	0.309	0.463	0.617	0.772	0.926	1.081	1.235	1.389	1.544	1.698	1.852	2.007	2.161	2.315	2.470	2.624	2.778	2.933	3.087	3.242	3.396
350.	0.318	0.477	0.636	0.795	0.953	1.112	1.271	1.430	1.589	1.748	1.907	2.066	2.225	2.384	2.543	2.702	2.860	3.019	3.178	3.337	3.496
360.	0.327	0.490	0.654	0.817	0.981	1.144	1.308	1.471	1.635	1.798	1.962	2.125	2.289	2.452	2.615	2.779	2.942	3.106	3.269	3.433	3.596
370.	0.336	0.504	0.672	0.840	1.008	1.176	1.344	1.512	1.680	1.848	2.016	2.184	2.352	2.520	2.688	2.856	3.024	3.192	3.360	3.528	3.697

Table 2.7c Torispherical Head Thickness Tables in Inches: Joint Efficiency = 0.85

TORISPHERICAL HEAD THICKNESS TABLES IN INCHES

JOINT EFF.=0.80 ALLOWABLE STRESS VALUE =13800. PSI

	INSIDE CROWN RADIUS,INCHES																				
PSI	12.	18.	24.	30.	36.	42.	48.	54.	60.	66.	72.	78.	84.	90.	96.	102.	108.	114.	120.	126.	132.
15.	0.014	0.022	0.029	0.036	0.043	0.051	0.058	0.065	0.072	0.079	0.087	0.094	0.101	0.108	0.115	0.123	0.130	0.137	0.144	0.152	0.159
20.	0.019	0.029	0.038	0.048	0.058	0.067	0.077	0.087	0.096	0.106	0.115	0.125	0.135	0.144	0.154	0.164	0.173	0.183	0.192	0.202	0.212
25.	0.024	0.036	0.048	0.060	0.072	0.084	0.096	0.108	0.120	0.132	0.144	0.156	0.168	0.180	0.192	0.204	0.216	0.229	0.241	0.253	0.265
30.	0.029	0.043	0.058	0.072	0.087	0.101	0.115	0.130	0.144	0.159	0.173	0.188	0.202	0.216	0.231	0.245	0.260	0.274	0.289	0.303	0.318
35.	0.034	0.051	0.067	0.084	0.101	0.118	0.135	0.152	0.168	0.185	0.202	0.219	0.236	0.253	0.269	0.286	0.303	0.320	0.337	0.354	0.370
40.	0.038	0.058	0.077	0.096	0.115	0.135	0.154	0.173	0.192	0.212	0.231	0.250	0.269	0.289	0.308	0.327	0.346	0.366	0.385	0.404	0.423
45.	0.043	0.065	0.087	0.108	0.130	0.152	0.173	0.195	0.217	0.238	0.260	0.281	0.303	0.325	0.346	0.368	0.390	0.411	0.433	0.455	0.476
50.	0.048	0.072	0.096	0.120	0.144	0.168	0.192	0.217	0.241	0.265	0.289	0.313	0.337	0.361	0.385	0.409	0.433	0.457	0.481	0.505	0.529
55.	0.053	0.079	0.106	0.132	0.159	0.185	0.212	0.238	0.265	0.291	0.318	0.344	0.371	0.397	0.423	0.450	0.476	0.503	0.529	0.556	0.582
60.	0.058	0.087	0.115	0.144	0.173	0.202	0.231	0.260	0.289	0.318	0.346	0.375	0.404	0.433	0.462	0.491	0.520	0.549	0.577	0.606	0.635
65.	0.063	0.094	0.125	0.156	0.188	0.219	0.250	0.282	0.313	0.344	0.375	0.407	0.438	0.469	0.501	0.532	0.563	0.594	0.626	0.657	0.688
70.	0.067	0.101	0.135	0.168	0.202	0.236	0.270	0.303	0.337	0.371	0.404	0.438	0.472	0.505	0.539	0.573	0.606	0.640	0.674	0.707	0.741
75.	0.072	0.108	0.144	0.180	0.217	0.253	0.289	0.325	0.361	0.397	0.433	0.469	0.505	0.541	0.578	0.614	0.650	0.686	0.722	0.758	0.794
80.	0.077	0.116	0.154	0.193	0.231	0.270	0.308	0.347	0.385	0.424	0.462	0.501	0.539	0.578	0.616	0.655	0.693	0.732	0.770	0.809	0.847
85.	0.082	0.123	0.164	0.205	0.245	0.286	0.327	0.368	0.409	0.450	0.491	0.532	0.573	0.614	0.655	0.696	0.736	0.777	0.818	0.859	0.900
90.	0.087	0.130	0.173	0.217	0.260	0.303	0.347	0.390	0.433	0.477	0.520	0.563	0.607	0.650	0.693	0.736	0.780	0.823	0.866	0.910	0.953
95.	0.091	0.137	0.183	0.229	0.274	0.320	0.366	0.412	0.457	0.503	0.549	0.595	0.640	0.686	0.732	0.777	0.823	0.869	0.915	0.960	1.006
100.	0.096	0.144	0.193	0.241	0.289	0.337	0.385	0.433	0.481	0.530	0.578	0.626	0.674	0.722	0.770	0.818	0.867	0.915	0.963	1.011	1.059
105.	0.101	0.152	0.202	0.253	0.303	0.354	0.404	0.455	0.506	0.556	0.607	0.657	0.708	0.758	0.809	0.859	0.910	0.960	1.011	1.062	1.112
110.	0.106	0.159	0.212	0.265	0.318	0.371	0.424	0.477	0.530	0.583	0.636	0.688	0.741	0.794	0.847	0.900	0.953	1.006	1.059	1.112	1.165
115.	0.111	0.166	0.221	0.277	0.332	0.388	0.443	0.498	0.554	0.609	0.664	0.720	0.775	0.831	0.886	0.941	0.997	1.052	1.107	1.163	1.218
120.	0.116	0.173	0.231	0.289	0.347	0.404	0.462	0.520	0.578	0.636	0.693	0.751	0.809	0.867	0.924	0.982	1.040	1.098	1.156	1.213	1.271
125.	0.120	0.181	0.241	0.301	0.361	0.421	0.482	0.542	0.602	0.662	0.722	0.782	0.843	0.903	0.963	1.023	1.083	1.144	1.204	1.264	1.324
130.	0.125	0.188	0.250	0.313	0.376	0.438	0.501	0.563	0.626	0.689	0.751	0.814	0.876	0.939	1.002	1.064	1.127	1.189	1.252	1.315	1.377
135.	0.130	0.195	0.260	0.325	0.390	0.455	0.520	0.585	0.650	0.715	0.780	0.845	0.910	0.975	1.040	1.105	1.170	1.235	1.300	1.365	1.430
140.	0.135	0.202	0.270	0.337	0.405	0.472	0.539	0.607	0.674	0.742	0.809	0.876	0.944	1.011	1.079	1.146	1.214	1.281	1.348	1.416	1.483
145.	0.140	0.210	0.279	0.349	0.419	0.489	0.559	0.629	0.698	0.768	0.838	0.908	0.978	1.048	1.117	1.187	1.257	1.327	1.397	1.467	1.536
150.	0.144	0.217	0.289	0.361	0.433	0.506	0.578	0.650	0.722	0.795	0.867	0.939	1.011	1.084	1.156	1.228	1.300	1.373	1.445	1.517	1.589
160.	0.154	0.231	0.308	0.385	0.462	0.539	0.617	0.694	0.771	0.848	0.925	1.002	1.079	1.156	1.233	1.310	1.387	1.464	1.541	1.618	1.696
170.	0.164	0.246	0.328	0.409	0.491	0.573	0.655	0.737	0.819	0.901	0.983	1.065	1.146	1.228	1.310	1.392	1.474	1.556	1.638	1.720	1.802
180.	0.173	0.260	0.347	0.434	0.520	0.607	0.694	0.780	0.867	0.954	1.041	1.127	1.214	1.301	1.387	1.474	1.561	1.648	1.734	1.821	1.908
190.	0.183	0.275	0.366	0.458	0.549	0.641	0.732	0.824	0.915	1.007	1.099	1.190	1.282	1.373	1.465	1.556	1.648	1.739	1.831	1.922	2.014
200.	0.193	0.289	0.385	0.482	0.578	0.675	0.771	0.867	0.964	1.060	1.156	1.253	1.349	1.446	1.542	1.638	1.735	1.831	1.927	2.024	2.120
210.	0.202	0.304	0.405	0.506	0.607	0.708	0.810	0.911	1.012	1.113	1.214	1.316	1.417	1.518	1.619	1.720	1.822	1.923	2.024	2.125	2.226
220.	0.212	0.318	0.424	0.530	0.636	0.742	0.848	0.954	1.060	1.166	1.272	1.378	1.484	1.590	1.696	1.802	1.908	2.015	2.121	2.227	2.333
230.	0.222	0.333	0.443	0.554	0.665	0.776	0.887	0.998	1.109	1.219	1.330	1.441	1.552	1.663	1.774	1.885	1.995	2.106	2.217	2.328	2.439
240.	0.231	0.347	0.463	0.578	0.694	0.810	0.925	1.041	1.157	1.273	1.388	1.504	1.620	1.735	1.851	1.967	2.082	2.198	2.316	2.429	2.545
250.	0.241	0.362	0.482	0.603	0.723	0.844	0.964	1.085	1.205	1.326	1.446	1.567	1.687	1.808	1.928	2.049	2.169	2.290	2.410	2.531	2.651
260.	0.251	0.376	0.501	0.627	0.752	0.877	1.003	1.128	1.253	1.379	1.504	1.630	1.755	1.880	2.006	2.131	2.256	2.382	2.507	2.632	2.758
270.	0.260	0.391	0.521	0.651	0.781	0.911	1.041	1.172	1.302	1.432	1.562	1.692	1.823	1.953	2.083	2.213	2.343	2.473	2.604	2.734	2.864
280.	0.270	0.405	0.540	0.675	0.810	0.945	1.080	1.215	1.350	1.485	1.620	1.755	1.890	2.025	2.160	2.295	2.430	2.565	2.700	2.835	2.970
290.	0.280	0.420	0.559	0.699	0.839	0.979	1.119	1.259	1.399	1.538	1.678	1.818	1.958	2.098	2.238	2.377	2.517	2.657	2.797	2.937	3.077
300.	0.289	0.434	0.579	0.723	0.868	1.013	1.157	1.302	1.447	1.592	1.736	1.881	2.026	2.170	2.315	2.460	2.604	2.749	2.894	3.038	3.183
310.	0.299	0.449	0.598	0.748	0.897	1.047	1.196	1.346	1.495	1.645	1.794	1.944	2.093	2.243	2.392	2.542	2.691	2.841	2.990	3.140	3.290
320.	0.309	0.463	0.617	0.772	0.926	1.081	1.235	1.389	1.544	1.698	1.852	2.007	2.161	2.315	2.470	2.624	2.778	2.933	3.087	3.242	3.396
330.	0.318	0.478	0.637	0.796	0.955	1.114	1.274	1.433	1.592	1.751	1.910	2.070	2.229	2.388	2.547	2.706	2.866	3.025	3.184	3.343	3.502
340.	0.328	0.492	0.656	0.820	0.984	1.148	1.312	1.476	1.640	1.804	1.968	2.132	2.297	2.461	2.625	2.789	2.953	3.117	3.281	3.445	3.609
350.	0.338	0.507	0.676	0.844	1.013	1.182	1.351	1.520	1.689	1.858	2.027	2.195	2.364	2.533	2.702	2.871	3.040	3.209	3.378	3.546	3.715
360.	0.347	0.521	0.695	0.869	1.042	1.216	1.390	1.563	1.737	1.911	2.085	2.258	2.432	2.606	2.779	2.953	3.127	3.301	3.474	3.648	3.822
370.	0.357	0.536	0.714	0.893	1.071	1.250	1.428	1.607	1.786	1.964	2.143	2.321	2.500	2.678	2.857	3.036	3.214	3.393	3.571	3.750	3.928

TABLE 2.7d Torispherical Head Thickness Tables in Inches: Joint Efficiency = 0.80

TORISPHERICAL HEAD THICKNESS TABLES IN INCHES

JOINT EFF.=0.70 ALLOWABLE STRESS VALUE =13800. PSI

PSI	12.	18.	24.	30.	36.	42.	48.	54.	60.	66.	72.	78.	84.	90.	96.	102.	108.	114.	120.	126.	132.
	INSIDE CROWN RADIUS,INCHES																				
15.	0.016	0.025	0.033	0.041	0.049	0.058	0.066	0.074	0.082	0.091	0.099	0.107	0.115	0.124	0.132	0.140	0.148	0.157	0.165	0.173	0.181
20.	0.022	0.033	0.044	0.055	0.066	0.077	0.088	0.099	0.110	0.121	0.132	0.143	0.154	0.165	0.176	0.187	0.198	0.209	0.220	0.231	0.242
25.	0.027	0.041	0.055	0.069	0.082	0.096	0.110	0.124	0.137	0.151	0.165	0.179	0.192	0.206	0.220	0.234	0.247	0.261	0.275	0.289	0.302
30.	0.033	0.049	0.066	0.082	0.099	0.115	0.132	0.148	0.165	0.181	0.198	0.214	0.231	0.247	0.264	0.280	0.297	0.313	0.330	0.346	0.363
35.	0.038	0.058	0.077	0.096	0.115	0.135	0.154	0.173	0.192	0.212	0.231	0.250	0.269	0.289	0.308	0.327	0.346	0.366	0.385	0.404	0.423
40.	0.044	0.066	0.088	0.110	0.132	0.154	0.176	0.198	0.220	0.242	0.264	0.286	0.308	0.330	0.352	0.374	0.396	0.418	0.440	0.462	0.484
45.	0.049	0.074	0.099	0.124	0.148	0.173	0.198	0.223	0.247	0.272	0.297	0.322	0.346	0.371	0.396	0.421	0.445	0.470	0.495	0.520	0.544
50.	0.055	0.082	0.110	0.137	0.165	0.192	0.220	0.247	0.275	0.302	0.330	0.357	0.385	0.412	0.440	0.467	0.495	0.522	0.550	0.577	0.605
55.	0.061	0.091	0.121	0.151	0.182	0.212	0.242	0.272	0.303	0.333	0.363	0.393	0.424	0.454	0.484	0.514	0.545	0.575	0.605	0.635	0.666
60.	0.066	0.099	0.132	0.165	0.198	0.231	0.264	0.297	0.330	0.363	0.396	0.429	0.462	0.495	0.528	0.561	0.594	0.627	0.660	0.693	0.726
65.	0.072	0.107	0.143	0.179	0.215	0.250	0.286	0.322	0.358	0.393	0.429	0.465	0.501	0.536	0.572	0.608	0.644	0.679	0.715	0.751	0.787
70.	0.077	0.116	0.154	0.193	0.231	0.270	0.308	0.347	0.385	0.424	0.462	0.501	0.539	0.578	0.616	0.655	0.693	0.732	0.770	0.809	0.847
75.	0.083	0.124	0.165	0.206	0.248	0.289	0.330	0.371	0.413	0.454	0.495	0.536	0.578	0.619	0.660	0.701	0.743	0.784	0.825	0.866	0.908
80.	0.088	0.132	0.176	0.220	0.264	0.308	0.352	0.396	0.440	0.484	0.528	0.572	0.616	0.660	0.704	0.748	0.792	0.836	0.880	0.924	0.968
85.	0.094	0.140	0.187	0.234	0.281	0.327	0.374	0.421	0.468	0.514	0.561	0.608	0.655	0.701	0.748	0.795	0.842	0.889	0.935	0.982	1.029
90.	0.099	0.149	0.198	0.248	0.297	0.347	0.396	0.446	0.495	0.545	0.594	0.644	0.693	0.743	0.792	0.842	0.891	0.941	0.990	1.040	1.089
95.	0.105	0.157	0.209	0.261	0.314	0.366	0.418	0.470	0.523	0.575	0.627	0.680	0.732	0.784	0.836	0.889	0.941	0.993	1.045	1.098	1.150
100.	0.110	0.165	0.220	0.275	0.330	0.385	0.440	0.495	0.550	0.605	0.660	0.715	0.770	0.825	0.880	0.935	0.990	1.045	1.101	1.156	1.211
105.	0.116	0.173	0.231	0.289	0.347	0.404	0.462	0.520	0.578	0.636	0.693	0.751	0.809	0.867	0.924	0.982	1.040	1.098	1.156	1.213	1.271
110.	0.121	0.182	0.242	0.303	0.363	0.424	0.484	0.545	0.605	0.666	0.726	0.787	0.847	0.908	0.969	1.029	1.090	1.150	1.211	1.271	1.332
115.	0.127	0.190	0.253	0.316	0.380	0.443	0.506	0.570	0.633	0.696	0.759	0.823	0.886	0.949	1.013	1.076	1.139	1.203	1.266	1.329	1.392
120.	0.132	0.198	0.264	0.330	0.396	0.462	0.528	0.594	0.660	0.726	0.793	0.859	0.925	0.991	1.057	1.123	1.189	1.255	1.321	1.387	1.453
125.	0.138	0.206	0.275	0.344	0.413	0.482	0.550	0.619	0.688	0.757	0.826	0.894	0.963	1.032	1.101	1.170	1.238	1.307	1.376	1.445	1.514
130.	0.143	0.215	0.286	0.358	0.429	0.501	0.572	0.644	0.716	0.787	0.859	0.930	1.002	1.073	1.145	1.216	1.288	1.360	1.431	1.503	1.574
135.	0.149	0.223	0.297	0.372	0.446	0.520	0.594	0.669	0.743	0.817	0.892	0.966	1.040	1.115	1.189	1.263	1.338	1.412	1.486	1.561	1.635
140.	0.154	0.231	0.308	0.385	0.462	0.539	0.617	0.694	0.771	0.848	0.925	1.002	1.079	1.156	1.233	1.310	1.387	1.464	1.541	1.618	1.696
145.	0.160	0.239	0.319	0.399	0.479	0.559	0.639	0.718	0.798	0.878	0.958	1.038	1.118	1.197	1.277	1.357	1.437	1.517	1.596	1.676	1.756
150.	0.165	0.248	0.330	0.413	0.495	0.578	0.661	0.743	0.826	0.908	0.991	1.074	1.156	1.239	1.321	1.404	1.486	1.569	1.652	1.734	1.817
160.	0.176	0.264	0.352	0.440	0.529	0.617	0.705	0.793	0.881	0.969	1.057	1.145	1.233	1.321	1.410	1.498	1.586	1.674	1.762	1.850	1.938
170.	0.187	0.281	0.374	0.468	0.562	0.655	0.749	0.843	0.936	1.030	1.123	1.217	1.311	1.404	1.498	1.591	1.685	1.779	1.872	1.966	2.059
180.	0.198	0.297	0.397	0.496	0.595	0.694	0.793	0.892	0.991	1.090	1.190	1.289	1.388	1.487	1.586	1.685	1.784	1.883	1.983	2.082	2.181
190.	0.209	0.314	0.419	0.523	0.628	0.733	0.837	0.942	1.046	1.151	1.256	1.360	1.465	1.570	1.674	1.779	1.884	1.988	2.093	2.198	2.302
200.	0.220	0.330	0.441	0.551	0.661	0.771	0.881	0.991	1.102	1.212	1.322	1.432	1.542	1.652	1.763	1.873	1.983	2.093	2.203	2.313	2.424
210.	0.231	0.347	0.463	0.578	0.694	0.810	0.925	1.041	1.157	1.273	1.388	1.504	1.620	1.735	1.851	1.967	2.082	2.198	2.314	2.429	2.545
220.	0.242	0.364	0.485	0.606	0.727	0.848	0.970	1.091	1.212	1.333	1.454	1.576	1.697	1.818	1.939	2.061	2.182	2.303	2.424	2.545	2.667
230.	0.253	0.380	0.507	0.634	0.760	0.887	1.014	1.141	1.267	1.394	1.521	1.647	1.774	1.901	2.028	2.154	2.281	2.408	2.535	2.661	2.788
240.	0.265	0.397	0.529	0.661	0.794	0.926	1.058	1.190	1.323	1.455	1.587	1.719	1.852	1.984	2.116	2.248	2.381	2.513	2.645	2.777	2.910
250.	0.276	0.413	0.551	0.689	0.827	0.964	1.102	1.240	1.378	1.516	1.653	1.791	1.929	2.067	2.204	2.342	2.480	2.618	2.756	2.893	3.031
260.	0.287	0.430	0.573	0.717	0.860	1.003	1.146	1.290	1.433	1.576	1.720	1.863	2.006	2.150	2.293	2.436	2.579	2.723	2.866	3.009	3.153
270.	0.298	0.446	0.595	0.744	0.893	1.042	1.191	1.339	1.488	1.637	1.786	1.935	2.084	2.232	2.381	2.530	2.679	2.828	2.977	3.125	3.274
280.	0.309	0.463	0.617	0.772	0.926	1.081	1.235	1.389	1.544	1.698	1.852	2.007	2.161	2.315	2.470	2.624	2.778	2.933	3.087	3.242	3.396
290.	0.320	0.480	0.640	0.799	0.959	1.119	1.279	1.439	1.599	1.759	1.919	2.079	2.238	2.398	2.558	2.718	2.878	3.038	3.198	3.358	3.518
300.	0.331	0.496	0.662	0.827	0.993	1.158	1.323	1.489	1.654	1.820	1.985	2.150	2.316	2.481	2.647	2.812	2.978	3.143	3.308	3.474	3.639
310.	0.342	0.513	0.684	0.855	1.026	1.197	1.368	1.539	1.710	1.880	2.051	2.222	2.393	2.564	2.735	2.906	3.077	3.248	3.419	3.590	3.761
320.	0.353	0.529	0.706	0.882	1.059	1.235	1.412	1.588	1.765	1.941	2.118	2.294	2.471	2.647	2.824	3.000	3.177	3.353	3.530	3.706	3.883
330.	0.364	0.546	0.728	0.910	1.092	1.274	1.456	1.638	1.820	2.002	2.184	2.366	2.548	2.730	2.912	3.094	3.276	3.458	3.640	3.822	4.004
340.	0.375	0.563	0.750	0.938	1.125	1.313	1.500	1.688	1.876	2.063	2.251	2.438	2.626	2.813	3.001	3.188	3.376	3.564	3.751	3.939	4.126
350.	0.386	0.579	0.772	0.965	1.159	1.352	1.545	1.738	1.931	2.124	2.317	2.510	2.703	2.896	3.089	3.283	3.476	3.669	3.862	4.055	4.248
360.	0.397	0.596	0.795	0.993	1.192	1.390	1.589	1.788	1.986	2.185	2.384	2.582	2.781	2.979	3.178	3.377	3.575	3.774	3.973	4.171	4.370
370.	0.408	0.613	0.817	1.021	1.225	1.429	1.633	1.838	2.042	2.246	2.450	2.654	2.858	3.063	3.267	3.471	3.675	3.879	4.083	4.288	4.492

TABLE 2.7e Torispherical Head Thickness Tables in Inches: Joint Efficiency = 0.70

TORISPHERICAL HEAD THICKNESS TABLES IN INCHES

JOINT EFF.=0.65 ALLOWABLE STRESS VALUE =13800. PSI

	INSIDE CROWN RADIUS,INCHES																				
PSI	12.	18.	24.	30.	36.	42.	48.	54.	60.	66.	72.	78.	84.	90.	96.	102.	108.	114.	120.	126.	132.
15.	0.018	0.027	0.036	0.044	0.053	0.062	0.071	0.080	0.089	0.098	0.107	0.115	0.124	0.133	0.142	0.151	0.160	0.169	0.178	0.187	0.195
20.	0.024	0.036	0.047	0.059	0.071	0.083	0.095	0.107	0.118	0.130	0.142	0.154	0.166	0.178	0.189	0.201	0.213	0.225	0.237	0.249	0.261
25.	0.030	0.044	0.059	0.074	0.089	0.104	0.118	0.133	0.148	0.163	0.178	0.192	0.207	0.222	0.237	0.252	0.266	0.281	0.296	0.311	0.326
30.	0.036	0.053	0.071	0.089	0.107	0.124	0.142	0.160	0.178	0.195	0.213	0.231	0.249	0.266	0.284	0.302	0.320	0.338	0.355	0.373	0.391
35.	0.041	0.062	0.083	0.104	0.124	0.145	0.166	0.187	0.207	0.228	0.249	0.269	0.290	0.311	0.332	0.352	0.373	0.394	0.415	0.435	0.456
40.	0.047	0.071	0.095	0.118	0.142	0.166	0.190	0.213	0.237	0.261	0.284	0.308	0.332	0.355	0.379	0.403	0.426	0.450	0.474	0.497	0.521
45.	0.053	0.080	0.107	0.133	0.160	0.187	0.213	0.240	0.267	0.293	0.320	0.346	0.373	0.400	0.426	0.453	0.480	0.506	0.533	0.560	0.586
50.	0.059	0.089	0.118	0.148	0.178	0.207	0.237	0.267	0.296	0.326	0.355	0.385	0.415	0.444	0.474	0.503	0.533	0.563	0.592	0.622	0.652
55.	0.065	0.098	0.130	0.163	0.195	0.228	0.261	0.293	0.326	0.358	0.391	0.424	0.456	0.489	0.521	0.554	0.586	0.619	0.652	0.684	0.717
60.	0.071	0.107	0.142	0.178	0.213	0.249	0.284	0.320	0.355	0.391	0.427	0.462	0.498	0.533	0.569	0.604	0.640	0.675	0.711	0.746	0.782
65.	0.077	0.116	0.154	0.193	0.231	0.270	0.308	0.347	0.385	0.424	0.462	0.501	0.539	0.578	0.616	0.655	0.693	0.732	0.770	0.809	0.847
70.	0.083	0.124	0.166	0.207	0.249	0.290	0.332	0.373	0.415	0.456	0.498	0.539	0.581	0.622	0.664	0.705	0.746	0.788	0.829	0.871	0.912
75.	0.089	0.133	0.178	0.222	0.267	0.311	0.355	0.400	0.444	0.489	0.533	0.578	0.622	0.667	0.711	0.755	0.800	0.844	0.889	0.933	0.978
80.	0.095	0.142	0.190	0.237	0.284	0.332	0.379	0.427	0.474	0.521	0.569	0.616	0.664	0.711	0.758	0.806	0.853	0.901	0.948	0.995	1.043
85.	0.101	0.151	0.201	0.252	0.302	0.353	0.403	0.453	0.504	0.554	0.604	0.655	0.705	0.755	0.806	0.856	0.907	0.957	1.007	1.058	1.108
90.	0.107	0.160	0.213	0.267	0.320	0.373	0.427	0.480	0.533	0.587	0.640	0.693	0.747	0.800	0.853	0.907	0.960	1.013	1.067	1.120	1.173
95.	0.113	0.169	0.225	0.281	0.338	0.394	0.450	0.507	0.563	0.619	0.676	0.732	0.788	0.844	0.901	0.957	1.013	1.070	1.126	1.182	1.239
100.	0.119	0.178	0.237	0.296	0.356	0.415	0.474	0.533	0.593	0.652	0.711	0.770	0.830	0.889	0.948	1.007	1.067	1.126	1.185	1.245	1.304
105.	0.124	0.187	0.249	0.311	0.373	0.436	0.498	0.560	0.622	0.685	0.747	0.809	0.871	0.933	0.996	1.058	1.120	1.182	1.245	1.307	1.369
110.	0.130	0.196	0.261	0.326	0.391	0.456	0.522	0.587	0.652	0.717	0.782	0.848	0.913	0.978	1.043	1.108	1.174	1.239	1.304	1.369	1.434
115.	0.136	0.204	0.273	0.341	0.409	0.477	0.545	0.613	0.682	0.750	0.818	0.886	0.954	1.022	1.091	1.159	1.227	1.295	1.363	1.431	1.500
120.	0.142	0.213	0.285	0.356	0.427	0.498	0.569	0.640	0.711	0.782	0.854	0.925	0.996	1.067	1.138	1.209	1.280	1.352	1.423	1.494	1.565
125.	0.148	0.222	0.296	0.370	0.445	0.519	0.593	0.667	0.741	0.815	0.889	0.963	1.037	1.111	1.186	1.260	1.334	1.408	1.482	1.556	1.630
130.	0.154	0.231	0.308	0.385	0.462	0.539	0.617	0.694	0.771	0.848	0.925	1.002	1.079	1.156	1.233	1.310	1.387	1.464	1.541	1.618	1.696
135.	0.160	0.240	0.320	0.400	0.480	0.560	0.640	0.720	0.800	0.880	0.960	1.040	1.121	1.201	1.281	1.361	1.441	1.521	1.601	1.681	1.761
140.	0.166	0.249	0.332	0.415	0.498	0.581	0.664	0.747	0.830	0.913	0.996	1.079	1.162	1.245	1.328	1.411	1.494	1.577	1.660	1.743	1.826
145.	0.172	0.258	0.344	0.430	0.516	0.602	0.688	0.774	0.860	0.946	1.032	1.118	1.204	1.290	1.376	1.462	1.548	1.634	1.720	1.805	1.891
150.	0.178	0.267	0.356	0.445	0.534	0.623	0.712	0.801	0.889	0.978	1.067	1.156	1.245	1.334	1.423	1.512	1.601	1.690	1.779	1.868	1.957
160.	0.190	0.285	0.380	0.474	0.569	0.664	0.759	0.854	0.949	1.044	1.139	1.234	1.328	1.423	1.518	1.613	1.708	1.803	1.898	1.993	2.087
170.	0.202	0.302	0.403	0.504	0.605	0.706	0.807	0.907	1.008	1.109	1.210	1.311	1.412	1.512	1.613	1.714	1.815	1.916	2.017	2.117	2.218
180.	0.214	0.320	0.427	0.534	0.641	0.747	0.854	0.961	1.068	1.174	1.281	1.388	1.495	1.602	1.708	1.815	1.922	2.029	2.135	2.242	2.349
190.	0.225	0.338	0.451	0.564	0.676	0.789	0.902	1.014	1.127	1.240	1.353	1.465	1.578	1.691	1.803	1.916	2.029	2.142	2.254	2.367	2.480
200.	0.237	0.356	0.475	0.593	0.712	0.831	0.949	1.068	1.187	1.305	1.424	1.543	1.661	1.780	1.899	2.017	2.136	2.255	2.373	2.492	2.611
210.	0.249	0.374	0.498	0.623	0.748	0.872	0.997	1.121	1.246	1.371	1.495	1.620	1.744	1.869	1.994	2.118	2.243	2.368	2.492	2.617	2.741
220.	0.261	0.392	0.522	0.653	0.783	0.914	1.044	1.175	1.306	1.436	1.567	1.697	1.828	1.958	2.089	2.219	2.350	2.481	2.611	2.742	2.872
230.	0.273	0.410	0.546	0.683	0.819	0.956	1.092	1.229	1.365	1.502	1.638	1.775	1.911	2.048	2.184	2.321	2.457	2.594	2.730	2.867	3.003
240.	0.285	0.427	0.570	0.712	0.855	0.997	1.140	1.282	1.425	1.567	1.709	1.852	1.994	2.137	2.279	2.422	2.564	2.707	2.849	2.992	3.134
250.	0.297	0.445	0.594	0.742	0.890	1.039	1.187	1.336	1.484	1.632	1.781	1.929	2.078	2.226	2.375	2.523	2.671	2.820	2.968	3.117	3.265
260.	0.309	0.463	0.617	0.772	0.926	1.081	1.235	1.389	1.544	1.698	1.852	2.007	2.161	2.315	2.470	2.624	2.778	2.933	3.087	3.242	3.396
270.	0.321	0.481	0.641	0.802	0.962	1.122	1.283	1.443	1.603	1.763	1.924	2.084	2.244	2.405	2.565	2.725	2.886	3.046	3.206	3.367	3.527
280.	0.333	0.499	0.665	0.831	0.998	1.164	1.330	1.496	1.663	1.829	1.995	2.162	2.328	2.494	2.660	2.827	2.993	3.159	3.325	3.492	3.658
290.	0.344	0.517	0.689	0.861	1.033	1.206	1.378	1.550	1.722	1.895	2.067	2.239	2.411	2.583	2.756	2.928	3.100	3.272	3.445	3.617	3.789
300.	0.356	0.535	0.713	0.891	1.069	1.247	1.426	1.604	1.782	1.960	2.138	2.316	2.495	2.673	2.851	3.029	3.207	3.386	3.564	3.742	3.920
310.	0.368	0.552	0.737	0.921	1.105	1.289	1.473	1.657	1.841	2.026	2.210	2.394	2.578	2.762	2.946	3.131	3.315	3.499	3.683	3.867	4.051
320.	0.380	0.570	0.760	0.951	1.141	1.331	1.521	1.711	1.901	2.091	2.281	2.471	2.662	2.852	3.042	3.232	3.422	3.612	3.802	3.992	4.182
330.	0.392	0.588	0.784	0.980	1.176	1.373	1.569	1.765	1.961	2.157	2.353	2.549	2.745	2.941	3.137	3.333	3.529	3.725	3.921	4.118	4.314
340.	0.404	0.606	0.808	1.010	1.212	1.414	1.616	1.818	2.020	2.222	2.424	2.626	2.829	3.031	3.233	3.435	3.637	3.839	4.041	4.243	4.445
350.	0.416	0.624	0.832	1.040	1.248	1.456	1.664	1.872	2.080	2.288	2.496	2.704	2.912	3.120	3.328	3.536	3.744	3.952	4.160	4.368	4.576
360.	0.428	0.642	0.856	1.070	1.284	1.498	1.712	1.926	2.140	2.354	2.568	2.782	2.996	3.210	3.424	3.637	3.851	4.065	4.279	4.493	4.707
370.	0.440	0.660	0.880	1.100	1.320	1.540	1.759	1.979	2.199	2.419	2.639	2.859	3.079	3.299	3.519	3.739	3.959	4.179	4.399	4.619	4.839

TABLE 2.7f Torispherical Head Thickness Tables in Inches: Joint Efficiency = 0.65

ELLIPSOIDAL HEAD THICKNESS TABLES IN INCHES MAJOR TO MINOR AXIS RATIO=2:1

JOINT EFF.=1.00 ALLOWABLE STRESS VALUE =13800. PSI

PSI	INSIDE DIAMETER, IN. 12.	18.	24.	30.	36.	42.	48.	54.	60.	66.	72.	78.	84.	90.	96.	102.	108.	114.	120.	126.	132.
15.	0.007	0.010	0.013	0.016	0.020	0.023	0.026	0.029	0.033	0.036	0.039	0.042	0.046	0.049	0.052	0.055	0.059	0.062	0.065	0.068	0.072
20.	0.009	0.013	0.017	0.022	0.026	0.030	0.035	0.039	0.043	0.048	0.052	0.057	0.061	0.065	0.070	0.074	0.078	0.083	0.087	0.091	0.096
25.	0.011	0.016	0.022	0.027	0.033	0.038	0.043	0.049	0.054	0.060	0.065	0.071	0.076	0.082	0.087	0.092	0.098	0.103	0.109	0.114	0.120
30.	0.013	0.020	0.026	0.033	0.039	0.046	0.052	0.059	0.065	0.072	0.078	0.085	0.091	0.098	0.104	0.111	0.117	0.124	0.130	0.137	0.144
35.	0.015	0.023	0.030	0.038	0.046	0.053	0.061	0.068	0.076	0.084	0.091	0.099	0.107	0.114	0.122	0.129	0.137	0.145	0.152	0.160	0.167
40.	0.017	0.026	0.035	0.043	0.052	0.061	0.070	0.078	0.087	0.096	0.104	0.113	0.122	0.130	0.139	0.148	0.157	0.165	0.174	0.183	0.191
45.	0.020	0.029	0.039	0.049	0.059	0.069	0.078	0.088	0.098	0.108	0.117	0.127	0.137	0.147	0.157	0.166	0.176	0.186	0.196	0.206	0.215
50.	0.022	0.033	0.043	0.054	0.065	0.076	0.087	0.098	0.109	0.120	0.130	0.141	0.152	0.163	0.174	0.185	0.196	0.207	0.217	0.228	0.239
55.	0.024	0.036	0.048	0.060	0.072	0.084	0.096	0.108	0.120	0.132	0.144	0.155	0.167	0.179	0.191	0.203	0.215	0.227	0.239	0.251	0.263
60.	0.026	0.039	0.052	0.065	0.078	0.091	0.104	0.117	0.130	0.144	0.157	0.170	0.183	0.196	0.209	0.222	0.235	0.248	0.261	0.274	0.287
65.	0.028	0.042	0.057	0.071	0.085	0.099	0.113	0.127	0.141	0.156	0.170	0.184	0.198	0.212	0.226	0.240	0.254	0.269	0.283	0.297	0.311
70.	0.030	0.046	0.061	0.076	0.091	0.107	0.122	0.137	0.152	0.167	0.183	0.198	0.213	0.228	0.244	0.259	0.274	0.289	0.305	0.320	0.335
75.	0.033	0.049	0.065	0.082	0.098	0.114	0.131	0.147	0.163	0.179	0.196	0.212	0.228	0.245	0.261	0.277	0.294	0.310	0.326	0.343	0.359
80.	0.035	0.052	0.070	0.087	0.104	0.122	0.139	0.157	0.174	0.191	0.209	0.226	0.244	0.261	0.278	0.296	0.313	0.331	0.348	0.365	0.383
85.	0.037	0.055	0.074	0.092	0.111	0.129	0.148	0.166	0.185	0.203	0.222	0.240	0.259	0.277	0.296	0.314	0.333	0.351	0.370	0.388	0.407
90.	0.039	0.059	0.078	0.098	0.117	0.137	0.157	0.176	0.196	0.215	0.235	0.255	0.274	0.294	0.313	0.333	0.352	0.372	0.392	0.411	0.431
95.	0.041	0.062	0.083	0.103	0.124	0.145	0.165	0.186	0.207	0.227	0.248	0.269	0.289	0.310	0.331	0.351	0.372	0.393	0.413	0.434	0.455
100.	0.044	0.065	0.087	0.109	0.131	0.152	0.174	0.196	0.218	0.239	0.261	0.283	0.305	0.326	0.348	0.370	0.392	0.413	0.435	0.457	0.479
105.	0.046	0.069	0.091	0.114	0.137	0.160	0.183	0.206	0.228	0.251	0.274	0.297	0.320	0.343	0.365	0.388	0.411	0.434	0.457	0.480	0.503
110.	0.048	0.072	0.096	0.120	0.144	0.168	0.191	0.215	0.239	0.263	0.287	0.311	0.335	0.359	0.383	0.407	0.431	0.455	0.479	0.503	0.527
115.	0.050	0.075	0.100	0.125	0.150	0.175	0.200	0.225	0.250	0.275	0.300	0.325	0.350	0.375	0.400	0.425	0.450	0.475	0.500	0.525	0.550
120.	0.052	0.078	0.104	0.131	0.157	0.183	0.209	0.235	0.261	0.287	0.313	0.339	0.366	0.392	0.418	0.444	0.470	0.496	0.522	0.548	0.574
125.	0.054	0.082	0.109	0.136	0.163	0.190	0.218	0.245	0.272	0.299	0.326	0.354	0.381	0.408	0.435	0.462	0.490	0.517	0.544	0.571	0.598
130.	0.057	0.085	0.113	0.141	0.170	0.198	0.226	0.255	0.283	0.311	0.339	0.368	0.396	0.424	0.453	0.481	0.509	0.537	0.566	0.594	0.622
135.	0.059	0.088	0.118	0.147	0.176	0.206	0.235	0.264	0.294	0.323	0.353	0.382	0.411	0.441	0.470	0.499	0.529	0.558	0.588	0.617	0.646
140.	0.061	0.091	0.122	0.152	0.183	0.213	0.244	0.274	0.305	0.335	0.366	0.396	0.427	0.457	0.487	0.518	0.548	0.579	0.609	0.640	0.670
145.	0.063	0.095	0.126	0.158	0.189	0.221	0.252	0.284	0.316	0.347	0.379	0.410	0.442	0.473	0.505	0.536	0.568	0.600	0.631	0.663	0.694
150.	0.065	0.098	0.131	0.163	0.196	0.229	0.261	0.294	0.326	0.359	0.392	0.424	0.457	0.490	0.522	0.555	0.588	0.620	0.653	0.686	0.718
160.	0.070	0.104	0.139	0.174	0.209	0.244	0.279	0.313	0.348	0.383	0.418	0.453	0.488	0.522	0.557	0.592	0.627	0.662	0.696	0.731	0.766
170.	0.074	0.111	0.148	0.185	0.222	0.259	0.296	0.333	0.370	0.407	0.444	0.481	0.518	0.555	0.592	0.629	0.666	0.703	0.740	0.777	0.814
180.	0.078	0.118	0.157	0.196	0.235	0.274	0.313	0.353	0.392	0.431	0.470	0.509	0.549	0.588	0.627	0.666	0.705	0.744	0.784	0.823	0.862
190.	0.083	0.124	0.165	0.207	0.248	0.290	0.331	0.372	0.414	0.455	0.496	0.538	0.579	0.620	0.662	0.703	0.745	0.786	0.827	0.869	0.910
200.	0.087	0.131	0.174	0.218	0.261	0.305	0.348	0.392	0.435	0.479	0.522	0.566	0.610	0.653	0.697	0.740	0.784	0.827	0.871	0.914	0.958
210.	0.091	0.137	0.183	0.229	0.274	0.320	0.366	0.411	0.457	0.503	0.549	0.594	0.640	0.686	0.732	0.777	0.823	0.869	0.914	0.960	1.006
220.	0.096	0.144	0.192	0.240	0.287	0.335	0.383	0.431	0.479	0.527	0.575	0.623	0.671	0.719	0.766	0.814	0.862	0.910	0.958	1.006	1.054
230.	0.100	0.150	0.200	0.250	0.301	0.351	0.401	0.451	0.501	0.551	0.601	0.651	0.701	0.751	0.801	0.851	0.902	0.952	1.002	1.052	1.102
240.	0.105	0.157	0.209	0.261	0.314	0.366	0.418	0.470	0.523	0.575	0.627	0.679	0.732	0.784	0.836	0.889	0.941	0.993	1.045	1.098	1.150
250.	0.109	0.163	0.218	0.272	0.327	0.381	0.436	0.490	0.544	0.599	0.653	0.708	0.762	0.817	0.871	0.926	0.980	1.034	1.089	1.143	1.198
260.	0.113	0.170	0.227	0.283	0.340	0.396	0.453	0.510	0.566	0.623	0.680	0.736	0.793	0.849	0.906	0.963	1.019	1.076	1.133	1.189	1.246
270.	0.118	0.176	0.235	0.294	0.353	0.412	0.470	0.529	0.588	0.647	0.706	0.765	0.823	0.882	0.941	1.000	1.059	1.117	1.176	1.235	1.294
280.	0.122	0.183	0.244	0.305	0.366	0.427	0.488	0.549	0.610	0.671	0.732	0.793	0.854	0.915	0.976	1.037	1.098	1.159	1.220	1.281	1.342
290.	0.126	0.190	0.253	0.316	0.379	0.442	0.505	0.569	0.632	0.695	0.758	0.821	0.884	0.948	1.011	1.074	1.137	1.200	1.264	1.327	1.390
300.	0.131	0.196	0.261	0.327	0.392	0.458	0.523	0.588	0.654	0.719	0.784	0.850	0.915	0.980	1.046	1.111	1.176	1.242	1.307	1.373	1.438
310.	0.135	0.203	0.270	0.338	0.405	0.473	0.540	0.608	0.675	0.743	0.811	0.878	0.946	1.013	1.081	1.148	1.216	1.283	1.351	1.418	1.486
320.	0.139	0.209	0.279	0.349	0.418	0.488	0.558	0.628	0.697	0.767	0.837	0.906	0.976	1.046	1.116	1.185	1.255	1.325	1.395	1.464	1.534
330.	0.144	0.216	0.288	0.360	0.431	0.503	0.575	0.647	0.719	0.791	0.863	0.935	1.007	1.079	1.151	1.222	1.294	1.366	1.438	1.510	1.582
340.	0.148	0.222	0.296	0.370	0.445	0.519	0.593	0.667	0.741	0.815	0.889	0.963	1.037	1.111	1.186	1.260	1.334	1.408	1.482	1.556	1.630
350.	0.153	0.229	0.305	0.381	0.458	0.534	0.610	0.687	0.763	0.839	0.915	0.992	1.068	1.144	1.220	1.297	1.373	1.449	1.526	1.602	1.678
360.	0.157	0.235	0.314	0.392	0.471	0.549	0.628	0.706	0.785	0.863	0.942	1.020	1.099	1.177	1.255	1.334	1.412	1.491	1.569	1.648	1.726
370.	0.161	0.242	0.323	0.403	0.484	0.565	0.645	0.726	0.807	0.887	0.968	1.048	1.129	1.210	1.290	1.371	1.452	1.532	1.613	1.694	1.774

Table 2.7g Ellipsoidal Head Thickness Tables in Inches: Joint Efficiency = 1.0

ELLIPSOIDAL HEAD THICKNESS TABLES IN INCHES   MAJOR TO MINOR AXIS RATIO=2:1

JOINT EFF.=0.90 ALLOWABLE STRESS VALUE =13800. PSI

PSI	INSIDE DIAMETER, IN. 12.	18.	24.	30.	36.	42.	48.	54.	60.	66.	72.	78.	84.	90.	96.	102.	108.	114.	120.	126.	132.
15.	0.007	0.011	0.014	0.018	0.022	0.025	0.029	0.033	0.036	0.040	0.043	0.047	0.051	0.054	0.058	0.062	0.065	0.069	0.072	0.076	0.080
20.	0.010	0.014	0.019	0.024	0.029	0.034	0.039	0.043	0.048	0.053	0.058	0.063	0.068	0.072	0.077	0.082	0.087	0.092	0.097	0.101	0.106
25.	0.012	0.018	0.024	0.030	0.036	0.042	0.048	0.054	0.060	0.066	0.072	0.079	0.085	0.091	0.097	0.103	0.109	0.115	0.121	0.127	0.133
30.	0.014	0.022	0.029	0.036	0.043	0.051	0.058	0.065	0.072	0.080	0.087	0.094	0.101	0.109	0.116	0.123	0.130	0.138	0.145	0.152	0.159
35.	0.017	0.025	0.034	0.042	0.051	0.059	0.068	0.076	0.085	0.093	0.101	0.110	0.118	0.127	0.135	0.144	0.152	0.161	0.169	0.178	0.186
40.	0.019	0.029	0.039	0.048	0.058	0.068	0.077	0.087	0.097	0.106	0.116	0.126	0.135	0.145	0.155	0.164	0.174	0.184	0.193	0.203	0.213
45.	0.022	0.033	0.043	0.054	0.065	0.076	0.087	0.098	0.109	0.120	0.130	0.141	0.152	0.163	0.174	0.185	0.196	0.207	0.217	0.228	0.239
50.	0.024	0.036	0.048	0.060	0.072	0.085	0.097	0.109	0.121	0.133	0.145	0.157	0.169	0.181	0.193	0.205	0.217	0.230	0.242	0.254	0.266
55.	0.027	0.040	0.053	0.066	0.080	0.093	0.106	0.120	0.133	0.146	0.159	0.173	0.186	0.199	0.213	0.226	0.239	0.253	0.266	0.279	0.292
60.	0.029	0.043	0.058	0.072	0.087	0.101	0.116	0.130	0.145	0.159	0.174	0.188	0.203	0.217	0.232	0.246	0.261	0.275	0.290	0.304	0.319
65.	0.031	0.047	0.063	0.079	0.094	0.110	0.126	0.141	0.157	0.173	0.189	0.204	0.220	0.236	0.251	0.267	0.283	0.298	0.314	0.330	0.346
70.	0.034	0.051	0.068	0.085	0.102	0.118	0.135	0.152	0.169	0.186	0.203	0.220	0.237	0.254	0.271	0.288	0.305	0.321	0.338	0.355	0.372
75.	0.036	0.054	0.073	0.091	0.109	0.127	0.145	0.163	0.181	0.199	0.218	0.236	0.254	0.272	0.290	0.308	0.326	0.344	0.363	0.381	0.399
80.	0.039	0.058	0.077	0.097	0.116	0.135	0.155	0.174	0.193	0.213	0.232	0.251	0.271	0.290	0.309	0.329	0.348	0.367	0.387	0.406	0.425
85.	0.041	0.062	0.082	0.103	0.123	0.144	0.164	0.185	0.205	0.226	0.247	0.267	0.288	0.308	0.329	0.349	0.370	0.390	0.411	0.431	0.452
90.	0.044	0.065	0.087	0.109	0.131	0.152	0.174	0.196	0.218	0.239	0.261	0.283	0.305	0.326	0.348	0.370	0.392	0.413	0.435	0.457	0.479
95.	0.046	0.069	0.092	0.115	0.138	0.161	0.184	0.207	0.230	0.253	0.276	0.299	0.322	0.344	0.367	0.390	0.413	0.436	0.459	0.482	0.505
100.	0.048	0.073	0.097	0.121	0.145	0.169	0.193	0.218	0.242	0.266	0.290	0.314	0.338	0.363	0.387	0.411	0.435	0.459	0.483	0.508	0.532
105.	0.051	0.076	0.102	0.127	0.152	0.178	0.203	0.228	0.254	0.279	0.305	0.330	0.355	0.381	0.406	0.432	0.457	0.482	0.508	0.533	0.558
110.	0.053	0.080	0.106	0.133	0.160	0.186	0.213	0.239	0.266	0.293	0.319	0.346	0.372	0.399	0.425	0.452	0.479	0.505	0.532	0.558	0.585
115.	0.056	0.083	0.111	0.139	0.167	0.195	0.222	0.250	0.278	0.306	0.334	0.361	0.389	0.417	0.445	0.473	0.500	0.528	0.556	0.584	0.612
120.	0.058	0.087	0.116	0.145	0.174	0.203	0.232	0.261	0.290	0.319	0.348	0.377	0.406	0.435	0.464	0.493	0.522	0.551	0.580	0.609	0.638
125.	0.060	0.091	0.121	0.151	0.181	0.212	0.242	0.272	0.302	0.332	0.363	0.393	0.423	0.453	0.484	0.514	0.544	0.574	0.604	0.635	0.665
130.	0.063	0.094	0.126	0.157	0.189	0.220	0.251	0.283	0.314	0.346	0.377	0.409	0.440	0.472	0.503	0.534	0.566	0.597	0.629	0.660	0.692
135.	0.065	0.098	0.131	0.163	0.196	0.229	0.261	0.294	0.326	0.359	0.392	0.424	0.457	0.490	0.522	0.555	0.588	0.620	0.653	0.686	0.718
140.	0.068	0.102	0.135	0.169	0.203	0.237	0.271	0.305	0.339	0.372	0.406	0.440	0.474	0.508	0.542	0.576	0.609	0.643	0.677	0.711	0.745
145.	0.070	0.105	0.140	0.175	0.210	0.245	0.281	0.316	0.351	0.386	0.421	0.456	0.491	0.526	0.561	0.596	0.631	0.666	0.701	0.736	0.771
150.	0.073	0.109	0.145	0.181	0.218	0.254	0.290	0.326	0.363	0.399	0.435	0.472	0.508	0.544	0.580	0.617	0.653	0.689	0.726	0.762	0.798
160.	0.077	0.116	0.155	0.193	0.232	0.271	0.310	0.348	0.387	0.426	0.464	0.503	0.542	0.580	0.619	0.658	0.697	0.735	0.774	0.813	0.851
170.	0.082	0.123	0.164	0.206	0.247	0.288	0.329	0.370	0.411	0.452	0.493	0.535	0.576	0.617	0.658	0.699	0.740	0.781	0.822	0.864	0.905
180.	0.087	0.131	0.174	0.218	0.261	0.305	0.348	0.392	0.435	0.479	0.522	0.566	0.610	0.653	0.697	0.740	0.784	0.827	0.871	0.914	0.958
190.	0.092	0.138	0.184	0.230	0.276	0.322	0.368	0.414	0.460	0.506	0.552	0.598	0.643	0.689	0.735	0.781	0.827	0.873	0.919	0.965	1.011
200.	0.097	0.145	0.194	0.242	0.290	0.339	0.387	0.435	0.484	0.532	0.581	0.629	0.677	0.726	0.774	0.823	0.871	0.919	0.968	1.016	1.065
210.	0.102	0.152	0.203	0.254	0.305	0.356	0.406	0.457	0.508	0.559	0.610	0.661	0.711	0.762	0.813	0.864	0.915	0.965	1.016	1.067	1.118
220.	0.106	0.160	0.213	0.266	0.319	0.373	0.426	0.479	0.532	0.586	0.639	0.692	0.745	0.799	0.852	0.905	0.958	1.011	1.065	1.118	1.171
230.	0.111	0.167	0.223	0.278	0.334	0.390	0.445	0.501	0.557	0.612	0.668	0.724	0.779	0.835	0.891	0.946	1.002	1.058	1.113	1.169	1.224
240.	0.116	0.174	0.232	0.290	0.348	0.407	0.465	0.523	0.581	0.639	0.697	0.755	0.813	0.871	0.929	0.987	1.045	1.104	1.162	1.220	1.278
250.	0.121	0.182	0.242	0.303	0.363	0.424	0.484	0.545	0.605	0.666	0.726	0.787	0.847	0.908	0.968	1.029	1.089	1.150	1.210	1.271	1.331
260.	0.126	0.189	0.252	0.315	0.378	0.441	0.503	0.566	0.629	0.692	0.755	0.818	0.881	0.944	1.007	1.070	1.133	1.196	1.259	1.322	1.385
270.	0.131	0.196	0.261	0.327	0.392	0.458	0.523	0.588	0.654	0.719	0.784	0.850	0.915	0.980	1.046	1.111	1.176	1.242	1.307	1.373	1.438
280.	0.136	0.203	0.271	0.339	0.407	0.474	0.542	0.610	0.678	0.746	0.813	0.881	0.949	1.017	1.085	1.152	1.220	1.288	1.356	1.423	1.491
290.	0.140	0.211	0.281	0.351	0.421	0.491	0.562	0.632	0.702	0.772	0.843	0.913	0.983	1.053	1.123	1.194	1.264	1.334	1.404	1.474	1.545
300.	0.145	0.218	0.291	0.363	0.436	0.508	0.581	0.654	0.726	0.799	0.872	0.944	1.017	1.090	1.162	1.235	1.308	1.380	1.453	1.525	1.598
310.	0.150	0.225	0.300	0.375	0.450	0.525	0.601	0.676	0.751	0.826	0.901	0.976	1.051	1.126	1.201	1.276	1.351	1.426	1.501	1.576	1.651
320.	0.155	0.232	0.310	0.387	0.465	0.542	0.620	0.697	0.775	0.852	0.930	1.007	1.085	1.162	1.240	1.317	1.395	1.472	1.550	1.627	1.705
330.	0.160	0.240	0.320	0.400	0.480	0.559	0.639	0.719	0.799	0.879	0.959	1.039	1.119	1.199	1.279	1.359	1.439	1.519	1.598	1.678	1.758
340.	0.165	0.247	0.329	0.412	0.494	0.576	0.659	0.741	0.824	0.906	0.988	1.071	1.153	1.235	1.318	1.400	1.482	1.565	1.647	1.729	1.812
350.	0.170	0.254	0.339	0.424	0.509	0.593	0.678	0.763	0.848	0.933	1.017	1.102	1.187	1.272	1.356	1.441	1.526	1.611	1.696	1.780	1.865
360.	0.174	0.262	0.349	0.436	0.523	0.610	0.698	0.785	0.872	0.959	1.047	1.134	1.221	1.308	1.395	1.483	1.570	1.657	1.744	1.831	1.919
370.	0.179	0.269	0.359	0.448	0.538	0.627	0.717	0.807	0.896	0.986	1.076	1.165	1.255	1.345	1.434	1.524	1.614	1.703	1.793	1.882	1.972

TABLE 2.7*h* Ellipsoidal Head Thickness Tables in Inches: Joint Efficiency = 0.90

ELLIPSOIDAL HEAD THICKNESS TABLES IN INCHES   MAJOR TO MINOR AXIS RATIO=2:1

JOINT EFF.=0.85 ALLOWABLE STRESS VALUE =13800. PSI

INSIDE DIAMETER,IN.

PSI	12.	18.	24.	30.	36.	42.	48.	54.	60.	66.	72.	78.	84.	90.	96.	102.	108.	114.	120.	126.	132.
15.	0.008	0.012	0.015	0.019	0.023	0.027	0.031	0.035	0.038	0.042	0.046	0.050	0.054	0.058	0.061	0.065	0.069	0.073	0.077	0.081	0.084
20.	0.010	0.015	0.020	0.026	0.031	0.036	0.041	0.046	0.051	0.056	0.061	0.067	0.072	0.077	0.082	0.087	0.092	0.097	0.102	0.107	0.113
25.	0.013	0.019	0.026	0.032	0.038	0.045	0.051	0.058	0.064	0.070	0.077	0.083	0.090	0.096	0.102	0.109	0.115	0.122	0.128	0.134	0.141
30.	0.015	0.023	0.031	0.038	0.046	0.054	0.061	0.069	0.077	0.084	0.092	0.100	0.107	0.115	0.123	0.130	0.138	0.146	0.153	0.161	0.169
35.	0.018	0.027	0.036	0.045	0.054	0.063	0.072	0.081	0.090	0.098	0.107	0.116	0.125	0.134	0.143	0.152	0.161	0.170	0.179	0.188	0.197
40.	0.020	0.031	0.041	0.051	0.061	0.072	0.082	0.092	0.102	0.113	0.123	0.133	0.143	0.154	0.164	0.174	0.184	0.194	0.205	0.215	0.225
45.	0.023	0.035	0.046	0.058	0.069	0.081	0.092	0.104	0.115	0.127	0.138	0.150	0.161	0.173	0.184	0.196	0.207	0.219	0.230	0.242	0.253
50.	0.026	0.038	0.051	0.064	0.077	0.090	0.102	0.115	0.128	0.141	0.154	0.166	0.179	0.192	0.205	0.217	0.230	0.243	0.256	0.269	0.281
55.	0.028	0.042	0.056	0.070	0.084	0.099	0.113	0.127	0.141	0.155	0.169	0.183	0.197	0.211	0.225	0.239	0.253	0.267	0.281	0.296	0.310
60.	0.031	0.046	0.061	0.077	0.092	0.107	0.123	0.138	0.154	0.169	0.184	0.200	0.215	0.230	0.246	0.261	0.276	0.292	0.307	0.322	0.338
65.	0.033	0.050	0.067	0.083	0.100	0.116	0.133	0.150	0.166	0.183	0.200	0.216	0.233	0.249	0.266	0.283	0.299	0.316	0.333	0.349	0.366
70.	0.036	0.054	0.072	0.090	0.107	0.125	0.143	0.161	0.179	0.197	0.215	0.233	0.251	0.269	0.287	0.305	0.322	0.340	0.358	0.376	0.394
75.	0.038	0.058	0.077	0.096	0.115	0.134	0.154	0.173	0.192	0.211	0.230	0.250	0.269	0.288	0.307	0.326	0.345	0.365	0.384	0.403	0.422
80.	0.041	0.061	0.082	0.102	0.123	0.143	0.164	0.184	0.205	0.225	0.246	0.266	0.287	0.307	0.328	0.348	0.369	0.389	0.409	0.430	0.450
85.	0.044	0.065	0.087	0.109	0.131	0.152	0.174	0.196	0.218	0.239	0.261	0.283	0.305	0.326	0.348	0.370	0.392	0.413	0.435	0.457	0.479
90.	0.046	0.069	0.092	0.115	0.138	0.161	0.184	0.207	0.230	0.253	0.276	0.299	0.322	0.346	0.369	0.392	0.415	0.438	0.461	0.484	0.507
95.	0.049	0.073	0.097	0.122	0.146	0.170	0.195	0.219	0.243	0.267	0.292	0.316	0.340	0.365	0.389	0.413	0.438	0.462	0.486	0.511	0.535
100.	0.051	0.077	0.102	0.128	0.154	0.179	0.205	0.230	0.256	0.282	0.307	0.333	0.358	0.384	0.410	0.435	0.461	0.486	0.512	0.538	0.563
105.	0.054	0.081	0.108	0.134	0.161	0.188	0.215	0.242	0.269	0.296	0.323	0.349	0.376	0.403	0.430	0.457	0.484	0.511	0.538	0.564	0.591
110.	0.056	0.084	0.113	0.141	0.169	0.197	0.225	0.253	0.282	0.310	0.338	0.366	0.394	0.422	0.451	0.479	0.507	0.535	0.563	0.591	0.620
115.	0.059	0.088	0.118	0.147	0.177	0.206	0.236	0.265	0.294	0.324	0.353	0.383	0.412	0.442	0.471	0.500	0.530	0.559	0.589	0.618	0.648
120.	0.061	0.092	0.123	0.154	0.184	0.215	0.246	0.276	0.307	0.338	0.369	0.399	0.430	0.461	0.492	0.522	0.553	0.584	0.614	0.645	0.676
125.	0.064	0.096	0.128	0.160	0.192	0.224	0.256	0.288	0.320	0.352	0.384	0.416	0.448	0.480	0.512	0.544	0.576	0.608	0.640	0.672	0.704
130.	0.067	0.100	0.133	0.166	0.200	0.233	0.266	0.300	0.333	0.366	0.399	0.433	0.466	0.499	0.533	0.566	0.599	0.632	0.666	0.699	0.732
135.	0.069	0.104	0.138	0.173	0.207	0.242	0.277	0.311	0.346	0.380	0.415	0.449	0.484	0.518	0.553	0.588	0.622	0.657	0.691	0.726	0.760
140.	0.072	0.108	0.143	0.179	0.215	0.251	0.287	0.323	0.358	0.394	0.430	0.466	0.502	0.538	0.574	0.609	0.645	0.681	0.717	0.753	0.789
145.	0.074	0.111	0.149	0.186	0.223	0.260	0.297	0.334	0.371	0.408	0.446	0.483	0.520	0.557	0.594	0.631	0.668	0.705	0.743	0.780	0.817
150.	0.077	0.115	0.154	0.192	0.230	0.269	0.307	0.346	0.384	0.423	0.461	0.499	0.538	0.576	0.615	0.653	0.691	0.730	0.768	0.807	0.845
160.	0.082	0.123	0.164	0.205	0.246	0.287	0.328	0.369	0.410	0.451	0.492	0.533	0.574	0.615	0.656	0.697	0.738	0.779	0.820	0.861	0.901
170.	0.087	0.131	0.174	0.218	0.261	0.305	0.348	0.392	0.435	0.479	0.522	0.566	0.610	0.653	0.697	0.740	0.784	0.827	0.871	0.914	0.958
180.	0.092	0.138	0.184	0.231	0.277	0.323	0.369	0.415	0.461	0.507	0.553	0.599	0.645	0.692	0.738	0.784	0.830	0.876	0.922	0.968	1.014
190.	0.097	0.146	0.195	0.243	0.292	0.341	0.389	0.438	0.487	0.535	0.584	0.633	0.681	0.730	0.779	0.827	0.876	0.925	0.973	1.022	1.071
200.	0.102	0.154	0.205	0.256	0.307	0.359	0.410	0.461	0.512	0.564	0.615	0.666	0.717	0.769	0.820	0.871	0.922	0.974	1.025	1.076	1.127
210.	0.108	0.161	0.215	0.269	0.323	0.377	0.430	0.484	0.538	0.592	0.646	0.699	0.753	0.807	0.861	0.915	0.968	1.022	1.076	1.130	1.184
220.	0.113	0.169	0.225	0.282	0.338	0.395	0.451	0.507	0.564	0.620	0.676	0.733	0.789	0.846	0.902	0.958	1.015	1.071	1.127	1.184	1.240
230.	0.118	0.177	0.236	0.295	0.354	0.413	0.472	0.530	0.589	0.648	0.707	0.766	0.825	0.884	0.943	1.002	1.061	1.120	1.179	1.238	1.297
240.	0.123	0.185	0.246	0.308	0.369	0.431	0.492	0.554	0.615	0.677	0.738	0.800	0.861	0.923	0.984	1.046	1.107	1.169	1.230	1.292	1.353
250.	0.128	0.192	0.256	0.320	0.384	0.449	0.513	0.577	0.641	0.705	0.769	0.833	0.897	0.961	1.025	1.089	1.153	1.217	1.282	1.346	1.410
260.	0.133	0.200	0.267	0.333	0.400	0.467	0.533	0.600	0.666	0.733	0.800	0.866	0.933	1.000	1.066	1.133	1.200	1.266	1.333	1.400	1.466
270.	0.138	0.208	0.277	0.346	0.415	0.484	0.554	0.623	0.692	0.761	0.831	0.900	0.969	1.038	1.107	1.177	1.246	1.315	1.384	1.453	1.523
280.	0.144	0.215	0.287	0.359	0.431	0.502	0.574	0.646	0.718	0.790	0.861	0.933	1.005	1.077	1.149	1.220	1.292	1.364	1.436	1.507	1.579
290.	0.149	0.223	0.297	0.372	0.446	0.520	0.595	0.669	0.744	0.818	0.892	0.967	1.041	1.115	1.190	1.264	1.338	1.413	1.487	1.561	1.636
300.	0.154	0.231	0.308	0.385	0.462	0.538	0.615	0.692	0.769	0.846	0.923	1.000	1.077	1.154	1.231	1.308	1.385	1.462	1.538	1.615	1.692
310.	0.159	0.238	0.318	0.397	0.477	0.556	0.636	0.715	0.795	0.874	0.954	1.033	1.113	1.192	1.272	1.351	1.431	1.510	1.590	1.669	1.749
320.	0.164	0.246	0.328	0.410	0.492	0.574	0.657	0.739	0.821	0.903	0.985	1.067	1.149	1.231	1.313	1.395	1.477	1.559	1.641	1.723	1.805
330.	0.169	0.254	0.339	0.423	0.508	0.592	0.677	0.762	0.846	0.931	1.016	1.100	1.185	1.270	1.354	1.439	1.523	1.608	1.693	1.777	1.862
340.	0.174	0.262	0.349	0.436	0.523	0.610	0.698	0.785	0.872	0.959	1.047	1.134	1.221	1.308	1.395	1.483	1.570	1.657	1.744	1.831	1.919
350.	0.180	0.269	0.359	0.449	0.539	0.628	0.718	0.808	0.898	0.988	1.077	1.167	1.257	1.347	1.437	1.526	1.616	1.706	1.796	1.885	1.975
360.	0.185	0.277	0.369	0.462	0.554	0.646	0.739	0.831	0.924	1.016	1.108	1.201	1.293	1.385	1.478	1.570	1.662	1.755	1.847	1.939	2.032
370.	0.190	0.285	0.380	0.475	0.570	0.665	0.759	0.854	0.949	1.044	1.139	1.234	1.329	1.424	1.519	1.614	1.709	1.804	1.899	1.994	2.088

TABLE 2.7i Ellipsoidal Head Thickness Tables in Inches: Joint Efficiency = 0.85

ELLIPSOIDAL HEAD THICKNESS TABLES IN INCHES MAJOR TO MINOR AXIS RATIO=2:1

JOINT EFF.=0.80 ALLOWABLE STRESS VALUE =13800. PSI

	INSIDE DIAMETER, IN.																				
PSI	12.	18.	24.	30.	36.	42.	48.	54.	60.	66.	72.	78.	84.	90.	96.	102.	108.	114.	120.	126.	132.
15.	0.008	0.012	0.016	0.020	0.024	0.029	0.033	0.037	0.041	0.045	0.049	0.053	0.057	0.061	0.065	0.069	0.073	0.077	0.082	0.086	0.090
20.	0.011	0.016	0.022	0.027	0.033	0.038	0.043	0.049	0.054	0.060	0.065	0.071	0.076	0.082	0.087	0.092	0.098	0.103	0.109	0.114	0.120
25.	0.014	0.020	0.027	0.034	0.041	0.048	0.054	0.061	0.068	0.075	0.082	0.088	0.095	0.102	0.109	0.116	0.122	0.129	0.136	0.143	0.149
30.	0.016	0.024	0.033	0.041	0.049	0.057	0.065	0.073	0.082	0.090	0.098	0.106	0.114	0.122	0.130	0.139	0.147	0.155	0.163	0.171	0.179
35.	0.019	0.029	0.038	0.048	0.057	0.067	0.076	0.086	0.095	0.105	0.114	0.124	0.133	0.143	0.152	0.162	0.171	0.181	0.190	0.200	0.209
40.	0.022	0.033	0.043	0.054	0.065	0.076	0.087	0.098	0.109	0.120	0.130	0.141	0.152	0.163	0.174	0.185	0.196	0.207	0.217	0.228	0.239
45.	0.024	0.037	0.049	0.061	0.073	0.086	0.098	0.110	0.122	0.135	0.147	0.159	0.171	0.183	0.196	0.208	0.220	0.232	0.245	0.257	0.269
50.	0.027	0.041	0.054	0.068	0.082	0.095	0.109	0.122	0.136	0.150	0.163	0.177	0.190	0.204	0.217	0.231	0.245	0.258	0.272	0.285	0.299
55.	0.030	0.045	0.060	0.075	0.090	0.105	0.120	0.135	0.150	0.164	0.179	0.194	0.209	0.224	0.239	0.254	0.269	0.284	0.299	0.314	0.329
60.	0.033	0.049	0.065	0.082	0.098	0.114	0.131	0.147	0.163	0.179	0.196	0.212	0.228	0.245	0.261	0.277	0.294	0.310	0.326	0.343	0.359
65.	0.035	0.053	0.071	0.088	0.106	0.124	0.141	0.159	0.177	0.194	0.212	0.230	0.247	0.265	0.283	0.300	0.318	0.336	0.353	0.371	0.389
70.	0.038	0.057	0.076	0.095	0.114	0.133	0.152	0.171	0.190	0.209	0.228	0.247	0.266	0.286	0.305	0.324	0.343	0.362	0.381	0.400	0.419
75.	0.041	0.061	0.082	0.102	0.122	0.143	0.163	0.184	0.204	0.224	0.245	0.265	0.286	0.306	0.326	0.347	0.367	0.387	0.408	0.428	0.449
80.	0.044	0.065	0.087	0.109	0.131	0.152	0.174	0.196	0.218	0.239	0.261	0.283	0.305	0.326	0.348	0.370	0.392	0.413	0.435	0.457	0.479
85.	0.046	0.069	0.092	0.116	0.139	0.162	0.185	0.208	0.231	0.254	0.277	0.301	0.324	0.347	0.370	0.393	0.416	0.439	0.462	0.485	0.509
90.	0.049	0.073	0.098	0.122	0.147	0.171	0.196	0.220	0.245	0.269	0.294	0.318	0.343	0.367	0.392	0.416	0.441	0.465	0.490	0.514	0.538
95.	0.052	0.078	0.103	0.129	0.155	0.181	0.207	0.233	0.258	0.284	0.310	0.336	0.362	0.388	0.413	0.439	0.465	0.491	0.517	0.543	0.568
100.	0.054	0.082	0.109	0.136	0.163	0.190	0.218	0.245	0.272	0.299	0.326	0.354	0.381	0.408	0.435	0.462	0.490	0.517	0.544	0.571	0.598
105.	0.057	0.086	0.114	0.143	0.171	0.200	0.228	0.257	0.286	0.314	0.343	0.371	0.400	0.428	0.457	0.486	0.514	0.543	0.571	0.600	0.628
110.	0.060	0.090	0.120	0.150	0.180	0.209	0.239	0.269	0.299	0.329	0.359	0.389	0.419	0.449	0.479	0.509	0.539	0.569	0.598	0.628	0.658
115.	0.063	0.094	0.125	0.156	0.188	0.219	0.250	0.282	0.313	0.344	0.375	0.407	0.438	0.469	0.501	0.532	0.563	0.594	0.626	0.657	0.688
120.	0.065	0.098	0.131	0.163	0.196	0.229	0.261	0.294	0.326	0.359	0.392	0.424	0.457	0.490	0.522	0.555	0.588	0.620	0.653	0.686	0.718
125.	0.068	0.102	0.136	0.170	0.204	0.238	0.272	0.306	0.340	0.374	0.408	0.442	0.476	0.510	0.544	0.578	0.612	0.646	0.680	0.714	0.748
130.	0.071	0.106	0.141	0.177	0.212	0.248	0.283	0.318	0.354	0.389	0.424	0.460	0.495	0.531	0.566	0.601	0.637	0.672	0.707	0.743	0.778
135.	0.073	0.110	0.147	0.184	0.220	0.257	0.294	0.331	0.367	0.404	0.441	0.477	0.514	0.551	0.588	0.624	0.661	0.698	0.735	0.771	0.808
140.	0.076	0.114	0.152	0.190	0.229	0.267	0.305	0.343	0.381	0.419	0.457	0.495	0.533	0.571	0.609	0.648	0.686	0.724	0.762	0.800	0.838
145.	0.079	0.118	0.158	0.197	0.237	0.276	0.316	0.355	0.395	0.434	0.473	0.513	0.552	0.592	0.631	0.671	0.710	0.750	0.789	0.829	0.868
150.	0.082	0.122	0.163	0.204	0.245	0.286	0.327	0.367	0.408	0.449	0.490	0.531	0.571	0.612	0.653	0.694	0.735	0.776	0.816	0.857	0.898
160.	0.087	0.131	0.174	0.218	0.261	0.305	0.348	0.392	0.435	0.479	0.522	0.566	0.610	0.653	0.697	0.740	0.784	0.827	0.871	0.914	0.958
170.	0.093	0.139	0.185	0.231	0.278	0.324	0.370	0.416	0.463	0.509	0.555	0.601	0.648	0.694	0.740	0.787	0.833	0.879	0.925	0.972	1.018
180.	0.098	0.147	0.196	0.245	0.294	0.343	0.392	0.441	0.490	0.539	0.588	0.637	0.686	0.735	0.784	0.833	0.882	0.931	0.980	1.029	1.078
190.	0.103	0.155	0.207	0.259	0.310	0.362	0.414	0.465	0.517	0.569	0.621	0.672	0.724	0.776	0.828	0.879	0.931	0.983	1.034	1.086	1.138
200.	0.109	0.163	0.218	0.272	0.327	0.381	0.436	0.490	0.544	0.599	0.653	0.708	0.762	0.817	0.871	0.926	0.980	1.034	1.089	1.143	1.198
210.	0.114	0.172	0.229	0.286	0.343	0.400	0.457	0.515	0.572	0.629	0.686	0.743	0.800	0.858	0.915	0.972	1.029	1.086	1.143	1.201	1.258
220.	0.120	0.180	0.240	0.300	0.359	0.419	0.479	0.539	0.599	0.659	0.719	0.779	0.839	0.899	0.958	1.018	1.078	1.138	1.198	1.258	1.318
230.	0.125	0.188	0.251	0.313	0.376	0.438	0.501	0.564	0.626	0.689	0.752	0.814	0.877	0.939	1.002	1.065	1.127	1.190	1.253	1.315	1.378
240.	0.131	0.196	0.261	0.327	0.392	0.458	0.523	0.588	0.654	0.719	0.784	0.850	0.915	0.980	1.046	1.111	1.176	1.242	1.307	1.373	1.438
250.	0.136	0.204	0.272	0.340	0.409	0.477	0.545	0.613	0.681	0.749	0.817	0.885	0.953	1.021	1.089	1.158	1.226	1.294	1.362	1.430	1.498
260.	0.142	0.212	0.283	0.354	0.425	0.496	0.567	0.637	0.708	0.779	0.850	0.921	0.991	1.062	1.133	1.204	1.275	1.346	1.416	1.487	1.558
270.	0.147	0.221	0.294	0.368	0.441	0.515	0.588	0.662	0.735	0.809	0.883	0.956	1.030	1.103	1.177	1.250	1.324	1.397	1.471	1.545	1.618
280.	0.153	0.229	0.305	0.381	0.458	0.534	0.610	0.687	0.763	0.839	0.915	0.992	1.068	1.144	1.220	1.297	1.373	1.449	1.526	1.602	1.678
290.	0.158	0.237	0.316	0.395	0.474	0.553	0.632	0.711	0.790	0.869	0.948	1.027	1.106	1.185	1.264	1.343	1.422	1.501	1.580	1.659	1.738
300.	0.163	0.245	0.327	0.409	0.490	0.572	0.654	0.736	0.817	0.899	0.981	1.063	1.144	1.226	1.308	1.390	1.471	1.553	1.635	1.717	1.798
310.	0.169	0.253	0.338	0.422	0.507	0.591	0.676	0.760	0.845	0.929	1.014	1.098	1.183	1.267	1.352	1.436	1.521	1.605	1.690	1.774	1.858
320.	0.174	0.262	0.349	0.436	0.523	0.610	0.698	0.785	0.872	0.959	1.047	1.134	1.221	1.308	1.395	1.483	1.570	1.657	1.744	1.831	1.919
330.	0.180	0.270	0.360	0.450	0.540	0.630	0.720	0.809	0.899	0.989	1.079	1.169	1.259	1.349	1.439	1.529	1.619	1.709	1.799	1.889	1.979
340.	0.185	0.278	0.371	0.463	0.556	0.649	0.741	0.834	0.927	1.019	1.112	1.205	1.297	1.390	1.483	1.576	1.668	1.761	1.854	1.946	2.039
350.	0.191	0.286	0.382	0.477	0.572	0.668	0.763	0.859	0.954	1.050	1.145	1.240	1.336	1.431	1.527	1.622	1.717	1.813	1.908	2.004	2.099
360.	0.196	0.294	0.393	0.491	0.589	0.687	0.785	0.883	0.981	1.080	1.178	1.276	1.374	1.472	1.570	1.668	1.767	1.865	1.963	2.061	2.159
370.	0.202	0.303	0.404	0.504	0.605	0.706	0.807	0.908	1.009	1.110	1.211	1.311	1.412	1.513	1.614	1.715	1.816	1.917	2.018	2.119	2.219

Table 2.7*j* Ellipsoidal Head Thickness Tables in Inches: Joint Efficiency = 0.80

ELLIPSOIDAL HEAD THICKNESS TABLES IN INCHES   MAJOR TO MINOR AXIS RATIO=2:1

JOINT EFF.=0.70 ALLOWABLE STRESS VALUE =13800. PSI

PSI	INSIDE DIAMETER, IN. 12.	18.	24.	30.	36.	42.	48.	54.	60.	66.	72.	78.	84.	90.	96.	102.	108.	114.	120.	126.	132.
15.	0.009	0.014	0.019	0.023	0.028	0.033	0.037	0.042	0.047	0.051	0.056	0.061	0.065	0.070	0.075	0.079	0.084	0.089	0.093	0.098	0.103
20.	0.012	0.019	0.025	0.031	0.037	0.043	0.050	0.056	0.062	0.068	0.075	0.081	0.087	0.093	0.099	0.106	0.112	0.118	0.124	0.130	0.137
25.	0.016	0.023	0.031	0.039	0.047	0.054	0.062	0.070	0.078	0.085	0.093	0.101	0.109	0.116	0.124	0.132	0.140	0.148	0.155	0.163	0.171
30.	0.019	0.028	0.037	0.047	0.056	0.065	0.075	0.084	0.093	0.103	0.112	0.121	0.130	0.140	0.149	0.158	0.168	0.177	0.186	0.196	0.205
35.	0.022	0.033	0.043	0.054	0.065	0.076	0.087	0.098	0.109	0.120	0.130	0.141	0.152	0.163	0.174	0.185	0.196	0.207	0.217	0.228	0.239
40.	0.025	0.037	0.050	0.062	0.075	0.087	0.099	0.112	0.124	0.137	0.149	0.162	0.174	0.186	0.199	0.211	0.224	0.236	0.249	0.261	0.273
45.	0.028	0.042	0.056	0.070	0.084	0.098	0.112	0.126	0.140	0.154	0.168	0.182	0.196	0.210	0.224	0.238	0.252	0.266	0.280	0.294	0.308
50.	0.031	0.047	0.062	0.078	0.093	0.109	0.124	0.140	0.155	0.171	0.186	0.202	0.218	0.233	0.249	0.264	0.280	0.295	0.311	0.326	0.342
55.	0.034	0.051	0.068	0.085	0.103	0.120	0.137	0.154	0.171	0.188	0.205	0.222	0.239	0.256	0.273	0.291	0.308	0.325	0.342	0.359	0.376
60.	0.037	0.056	0.075	0.093	0.112	0.131	0.149	0.168	0.186	0.205	0.224	0.242	0.261	0.280	0.298	0.317	0.336	0.354	0.373	0.392	0.410
65.	0.040	0.061	0.081	0.101	0.121	0.141	0.162	0.182	0.202	0.222	0.242	0.263	0.283	0.303	0.323	0.343	0.364	0.384	0.404	0.424	0.444
70.	0.044	0.065	0.087	0.109	0.131	0.152	0.174	0.196	0.218	0.239	0.261	0.283	0.305	0.326	0.348	0.370	0.392	0.413	0.435	0.457	0.479
75.	0.047	0.070	0.093	0.117	0.140	0.163	0.186	0.210	0.233	0.256	0.280	0.303	0.326	0.350	0.373	0.396	0.420	0.443	0.466	0.490	0.513
80.	0.050	0.075	0.099	0.124	0.149	0.174	0.199	0.224	0.249	0.274	0.298	0.323	0.348	0.373	0.398	0.423	0.448	0.472	0.497	0.522	0.547
85.	0.053	0.079	0.106	0.132	0.159	0.185	0.211	0.238	0.264	0.291	0.317	0.343	0.370	0.396	0.423	0.449	0.476	0.502	0.528	0.555	0.581
90.	0.056	0.084	0.112	0.140	0.168	0.196	0.224	0.252	0.280	0.308	0.336	0.364	0.392	0.420	0.448	0.476	0.504	0.532	0.560	0.588	0.615
95.	0.059	0.089	0.118	0.148	0.177	0.207	0.236	0.266	0.295	0.325	0.354	0.384	0.413	0.443	0.473	0.502	0.532	0.561	0.591	0.620	0.650
100.	0.062	0.093	0.124	0.155	0.187	0.218	0.249	0.280	0.311	0.342	0.373	0.404	0.435	0.466	0.497	0.528	0.560	0.591	0.622	0.653	0.684
105.	0.065	0.098	0.131	0.163	0.196	0.229	0.261	0.294	0.326	0.359	0.392	0.424	0.457	0.490	0.522	0.555	0.588	0.620	0.653	0.686	0.718
110.	0.068	0.103	0.137	0.171	0.205	0.239	0.274	0.308	0.342	0.376	0.410	0.445	0.479	0.513	0.547	0.581	0.616	0.650	0.684	0.718	0.752
115.	0.072	0.107	0.143	0.179	0.215	0.250	0.286	0.322	0.358	0.393	0.429	0.465	0.501	0.536	0.572	0.608	0.644	0.679	0.715	0.751	0.787
120.	0.075	0.112	0.149	0.187	0.224	0.261	0.299	0.336	0.373	0.410	0.448	0.485	0.522	0.560	0.597	0.634	0.672	0.709	0.746	0.784	0.821
125.	0.078	0.117	0.155	0.194	0.233	0.272	0.311	0.350	0.389	0.428	0.466	0.505	0.544	0.583	0.622	0.661	0.700	0.739	0.777	0.816	0.855
130.	0.081	0.121	0.162	0.202	0.243	0.283	0.323	0.364	0.404	0.445	0.485	0.526	0.566	0.606	0.647	0.687	0.728	0.768	0.809	0.849	0.889
135.	0.084	0.126	0.168	0.210	0.252	0.294	0.336	0.378	0.420	0.462	0.504	0.546	0.588	0.630	0.672	0.714	0.756	0.798	0.840	0.882	0.924
140.	0.087	0.131	0.174	0.218	0.261	0.305	0.348	0.392	0.435	0.479	0.522	0.566	0.610	0.653	0.697	0.740	0.784	0.827	0.871	0.914	0.958
145.	0.090	0.135	0.180	0.225	0.271	0.316	0.361	0.406	0.451	0.496	0.541	0.586	0.631	0.676	0.722	0.767	0.812	0.857	0.902	0.947	0.992
150.	0.093	0.140	0.187	0.233	0.280	0.327	0.373	0.420	0.467	0.513	0.560	0.607	0.653	0.700	0.747	0.793	0.840	0.886	0.933	0.980	1.026
160.	0.100	0.149	0.199	0.249	0.299	0.348	0.398	0.448	0.498	0.547	0.597	0.647	0.697	0.747	0.796	0.846	0.896	0.946	0.995	1.045	1.095
170.	0.106	0.159	0.212	0.264	0.317	0.370	0.423	0.476	0.529	0.582	0.635	0.688	0.740	0.793	0.846	0.899	0.952	1.005	1.058	1.111	1.164
180.	0.112	0.168	0.224	0.280	0.336	0.392	0.448	0.504	0.560	0.616	0.672	0.728	0.784	0.840	0.896	0.952	1.008	1.064	1.120	1.176	1.232
190.	0.118	0.177	0.236	0.296	0.355	0.414	0.473	0.532	0.591	0.650	0.709	0.769	0.828	0.887	0.946	1.005	1.064	1.123	1.182	1.242	1.301
200.	0.124	0.187	0.249	0.311	0.373	0.436	0.498	0.560	0.622	0.685	0.747	0.809	0.871	0.934	0.996	1.058	1.120	1.183	1.245	1.307	1.369
210.	0.131	0.196	0.261	0.327	0.392	0.458	0.523	0.588	0.654	0.719	0.784	0.850	0.915	0.980	1.046	1.111	1.176	1.242	1.307	1.373	1.438
220.	0.137	0.205	0.274	0.342	0.411	0.479	0.548	0.616	0.685	0.753	0.822	0.890	0.959	1.027	1.096	1.164	1.233	1.301	1.370	1.438	1.507
230.	0.143	0.215	0.286	0.358	0.430	0.501	0.573	0.644	0.716	0.788	0.859	0.931	1.002	1.074	1.146	1.217	1.289	1.360	1.432	1.504	1.575
240.	0.149	0.224	0.299	0.374	0.448	0.523	0.598	0.672	0.747	0.822	0.897	0.971	1.046	1.121	1.196	1.270	1.345	1.420	1.494	1.569	1.644
250.	0.156	0.234	0.311	0.389	0.467	0.545	0.623	0.701	0.778	0.856	0.934	1.012	1.090	1.168	1.245	1.323	1.401	1.479	1.557	1.635	1.713
260.	0.162	0.243	0.324	0.405	0.486	0.567	0.648	0.729	0.810	0.891	0.972	1.053	1.133	1.214	1.295	1.376	1.457	1.538	1.619	1.700	1.781
270.	0.168	0.252	0.336	0.420	0.505	0.589	0.673	0.757	0.841	0.925	1.009	1.093	1.177	1.261	1.345	1.429	1.514	1.598	1.682	1.766	1.850
280.	0.174	0.262	0.349	0.436	0.523	0.610	0.698	0.785	0.872	0.959	1.047	1.134	1.221	1.308	1.395	1.483	1.570	1.657	1.744	1.831	1.919
290.	0.181	0.271	0.361	0.452	0.542	0.632	0.723	0.813	0.903	0.994	1.084	1.174	1.265	1.355	1.445	1.536	1.626	1.716	1.807	1.897	1.987
300.	0.187	0.280	0.374	0.467	0.561	0.654	0.748	0.841	0.935	1.028	1.121	1.215	1.308	1.402	1.495	1.589	1.682	1.776	1.869	1.963	2.056
310.	0.193	0.290	0.386	0.483	0.579	0.676	0.773	0.869	0.966	1.062	1.159	1.256	1.352	1.449	1.545	1.642	1.738	1.835	1.932	2.028	2.125
320.	0.199	0.299	0.399	0.499	0.598	0.698	0.798	0.897	0.997	1.097	1.197	1.296	1.396	1.496	1.595	1.695	1.795	1.894	1.994	2.094	2.194
330.	0.206	0.309	0.411	0.514	0.617	0.720	0.823	0.926	1.028	1.131	1.234	1.337	1.440	1.543	1.645	1.748	1.851	1.954	2.057	2.160	2.262
340.	0.212	0.318	0.424	0.530	0.636	0.742	0.848	0.954	1.060	1.166	1.272	1.378	1.483	1.589	1.695	1.801	1.907	2.013	2.119	2.225	2.331
350.	0.218	0.327	0.436	0.545	0.655	0.764	0.873	0.982	1.091	1.200	1.309	1.418	1.527	1.636	1.745	1.855	1.964	2.073	2.182	2.291	2.400
360.	0.224	0.337	0.449	0.561	0.673	0.786	0.898	1.010	1.122	1.234	1.347	1.459	1.571	1.683	1.796	1.908	2.020	2.132	2.244	2.357	2.469
370.	0.231	0.346	0.461	0.577	0.692	0.807	0.923	1.038	1.153	1.269	1.384	1.500	1.615	1.730	1.846	1.961	2.076	2.192	2.307	2.422	2.538

TABLE 2.7k Ellipsoidal Head Thickness Tables in Inches: Joint Efficiency = 0.70

ELLIPSOIDAL HEAD THICKNESS TABLES IN INCHES   MAJOR TO MINOR AXIS RATIO=2:1

JOINT EFF.=0.65   ALLOWABLE STRESS VALUE =13800. PSI

INSIDE DIAMETER,IN.

PSI	12.	18.	24.	30.	36.	42.	48.	54.	60.	66.	72.	78.	84.	90.	96.	102.	108.	114.	120.	126.	132.
15.	0.010	0.015	0.020	0.025	0.030	0.035	0.040	0.045	0.050	0.055	0.060	0.065	0.070	0.075	0.080	0.085	0.090	0.095	0.100	0.105	0.110
20.	0.013	0.020	0.027	0.033	0.040	0.047	0.054	0.060	0.067	0.074	0.080	0.087	0.094	0.100	0.107	0.114	0.120	0.127	0.134	0.140	0.147
25.	0.017	0.025	0.033	0.042	0.050	0.059	0.067	0.075	0.084	0.092	0.100	0.109	0.117	0.125	0.134	0.142	0.151	0.159	0.167	0.176	0.184
30.	0.020	0.030	0.040	0.050	0.060	0.070	0.080	0.090	0.100	0.110	0.120	0.130	0.141	0.151	0.161	0.171	0.181	0.191	0.201	0.211	0.221
35.	0.023	0.035	0.047	0.059	0.070	0.082	0.094	0.105	0.117	0.129	0.141	0.152	0.164	0.176	0.187	0.199	0.211	0.222	0.234	0.246	0.258
40.	0.027	0.040	0.054	0.067	0.080	0.094	0.107	0.120	0.134	0.147	0.161	0.174	0.187	0.201	0.214	0.228	0.241	0.254	0.268	0.281	0.294
45.	0.030	0.045	0.060	0.075	0.090	0.105	0.120	0.136	0.151	0.166	0.181	0.196	0.211	0.226	0.241	0.256	0.271	0.286	0.301	0.316	0.331
50.	0.033	0.050	0.067	0.084	0.100	0.117	0.134	0.151	0.167	0.184	0.201	0.218	0.234	0.251	0.268	0.284	0.301	0.318	0.335	0.351	0.368
55.	0.037	0.055	0.074	0.092	0.110	0.129	0.147	0.166	0.184	0.202	0.221	0.239	0.258	0.276	0.294	0.313	0.331	0.350	0.368	0.387	0.405
60.	0.040	0.060	0.080	0.100	0.120	0.141	0.161	0.181	0.201	0.221	0.241	0.261	0.281	0.301	0.321	0.341	0.361	0.382	0.402	0.422	0.442
65.	0.044	0.065	0.087	0.109	0.131	0.152	0.174	0.196	0.218	0.239	0.261	0.283	0.305	0.326	0.348	0.370	0.392	0.413	0.435	0.457	0.479
70.	0.047	0.070	0.094	0.117	0.141	0.164	0.187	0.211	0.234	0.258	0.281	0.305	0.328	0.351	0.375	0.398	0.422	0.445	0.469	0.492	0.515
75.	0.050	0.075	0.100	0.126	0.151	0.176	0.201	0.226	0.251	0.276	0.301	0.326	0.351	0.377	0.402	0.427	0.452	0.477	0.502	0.527	0.552
80.	0.054	0.080	0.107	0.134	0.161	0.187	0.214	0.241	0.268	0.295	0.321	0.348	0.375	0.402	0.428	0.455	0.482	0.509	0.536	0.562	0.589
85.	0.057	0.085	0.114	0.142	0.171	0.199	0.228	0.256	0.285	0.313	0.341	0.370	0.398	0.427	0.455	0.484	0.512	0.541	0.569	0.598	0.626
90.	0.060	0.090	0.121	0.151	0.181	0.211	0.241	0.271	0.301	0.331	0.362	0.392	0.422	0.452	0.482	0.512	0.542	0.572	0.603	0.633	0.663
95.	0.064	0.095	0.127	0.159	0.191	0.223	0.254	0.286	0.318	0.350	0.382	0.413	0.445	0.477	0.509	0.541	0.573	0.604	0.636	0.668	0.700
100.	0.067	0.100	0.134	0.167	0.201	0.234	0.268	0.301	0.335	0.368	0.402	0.435	0.469	0.502	0.536	0.569	0.603	0.636	0.670	0.703	0.737
105.	0.070	0.105	0.141	0.176	0.211	0.246	0.281	0.316	0.352	0.387	0.422	0.457	0.492	0.527	0.563	0.598	0.633	0.668	0.703	0.738	0.773
110.	0.074	0.111	0.147	0.184	0.221	0.258	0.295	0.332	0.368	0.405	0.442	0.479	0.516	0.553	0.589	0.626	0.663	0.700	0.737	0.774	0.810
115.	0.077	0.116	0.154	0.193	0.231	0.270	0.308	0.347	0.385	0.424	0.462	0.501	0.539	0.578	0.616	0.655	0.693	0.732	0.770	0.809	0.847
120.	0.080	0.121	0.161	0.201	0.241	0.281	0.322	0.362	0.402	0.442	0.482	0.522	0.563	0.603	0.643	0.683	0.723	0.764	0.804	0.844	0.884
125.	0.084	0.126	0.167	0.209	0.251	0.293	0.335	0.377	0.419	0.461	0.502	0.544	0.586	0.628	0.670	0.712	0.754	0.795	0.837	0.879	0.921
130.	0.087	0.131	0.174	0.218	0.261	0.305	0.348	0.392	0.435	0.479	0.522	0.566	0.610	0.653	0.697	0.740	0.784	0.827	0.871	0.914	0.958
135.	0.090	0.136	0.181	0.226	0.271	0.317	0.362	0.407	0.452	0.497	0.543	0.588	0.633	0.678	0.723	0.769	0.814	0.859	0.904	0.950	0.995
140.	0.094	0.141	0.188	0.234	0.281	0.328	0.375	0.422	0.469	0.516	0.563	0.610	0.657	0.703	0.750	0.797	0.844	0.891	0.938	0.985	1.032
145.	0.097	0.146	0.194	0.243	0.291	0.340	0.389	0.437	0.486	0.534	0.583	0.631	0.680	0.729	0.777	0.826	0.874	0.923	0.971	1.020	1.069
150.	0.101	0.151	0.201	0.251	0.302	0.352	0.402	0.452	0.503	0.553	0.603	0.653	0.704	0.754	0.804	0.854	0.905	0.955	1.005	1.055	1.106
160.	0.107	0.161	0.214	0.268	0.322	0.375	0.429	0.482	0.536	0.590	0.643	0.697	0.751	0.804	0.858	0.911	0.965	1.019	1.072	1.126	1.179
170.	0.114	0.171	0.228	0.285	0.342	0.399	0.456	0.513	0.570	0.627	0.684	0.741	0.797	0.854	0.911	0.968	1.025	1.082	1.139	1.196	1.253
180.	0.121	0.181	0.241	0.302	0.362	0.422	0.483	0.543	0.603	0.664	0.724	0.784	0.845	0.905	0.965	1.025	1.086	1.146	1.206	1.267	1.327
190.	0.127	0.191	0.255	0.318	0.382	0.446	0.509	0.573	0.637	0.700	0.764	0.828	0.892	0.955	1.019	1.083	1.146	1.210	1.274	1.337	1.401
200.	0.134	0.201	0.268	0.335	0.402	0.469	0.536	0.603	0.670	0.737	0.804	0.872	0.939	1.006	1.073	1.140	1.207	1.274	1.341	1.408	1.475
210.	0.141	0.211	0.282	0.352	0.422	0.493	0.563	0.634	0.704	0.774	0.845	0.915	0.986	1.056	1.126	1.197	1.267	1.338	1.408	1.478	1.549
220.	0.148	0.221	0.295	0.369	0.443	0.516	0.590	0.664	0.738	0.811	0.885	0.959	1.033	1.106	1.180	1.254	1.328	1.401	1.475	1.549	1.623
230.	0.154	0.231	0.308	0.386	0.463	0.540	0.617	0.694	0.771	0.848	0.925	1.003	1.080	1.157	1.234	1.311	1.388	1.465	1.542	1.620	1.697
240.	0.161	0.241	0.322	0.402	0.483	0.563	0.644	0.724	0.805	0.885	0.966	1.046	1.127	1.207	1.288	1.368	1.449	1.529	1.610	1.690	1.771
250.	0.168	0.252	0.335	0.419	0.503	0.587	0.671	0.755	0.838	0.922	1.006	1.090	1.174	1.258	1.342	1.425	1.509	1.593	1.677	1.761	1.845
260.	0.174	0.262	0.349	0.436	0.523	0.610	0.698	0.785	0.872	0.959	1.047	1.134	1.221	1.308	1.395	1.483	1.570	1.657	1.744	1.831	1.919
270.	0.181	0.272	0.362	0.453	0.543	0.634	0.725	0.815	0.906	0.996	1.087	1.177	1.268	1.359	1.449	1.540	1.630	1.721	1.811	1.902	1.993
280.	0.188	0.282	0.376	0.470	0.564	0.658	0.752	0.845	0.939	1.033	1.127	1.221	1.315	1.409	1.503	1.597	1.691	1.785	1.879	1.973	2.067
290.	0.195	0.292	0.389	0.487	0.584	0.681	0.778	0.876	0.973	1.070	1.168	1.265	1.362	1.460	1.557	1.654	1.751	1.849	1.946	2.043	2.141
300.	0.201	0.302	0.403	0.503	0.604	0.705	0.805	0.906	1.007	1.107	1.208	1.309	1.409	1.510	1.611	1.711	1.812	1.913	2.013	2.114	2.215
310.	0.208	0.312	0.416	0.520	0.624	0.728	0.832	0.936	1.040	1.144	1.248	1.353	1.457	1.561	1.665	1.769	1.873	1.977	2.081	2.185	2.289
320.	0.215	0.322	0.430	0.537	0.644	0.752	0.859	0.967	1.074	1.181	1.289	1.396	1.504	1.611	1.719	1.826	1.933	2.041	2.148	2.256	2.363
330.	0.222	0.332	0.443	0.554	0.665	0.775	0.886	0.997	1.108	1.219	1.329	1.440	1.551	1.662	1.772	1.883	1.994	2.105	2.216	2.326	2.437
340.	0.228	0.342	0.457	0.571	0.685	0.799	0.913	1.027	1.141	1.256	1.370	1.484	1.598	1.712	1.826	1.940	2.055	2.169	2.283	2.397	2.511
350.	0.235	0.353	0.470	0.588	0.705	0.823	0.940	1.058	1.175	1.293	1.410	1.528	1.645	1.763	1.880	1.998	2.115	2.233	2.350	2.468	2.585
360.	0.242	0.363	0.484	0.604	0.725	0.846	0.967	1.088	1.209	1.330	1.451	1.572	1.692	1.813	1.934	2.055	2.176	2.297	2.418	2.539	2.660
370.	0.249	0.373	0.497	0.621	0.746	0.870	0.994	1.118	1.243	1.367	1.491	1.615	1.740	1.864	1.988	2.112	2.237	2.361	2.485	2.609	2.734

TABLE 2.7I Ellipsoidal Head Thickness Tables in Inches: Joint Efficiency = 0.65

2. Locate the desired diameter for ellipsoidal heads or the inside crown radius for torispherical heads. The diameter or inside crown radius values are printed across the top of the page for each table.
3. Locate the desired pressure. The pressure values are located in the left column.
4. Read the thickness corresponding to the diameter for ellipsoidal heads or the inside crown radius for torispherical heads and pressure.
5. Each table is based on an allowable stress value of 13,800 psi. For other stress values, multiply the thickness read from the table by the following *factor*.

$$\text{Factor} = \frac{14{,}178}{(\text{allowable stress})^{1.003}}$$

The user is cautioned that the thicknesses calculated using the tables with the above factor are more than adequate for the purpose of estimation, but these are not exact. Calculations with appropriate Code equations are required for all formal Code documentation.

Thinning due to the forming process and corrosion allowances are not included in these tables. The thickness equations for ellipsoidal and torispherical (flanged and dished) heads may be found in Table 2.8 (see also Code Par. UW-12). Table 2.9 sets forth the requirements and gives the location of the desired design calculations for various materials which may be planned to be used for Code construction.

Elements of Joint Design for Heads

As stated in the beginning of this chapter regarding the new design thickness stress multipliers, a number of important factors, such as stress values, quality factors (for castings), and joint efficiencies, must be considered in order to determine the thickness required for heads under internal pressure according to the degree of radiographic examination. Many designers have difficulty in determining these factors.

It should be remembered that the *ASME Boiler and Pressure Vessel Code* is a progressive, viable code and that, when necessary, changes are made. Therefore, the latest edition of the Code should always be used.

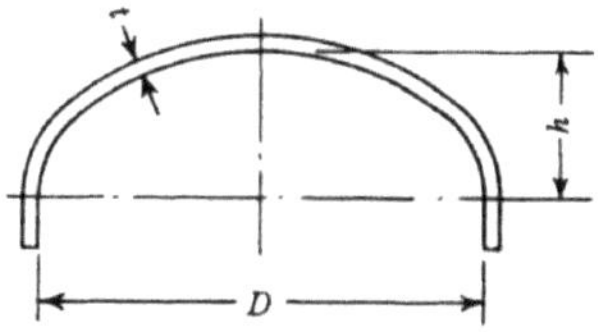

Ellipsoidal head

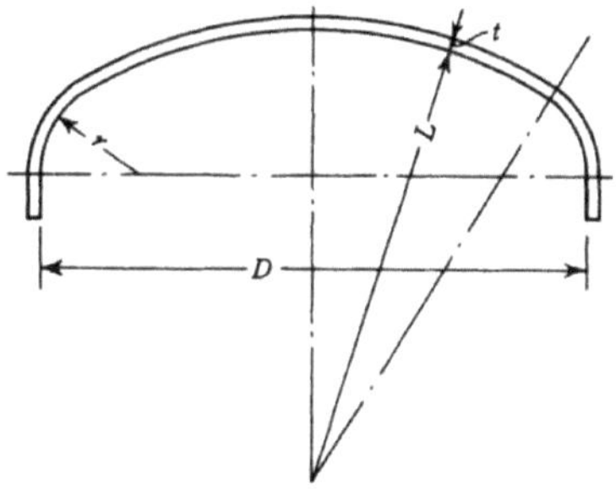

Torispherical head

Equations	**Remarks**	**Code references**
	Ellipsoidal heads under internal pressure	
$t = \dfrac{PD}{2SE - 0.2P}$ or $P = \dfrac{2SEt}{D + 0.2t}$	Equation is used when ratio of the major axis, *D*, to the minor axis, *h*, is 2:1 (see illustration above). For other ratios, see Appendix 1-4	Par. UG-32(d) Appendix 1 (also footnotes) Pars. UG-32(c), 1-4
	Minimum thickness (after forming)	Par. UG-32(a) and (b)
where *t* = minimum thickness, in *P* = internal pressure, psi *S* = allowable stress, psi *E* = minimum joint efficiency, percent *D* = inside diameter of head skirt, in	Allowable percentages of stress value and joint efficiencies for material depend upon type of radiography—full, spot, or no radiography	Pars. UW-11, UW-12 Pars. UW-51, UW-52
	Required skirt length	Pars. UG-32(1), and (m) Fig. UW-13.1
	Attachment of heads to shells	Pars. UG-32(K),
	Tolerance for formed heads	Par. UW-13
	Cold-formed heads of P-1, Group 1 and 2 material may require heat treatment	Par. UCS-79

TABLE 2.8 Calculations for Ellipsoidal and Torispherical Flanged and Dished Heads under Internal Pressure

Equations	Remarks	Code references
	Torispherical heads under internal pressure	
$t = \dfrac{0.885PL}{SE - 0.1P}$ or $P = \dfrac{SEt}{0.885L + 0.1t}$ in which L equals the inside crown radius (in inches) and where other unknown are same as above	ASME head is one in which the knuckle radius r is 6% of the inside crown radius L	Pars. UG-32(d),
	Minimum thickness (after forming)	Pars. UG-32(a) and (b); also footnotes
	Allowable percentages of stress value and joint efficiencies for material depend upon type of radiography—full, spot, or no radiography	Pars. UW-11, UW-12 Pars. UW-51, UW-52
	Required skirt length	Pars. UG-32(K),
	Attachment of heads to shells	Fig. UG-13
	Tolerance for formed heads	Par. UG-81
	Cold-formed heads may require heat treatment	Par. UCS-79

Table 2.8 (*Contd*)

Equations	Remarks	Code references
Compute thicknesses by appropriate procedure in Code Pars. UG-33(a), (d), and (e). Use greater value as Indicated	Carbon steel heads	Pars. UG-33, UCS-33 Code Section II, Part D, Subpart 3
	Nonferrous metal heads	Pars. UG-33, UNF-33 Code Section II, Part D, Subpart 3
	High-alloy steel heads	Pars. UG-33, UHA-31 Code Section II, Part D, Subpart 3
Thickness of cast-iron heads shall not be less than that for plus heads nor less than 1% of the inside diameter of head skirt	Cast-iron heads	Pars. UCI-32, UCI-33
	Clad steel heads	Pars. UCL-26, UCS-33
	Cast ductile iron heads	Pars. UCD-32, UCD-33
	Cladding may be included in design calculations	Pars. UCL-23(b), UCL-23(c)
	Stress values of materials (use appropriate table)	Code Section II, Part D, Subpart 3

Table 2.9 Calculation for Ellipsoidal and Torispherical Flanged and Dished Heads under External Pressure

Thickness of Flat Heads and Bolted Flat Cover Plates

The thickness equations for flat heads and bolted flat cover plates may be found in Table 2.10.

Equations	Remarks	Code references
	Unstayed flat heads	
Circular		
$t = d\sqrt{\frac{CP}{SE}}$	Values of C and d	Par. UG-34 Fig. UG-34
	With full face gasket	Par. UG-34(d)(11) Fig. UG-34 Sketch (p)
	Details of welded joints	Fig. UG-34
Noncircular		
$t = d\sqrt{\frac{ZCP}{SE}}$	Values of C, d, and Z	Par. UG-34 Fig. UG-34
Nomenclature	Modify formulas for unstayed flat heads and covers attached by bolts causing edge moments	Par. UG-34 App. 2
	Bolted heads (spherically dished covers)	
Use equations in App. 1		Par. 1-6 Fig. 1-6
Bolted Flange Connections		
Use equations in App. 2	General recommendations	Pars. UG-11, UG-44 App. 2
	Design considerations	App. 2
	Required loadings, moments, and stresses	App. 2
	Types of flanges	Fig. 2-4
	Limitations on optional type flanges	Par. 2-4(c)
	Limitations on hubbed flanges machined from plate	Par. 2-2(d)(2)
	Blind flanges conforming to an accepted standard may be used for the appropriate pressure–temperature rating	Table U-3
	Bolt loads	Par. 2-5
	Reverse flanges	Par. 2-13

Table 2.10 Calculations for Unstayed Flat and Bolted Heads and Bolted Flange Connections

Additional information needed for bolted flat cover plates is the allowable strength of the chosen bolts. The data required is the bolt root area and the bolt allowable stress. The designer is cautioned that in paragraph 2-3, Notation, the definitions for *Ab* and *Am* are based on the root diameter. (This is contrasted with standard ASME B1.1, Unified Inch Screw Threads, wherein the data for stress area is presented.) Table 2.11 presents the root area for commonly used bolt sizes in both UNC and 8-Thread series and is included as a handy reference guide for the designer. Note that the root area can also be calculated using the following equation:

$$A_{\text{root}} = \frac{\pi}{4}\left(D_b - \frac{1.296}{tpi}\right)^2$$

where

A_{root} = root area (in.^{2})
D_b = bolt nominal diameter (in.)
tpi = number of threads per inch

	Standard UNC Thread Series		8-Thread Series	
Bolt diameter inch	**Threads per inch**	**Root area square inch**	**Threads per inch**	**Root area square inch**
½	13	0.126		
5/8	11	0.202		
¾	10	0.302		
7/8	9	0.420		
1	8	0.552	8	0.552
1 1/8	7	0.694	8	0.728
1 ¼	7	0.891	8	0.930
1 3/8	6	1.055	8	1.156
1 ½	6	1.295	8	1.406
1 5/8			8	1.681
1 ¾	5	1.746	8	1.981
1 7/8			8	2.305
2	4 ½	2.302	8	2.653
2 ¼	4 ½	3.023	8	3.424
2 ½	4	3.719	8	4.293
2 ¾	4	4.622	8	5.260
3	4	5.624	8	6.326

TABLE 2.11 Root Area for Commonly Used Bolt Sizes (Standard UNC Thread and 8-Thread Series)

Various Other Designs for Heads

The ASME Code sets forth the requirements and allows the designer the latitude to design pressure vessels with heads of a "nonstandard" design. This is not to say that the average ASME Code shop will not see such designs; however, they should not rule out the possibility. Such designs include conical, toriconical, and hemispherical heads. Table 2.12 simplifies the Code design requirements of conical and toriconical heads under internal pressure and lists the Code design calculations and the Code references.

The hemispherical head is a more common design and is often used for gas pressure tanks such as propane tanks. Table 2.13 lists the equations and the Code references for the designs of such heads.

Allowable Pressure, Pitch, and Thickness

The areas of flat surfaces which are not supported under internal pressure have the tendency while under pressure to deform, bulge, and try to form a spherical or cylindrical shape. An example of this phenomenon can be seen by blowing air into a cereal box. On the application of pressure, the box becomes tubular in shape, and if one could blow hard enough, it would rupture. The Code recognizes this problem and has included requirements for adding bracing or stays between these flat surfaces. Table 2.14 states the equations for braced and stayed surfaces and gives the Code references that apply.

Calculations for Reinforcement of Openings

To determine whether an opening is adequately reinforced, it is first necessary to determine whether the areas of reinforcement available will be sufficient without the use of a pad. Figure 2.8 (taken from Code Fig. UG-37.1) includes a schematic diagram of an opening showing the involved areas and all the required calculations. The total cross-sectional area of reinforcement required is shown as A. The area of reinforcement available without a pad includes the following:

1. The area of excess thickness in the shell or head, A_1
2. The area of excess thickness in the nozzle wall, A_2
3. The area available in the nozzle projecting inward, A_3
4. The cross-sectional area of welds, $A_{41} + A_{43}$

If $A_1 + A_2 + A_3 + A_{41} + A_{43} \geq A$, the opening is adequately reinforced.

If $A_1 + A_2 + A_3 + A_{41} + A_{43} < A$, a pad is needed.

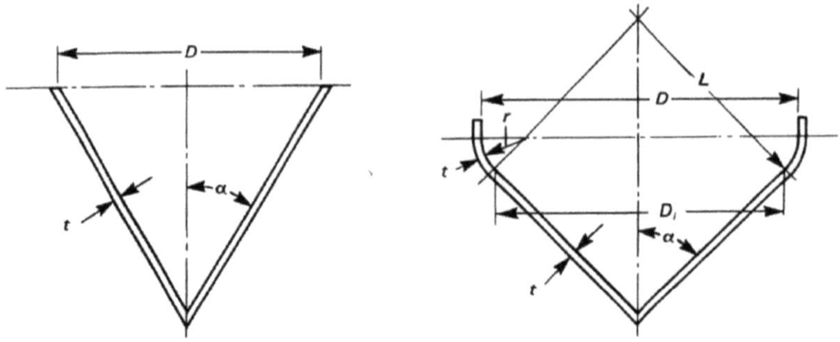

Conical

Toriconical (Cone Head With Knuckle)

Equations	Remarks	Code references
	Conical heads under internal pressure	
$t = \dfrac{PD}{2\cos\alpha(SE - 0.6P)}$	Used when one-half the apex angle (α) does not exceed 30°	Par. UG-32(f)
	Use toriconical head when angle α exceeds 30°	Par. UG-32(g)
	Requirements for reinforcing ring at large end	Par. 1-5(d)
	Requirements for reinforcing ring at small end	Par. 1-5(e)
	Requirements for reinforcing if also subjected to external pressure	Par. 1-8
	Toriconical heads under internal pressure	
$t_c = \dfrac{PD_i}{2\cos\alpha(SE - 0.6P)}$	Thickness of conical portion of toriconical head	Par. UG-32(g)
	Inside knuckle radius shall be neither less than 6% of the outside diameter of the head skirt nor less than three times the knuckle thickness	
$t_k = \dfrac{PLM}{2SE - 0.2P}$	Thickness of knuckle portion of toriconical head	Par. 1-4(d)
where $M = \frac{1}{4}\left(3 + \sqrt{L/r}\right)$		Par. 1-4(d)
$L = \dfrac{D_i}{2\cos\alpha}$		Par. UG-32(g)

Table 2.12 Calculations for Conical and Toriconical Heads

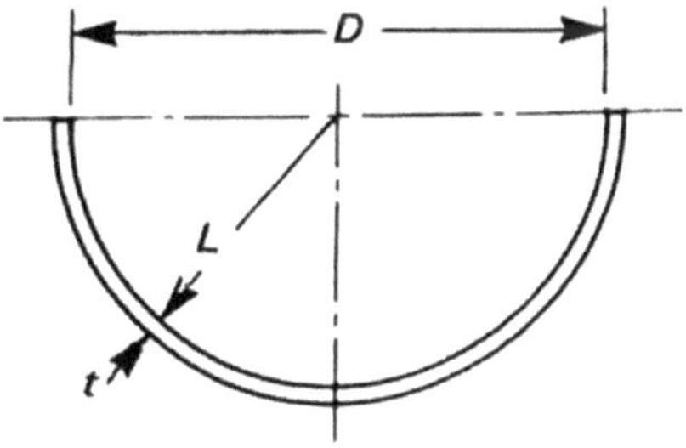

Hemispherical

Equations	Remarks	Code references
	Heads under internal pressure	
$t = \dfrac{PL}{2SE - 0.2P}$	Limited to: $t \leq 0.356L$, $P \leq 0.665SE$	Par. UG-32(e)
	Thick hemispherical heads ($t > 0.356L$, $P > 0.665SE$)	Par. 1-3
	Skirt required if the head is thicker than the adjacent shell	Par. UG-32(k)
	Heads under external pressure	
	Thickness determined in the same manner as a spherical shell	Pars. UG-32(c), UG-28(d)

TABLE 2.13 Calculations for Hemispherical Heads

If the reinforcement is found to be inadequate, then the area of pad needed (A_5) may be calculated as follows:

$$A_5 = A - \left(A_1 + A_2 + A_3 + A_4 \right)$$

If a pad is used, the calculations for A_2 and A_{41} must be revised. Then, if $A_1 + A_2 + A_3 + A_{41} + A_{42} + A_{43} + A_5 \geq A$, the opening is adequately reinforced.

Figure 2.9 (taken from Code Fig. UG-41.1) shows the required weld strength and the weld strength paths for consideration. Table 2.15 lists additional considerations along with Code references for openings and reinforcement.

Additional nomenclature for Figs. 2.8 and 2.9 is as follows:

c = corrosion allowance

D_p = outside diameter of reinforcing element (The actual size of reinforcing element may exceed the limits of reinforcement

Equations	Remarks	Code references
$t = p\sqrt{P/SC}$	Measurement of pitch *p* and values for *C*	Par. UG-47(a)
	Minimum thickness of stayed plate is 5/16 in. except in welded construction and cylindrical or spherical outer shell plates	Par. UG-47(b), UW-19 App. 17
	Thickness requirements for stayed jackets extending completely around a cylindrical shell, spherical vessel, or formed head	Pars. UG-47(a), UG-27(c), UG-27(d), UG-32
	Nozzles or other formed openings	Par. UG-37(d)(2)
	If two plates are stayed and only one requires staying, the value of *C* is governed by the thickness of the plate requiring staying	Par. UG-47(d)
	Welded stayed construction	Par. UW-19 Fig. UW-19.1
	Maximum pitch for welded and screwed stays with ends riveted over is 8½ in. For welded-in staybolts, the maximum pitch is 15 times the diameter of the staybolt	Pars. UG-47(f), UG-19(a)
	End requirements, location, and dimensions of staybolts	Pars. UG-48 to UG-50
	Hole sizes for screw stays	Par. UG-83
	Dimpled or embossed assemblies	App. 17

Table 2.14 Calculations for Braced and Stayed Surfaces

established by Code Par. UG-40; however, credit cannot be taken for any material outside these limits.)

d = finished diameter of circular opening or finished dimension (chord length at midsurface of thickness excluding excess thickness available for reinforcement) of nonradial opening in the plane under consideration [see Code Figs. UG-37.1 and UG-40]

E_1 = 1 when an opening is in the solid plate or in a Category B butt joint;

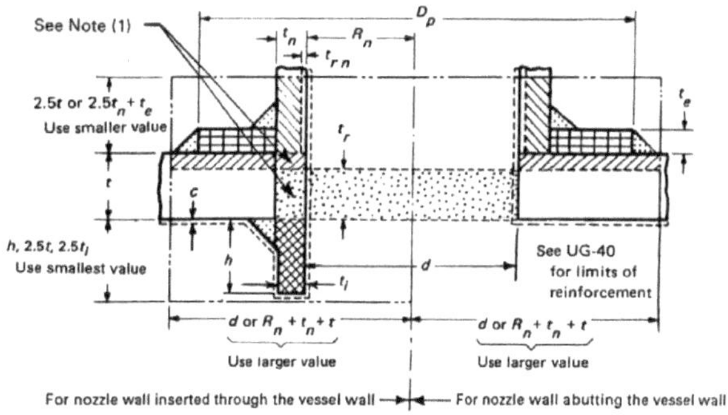

Without Reinforcing Element

$A = d\,t_r\,F + 2t_n\,t_r\,F(1 - f_{r1})$	Area required
$A_1 = d(E_1 t - Ft_r) - 2t_n(E_1 t - Ft_r)(1 - f_{r1})$ $= 2(t + t_n)(E_1 t - Ft_r) - 2t_n(E_1 t - Ft_r)(1 - f_{r1})$	Area available in shell; use larger value
$A_2 = 5(t_n - t_{rn})\,f_{r2}\,t$ $= 5(t_n - t_{rn})\,f_{r2}\,t_n$	Area available in nozzle projecting outward; use smaller value
$A_3 = 5t\,t_i\,f_{r2}$ $= 5t_i\,t_i\,f_{r2}$ $= 2h\,t_i\,f_{r2}$	Area available in inward nozzle; use smallest value
A_{41} = outward nozzle weld = $(\text{leg})^2 f_{r2}$	Area available in outward weld
A_{43} = inward nozzle weld = $(\text{leg})^2 f_{r2}$	Area available in inward weld
If $A_1 + A_2 + A_3 + A_{41} + A_{43} \geq A$	Opening is adequately reinforced
If $A_1 + A_2 + A_3 + A_{41} + A_{43} < A$	Opening is not adequately reinforced so reinforcing elements must be added and/or thicknesses must be increased

With Reinforcing Element Added

A = same as A, above	Area required
A_1 = same as A_1, above	Area available
$A_2 = 5(t_n - t_{rn})f_{r2}\,t$ $= 2(t_n - t_{rn})(2.5t_n + t_e)f_{r2}$	Area available in nozzle projecting outward; use smaller area
A_3 = same as A_3, above	Area available in inward nozzle
A_{41} = outward nozzle weld = $(\text{leg})^2 f_{r3}$	Area available in outward weld
A_{42} = outer element weld = $(\text{leg})^2 f_{r4}$	Area available in outer weld
A_{43} = inward nozzle weld = $(\text{leg})^2 f_{r2}$	Area available in inward weld
$A_5 = (D_p - d - 2t_n)\,t_e\,f_{r4}$ (Note (2))	Area available in element
If $A_1 + A_2 + A_3 + A_{41} + A_{42} + A_{43} + A_5 \geq A$	Opening is adequately reinforced

GENERAL NOTE: This figure illustrates a common nozzle configuration and is not intended to prohibit other configurations permitted by the Code.

NOTES:
(1) Includes consideration of these areas if $S_n/S_v < 1.0$ (both sides of centerline).
(2) This formula is applicable for a rectangular cross-sectional element that falls within the limits of reinforcement.

FIGURE 2.8 Nomenclature and formulas for reinforcement of openings (Code Fig. UG-37.1). (*Courtesy of the American Society of Mechanical Engineers.*)

= 0.85 when an opening is located in an ERW (electric resistance welding) or autogenously welded pipe or tube. If the ERW or autogenously welded joint is clearly identifiable and it can be shown that the opening does not pass through this weld joint, then E_1 may be determined using the other rules of this paragraph;

= joint efficiency obtained from Code Table UW-12 when any part of the opening passes through any other welded joint

F = correction factor that compensates for the variation in internal pressure stresses on different planes with respect to the axis

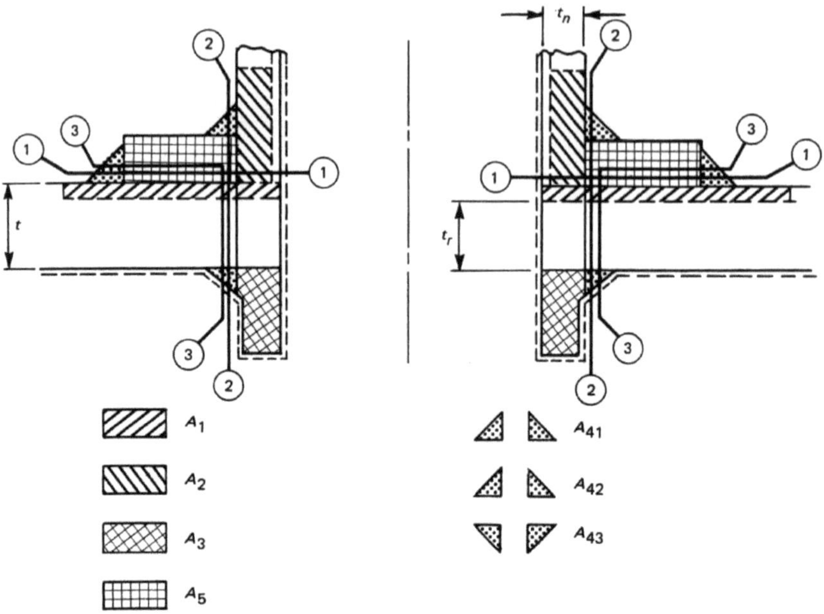

W = total weld load [UG-41(b)(2)]
$= [A - A_1 + 2t_n f_{r1}(E_1 t - F t_r)]S_v$
$W_{1\text{-}1}$ = weld load for strength path 1-1 [UG-41(b)(1)]
$= (A_2 + A_5 + A_{41} + A_{42})S_v$
$W_{2\text{-}2}$ = weld load for strength path 2-2 [UG-41(b)(1)]
$= (A_2 + A_3 + A_{41} + A_{43} + 2t_n t f_{r1})S_v$
$W_{3\text{-}3}$ = weld load for strength path 3-3 [UG-41(b)(1)]
$= (A_2 + A_3 + A_5 + A_{41} + A_{42} + A_{43} + 2t_n t f_{r1})S_v$

(a) Depicts Typical Nozzle Detail With Neck Inserted Through the Vessel Wall

FIGURE 2.9 Nozzle attachment weld loads and weld strength paths to be considered (Code Fig. UG-41.1). (*Courtesy of the American Society of Mechanical Engineers.*)

of a vessel. A value of 1.00 shall be used for all configurations except for integrally reinforced openings in cylindrical shells and cones where Code Fig. UG-37 may be used. [See Code Par. UW-16(c)(1).]

$f_{r1} = S_n/S_v$ for nozzle wall inserted through the vessel wall;
$= 1.0$ for nozzle wall abutting the vessel wall and for nozzles shown in Code Fig. UG-40, sketches (j), (k), (n), and (o)

$f_{r2} = S_n/S_v$

$f_{r3} =$ (lesser of S_n or S_p)$/S_v$

$f_{r4} = S_p/S_v$

R_n = inside radius of the nozzle under consideration

S = allowable stress value in tension (see Code Par. UG-23). For welded pipe or tubing, the allowable stress of the corresponding seamless product form. If there is no corresponding seamless product form, use the allowable stress for the welded product form divided by 0.85.

S_n = allowable stress in nozzle (see S above)

S_p = allowable stress in reinforcing element (plate) (see S above)

S_v = allowable stress in vessel (see S above)

t = specified vessel wall thickness (not including forming allowances). For pipe, it is the nominal thickness less the manufacturing undertolerance allowed in the pipe specification

t_e = thickness or height of reinforcing element (see Code Fig. UG-40)

t_i = nominal thickness of the internal projection of nozzle wall

t_n = nozzle wall thickness. Except for pipe, this is the wall thickness not including forming allowances. For pipe, use the nominal thickness [see Code Par. UG-16(d)]

t_r = required thickness of a seamless shell based on the circumferential stress, or of a formed head—computed by the Code rules for the designated pressure—using $E = 1$, and for shells fabricated from welded pipe or tubing, the allowable stress of the corresponding seamless product form. If there is no corresponding seamless product form, use the allowable stress for the welded product form divided by 0.85, except that:

(a) when the opening and its reinforcement are entirely within the spherical portion of a torispherical head, t_r is the thickness required by Code Par. 1-4(d), using $M = 1$;

(b) when the opening is in a cone, t_r is the thickness required for a seamless cone of diameter D measured where the nozzle axis pierces the inside wall of the cone;

(c) when the opening and its reinforcement are in an ellipsoidal head and are located entirely within a circle, the center of which coincides with the center of the head and the diameter of which is equal to 80% of the shell diameter, t_r is the thickness required for a seamless sphere of radius K_1D, where D is the shell diameter and K_1 is given by Code Table UG-37.

t_{rn} = required thickness of a seamless nozzle wall, using $E = 1$, and, for nozzles fabricated from welded pipe or tubing, the allowable stress of the corresponding seamless product form. If there is no corresponding seamless product form, use the allowable stress for the welded product form divided by 0.85

W = total load to be carried by attachment welds (see Code Par. UG-41)

Cryogenic and Low-Temperature Vessels

Cryogenic and low-temperature pressure vessels are an important segment of the industry. But low-temperature service requires

Remarks	Code references
To correct for corrosion, deduct the specified allowance from shell thickness, t and nozzle thickness, t_n, but add twice its value to the diameter of the opening, d	
The total load to be carried by attachment welds, W, must be equal to or greater than the lesser of W_{1-1}, W_{2-2}, and W_{3-3} (see Code Fig. 2.13). See Code Par. UG-41(b) for additional conditions	
Large openings in cylindrical shells	Par. 1-7
Openings in flat heads	Par. UG-39
Required weld sizes	Par. UW-16 Fig. UW-16.1
Reinforcement is considered integral when it is inherent in the shell plate or nozzle. Reinforcement built up by welding is also considered integral. The installation of a reinforcement pad is not considered to be integral	
For vessels under external pressure, reinforced area required need only be one-half that for internal pressure if the shell thickness satisfies the external-pressure requirements	Par. UG-37(d)
Limits of reinforcement	Par. UG-40
Strength of reinforcement	Par. UG-41
Reinforcement of multiple openings	Par. UG-42
Welded connections	Pars. UW-15, UW-16
Openings in or adjacent to welds	Par. UW-14
Large openings in cylindrical shells	Par. 1-7
Openings in flat heads	Par. UG-39
Single openings not requiring reinforcement— 3 in. pipe size, vessel thickness 3/8 in. or less 2 in. pipe size, vessel thickness over 3/8 in.	Par. UG-36
Openings in head or girth seam requiring radiography	Pars. UW-14, UG-37
Flued openings in formed heads	Par. UG-38 Fig. UG-38

TABLE 2.15 Openings and Reinforced Openings

particular attention to materials and design. Many materials exposed to low temperatures lose their toughness and ductility. As the temperature goes down, the metal has a tendency to become brittle. This may cause brittle fracture at full stress.

Most materials used for service at normal temperatures have a marked decrease in impact resistance as the temperatures decrease.

Materials in low-temperature service must have the proven ability to safely resist high-stress changes and shock loads. Only impact-tested materials able to withstand unavoidable stress concentrations must be used for cryogenic vessels. All sharp or abrupt transitions or changes of sections, corners, or notches should be eliminated in the design of the vessel and the fabrication of its parts.

The manufacturers, purchasers, and users of low-temperature pressure vessels often misunderstand certain paragraphs in the *ASME Boiler and Pressure Vessel Code.* The following are clarifications to various paragraphs applying to low-temperature vessels and to indicate when impact tests are required. When such tests are necessary, each part of the vessel, such as shells, heads, nozzles, and reinforcing pads, as well as other pressure parts, must be considered. Additionally, the non-pressure parts welded to the pressure parts shall also be considered. Table 2.16 lists the various Code requirements for low-temperature operations.

Remarks	Code references
Some materials and welds do not require impact test; consult the paragraphs in Code Subsection C to which material applies (see Table 2.1)	Pars. UG-20, UCS-66 thru UCS-68, UHA-51, UCL-27 Figs. UCS-66, UCS-66.1
When impact tests are required, weld metal and heat-affected zone must have impact tests	Pars. UG-84, UF-5, UNF-65, UHT-5
Low-temperature vessels	Part ULT
Welded carbon steel vessels must be postweld heat treated unless exempted from impact tests	Par. UCS-68
Vessels must be stamped according to the Code requirements (see Fig. 2.12)	Par. UG-116
All joints of Category A must be Type 1 of Code Table UW-12 [see UW-2(b)(1) for austenitic stainless steel exemptions]	Pars. UW-2(b), UW-3
All joints of Category B must be Type 1 or 2 of Code Table UW-12	Par. UW-2(b)
All joints of Categories C and D must be full-penetration welds [see UW-2(b)(4) for exemptions to Category D welds]	Par. UW-2(b)
Brazed vessels	Pars. UB-22, UG-84
High-alloy vessels	Par. UHA-51
Nonferrous vessels	Par. UNF-65

Table 2.16 Requirements for Low-Temperature Service

Impact-Test Exemptions

With cost a consideration in designing and manufacturing carbon and low-alloy steel pressure vessels, the design rules mandate strict requirements for the notch toughness of certain materials. The designer must evaluate each component material used in the design of the vessel separately. This section will assist the designer to better understand the methods to employ when seeking relief from impact test requirements for carbon steel vessel materials.

It should be noted that there have been several changes to Section VIII impact testing rules from the earlier editions of the Code, and they have become much more complex. The affected rules include the minimum design metal temperature (MDMT) calculation and the means of allowing a reduction to the MDMT for impact testing requirements.

The fundamental impact testing rule is that when the combination of thickness and MDMT of a material falls below the proper curve shown in Fig. 2.10 (Code Fig. UCS-66), impact testing is required, unless otherwise exempted in the Code. Note that the Code also includes a metric version of the exemption curves, shown as Code Fig. UCS-66M.

An important reason for impact testing when it is clearly required is to avoid brittle fracture—a disastrous type of failure. If impact testing is required, but barely, one should weigh the cost of testing versus going through the Code procedure to try to lift the material to the not-required position in Fig. 2.10. If the latter seems feasible, one should keep the Code close at hand because there are many variables involved. The tabular values for Fig. 2.10 included here as Table 2.17 are typically somewhat easier to work with than the graph.

The first step the designer should take is to check which vessel part may be exempted from the impact-test requirement. The method to claim exemptions is to use one of the following exemption models:

Exemption model 1. The designer should consult Code Par. UG-20(f). When all of the conditions of that paragraph are met, the impact-testing requirements of Code Par. UG-84 are no longer mandatory.

Exemption model 2. In the event that the designer cannot exempt all the remainder of the materials in exemption model 1, above, and depending on whether the stress in tension for the design conditions is less than the maximum allowable design stress, the Code allows a possible exemption from impact testing. Assuming that the example vessel uses the full allowable stress in its calculated thickness, lowering the design stress will increase the vessel thickness. If this is the case, the following sequence will give the reader an idea of how to use the chart in Fig. 2.11 to gain additional relief from impact testing. (See also Fig. 2.12.) *For this model, assume an MDMT of 10°F.* Note that for service temperatures below –50°F, impact testing is mandatory.

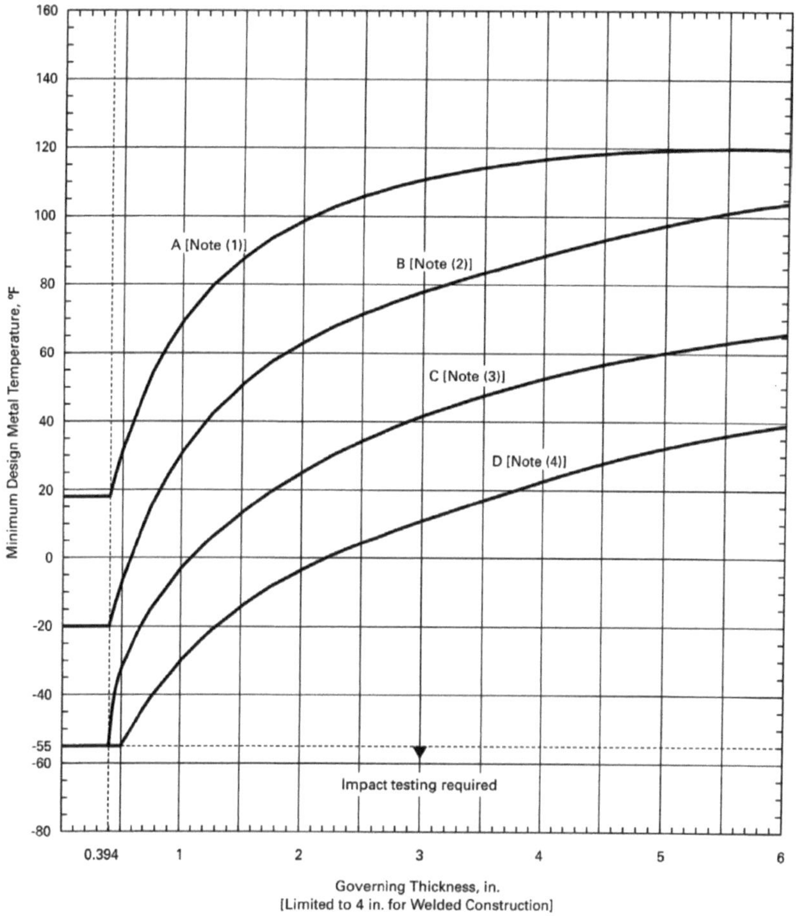

GENERAL NOTES:
(a) Tabular values for this figure are provided in Table UCS-66.
(b) See UCS-66(a).
(c) For bolting and nuts, the following impact test exemption temperatures shall apply:

Bolting

Spec. No.	Grade	Diameter, in. (mm)	Impact Test Exemption Temperature, °F (°C)
SA-193	B5	Up to 4 (100), incl.	-20 (-30)
SA-193	B7	Up to $2\frac{1}{2}$ in. (64), incl.	-55 (-48)
SA-193	...	Over $2\frac{1}{2}$ (64) to 7 (175), incl.	-40 (-40)
SA-193	B7M	Up to $2\frac{1}{2}$ (64), incl.	-55 (-48)

FIGURE 2.10 Impact-test exemption curves (Code Fig. UCS-66). (*Courtesy of the American Society of Mechanical Engineers.*)

1. Observe that for the material thickness of 1.5 in., curve C is about 5°F above the 10°F MDMT at the 1.5 in. thickness point.
2. The curve of Fig. 2.13 must be lowered by more than 5°F since the increased vessel thickness will result in a shift on curve C to a higher temperature; try, say, 15°F.

Bolting			
Spec. No.	Grade	Diameter, in. (mm)	Impact Test Exemption Temperature, °F (°C)
SA-193	B16	Up to 7 (175), incl.	-20 (-30)
SA-307	B	All	-20 (-30)
SA-320	L7, L7A, L7M	Up to 2½ (64), incl.	See General Note (c) of Figure UG-84.1
SA-320	L43	Up to 1 (25), incl.	See General Note (c) of Figure UG-84.1
SA-325	1	½ (13) to 1½ (38)	-20 (-30)
SA-354	BC	Up to 4 (100), incl.	0 (-18)
SA-354	BD	Up to 4 (100), incl.	+20 (-7)
SA-437	B4B, B4C	All diameters	See General Note (c) of Figure UG-84.1
SA-449	...	Up to 3 (75), incl.	-20 (-30)
SA-540	B21 Cl. All	All	Impact test required
SA-540	B22 Cl. 3	Up to 4 (100), incl.	Impact test required
SA-540	B23 Cl. 1, 2	All	Impact test required
SA-540	B23 Cl. 3, 4	Up to 6 (150), incl.	See General Note (c) of Figure UG-84.1
SA-540	B23 Cl. 3, 4	Over 6 (150) to 9½ (240), incl.	Impact test required
SA-540	B23 Cl. 5	Up to 8 (200), incl.	See General Note (c) of Figure UG-84.1
SA-540	B23 Cl. 5	Over 8 (200) to 9½ (240), incl.	Impact test required
SA-540	B24 Cl. 1	Up to 6 (150), incl.	See General Note (c) of Figure UG-84.1
SA-540	B24 Cl. 1	Over 6 (150) to 8 (200), incl.	Impact test required
SA-540	B24 Cl. 2	Up to 7 (175), incl.	See General Note (c) of Figure UG-84.1
SA-540	B24 Cl. 2	Over 7 (175) to 9½ (240), incl.	Impact test required
SA-540	B24 Cl. 3, 4	Up to 8 (200), incl.	See General Note (c) of Figure UG-84.1
SA-540	B24 Cl. 3, 4	Over 8 (200) to 9½ (240), incl.	Impact test required
SA-540	B24 Cl. 5	Up to 9½ (240), incl.	See General Note (c) of Figure UG-84.1
SA-540	B24V Cl. 3	All	See General Note (c) of Figure UG-84.1

Nuts		
Spec. No.	Grade	Impact Test Exemption Temperature, °F (°C)
SA-194	2, 2H, 2HM, 3, 4, 7, 7M, and 16	-55 (-48)
SA-540	B21/B22/B23/B24/B24V	-55 (-48)

(d) When no class or grade is shown, all classes or grades are included.
(e) The following shall apply to all material assignment notes:
(1) Cooling rates faster than those obtained by cooling in air, followed by tempering, as permitted by the material specification, are considered to be equivalent to normalizing or normalizing and tempering heat treatments.
(2) Fine grain practice is defined as the procedure necessary to obtain a fine austenitic grain size as described in SA-20.
(3) Normalized rolling condition is not considered as being equivalent to normalizing.
(f) Castings not listed in Notes (1) and (2) below shall be impact tested.

NOTES:
(1) Curve A applies to:
(a) all carbon and all low alloy steel plates, structural shapes, and bars not listed in Curves B, C, and D below;
(b) SA-216 Grades WCB and WCC if normalized and tempered or water-quenched and tempered; SA-217 Grade WC6 if normalized and tempered or water-quenched and tempered.
(2) Curve B applies to:
(a) see below:
SA-216 Grade WCA if normalized and tempered or water-quenched and tempered
SA-216 Grades WCB and WCC for thicknesses not exceeding 2 in. (50 mm) , if produced to fine grain practice and water-quenched and tempered
SA-217 Grade WC9 if normalized and tempered
SA-285 Grades A and B
SA-299
SA-414 Grade A
SA-515 Grade 60
SA-516 Grades 65 and 70 if not normalized
SA-612 if not normalized

FIGURE 2.10 *(Contd)*

3. On the curved line of Fig. 2.11, enter a "trial" temperature of 15°F on the horizontal temperature line between 10 and 20°F. Follow this line up vertically until it hits the curved line. Follow the horizontal line to the left and read a design stress to allowable tensile stress ratio of 0.85.
4. Calculate the new design stress by multiplying 0.85 by the allowable stress of 15,000 psi (the stress of the material found in Code Section II, Part D, Table 1A), yielding 12,750 psi.
5. Recalculate the vessel thickness using the design formula and substitute 1.76 in. as the new stress value (an approximation for a trial value is 1.50 in./0.85).

NOTES (CONT'D):

SA-662 Grade B if not normalized
SA/EN 10028-2 Grades P235GH, P265GH, P295GH, and P355GH as rolled
SA/AS 1548 Grades PT430NR and PT460NR

(b) except for cast steels, all materials of Curve A, if produced to fine grain practice and normalized, that are not listed in Curves C and D below;

(c) all pipe, fittings, forgings and tubing not listed for Curves C and D below;

(d) parts permitted under UG-11 shall be included in Curve B even when fabricated from plate that otherwise would be assigned to a different curve.

(3) Curve C applies to:

(a) see below:

SA-182 Grades F21 and F22 if normalized and tempered
SA-302 Grades C and D
SA-336 F21 and F22 if normalized and tempered, or liquid quenched and tempered
SA-387 Grades 21 and 22 if normalized and tempered, or liquid quenched and tempered
SA-516 Grades 55 and 60 if not normalized
SA-533 Types B and C Class 1
SA-662 Grade A
SA/EN 10028-2 Grade 10CrMo9–10 if normalized and tempered

(b) all materials listed in 2(a) and 2(c) for Curve B if produced to fine grain practice and normalized, normalized and tempered, or liquid quenched and tempered as permitted in the material specification, and not listed for Curve D below.

(4) Curve D applies to:

SA-203
SA-299 if normalized
SA-508 Grade 1
SA-516 if normalized or quenched and tempered
SA-524 Classes 1 and 2
SA-537 Classes 1, 2, and 3
SA-612 if normalized
SA-662 if normalized
SA-738 Grade A
SA-738 Grade A with Cb and V deliberately added in accordance with the provisions of the material specification, not colder than −20°F (−29°C)
SA-738 Grade B not colder than −20°F (−29°C)
SA/AS 1548 Grades PT430N and PT460N
SA/EN 10028-2 Grades P235GH, P265GH, P295GH, and P355GH if normalized
SA/EN 10028-3 Grade P275NH

Figure 2.10 *(Contd)*

6. Look at Fig. 2.13 again. Lower curve C by 15°F at a thickness of 1.76 in. to get a curve C MDMT of 15°F. Since the curve is now below the vessel MDMT of 10°F, 1.76 in. thick SA-516 grade 60 may be used without postweld heat treating (PWHT) normalizing or impact testing the materials. From the standpoint of cost and economics, this method is probably the best choice.
7. If PWHT is performed when it is not otherwise a requirement of Section VIII, Division 1, a 30°F reduction in impact-testing exemption temperature shall be given to the minimum permissible temperature from Fig. UCS-66 for P-1 material (see Code Par. UCS-68). Utilizing this method would be costly.
8. Further exemptions shall be determined as follows:

   a. If reinforcing pads are required, such as for the nozzle reinforcement, such reinforcement would be evaluated similarly with one of the exemption models above, using the pad thickness as the governing thickness.

   b. The 6 in. SA-106 grade B nozzle neck is a curve B material per Fig. 2.13, note (b)(3). The nozzle neck nominal

Customary Units					SI Units				
Thickness, in.	Curve A, °F	Curve B, °F	Curve C, °F	Curve D, °F	Thickness, mm	Curve A, °C	Curve B, °C	Curve C, °C	Curve D, °C
0.25	18	-20	-55	-55	6.4	-8	-29	-48	-48
0.3125	18	-20	-55	-55	7.9	-8	-29	-48	-48
0.375	18	-20	-55	-55	9.5	-8	-29	-48	-48
0.4375	25	-13	-40	-55	11.1	-4	-25	-40	-48
0.5	32	-7	-34	-55	12.7	0	-22	-37	-48
0.5625	37	-1	-26	-51	14.3	3	-18	-32	-46
0.625	43	5	-22	-48	15.9	6	-15	-30	-44
0.6875	48	10	-18	-45	17.5	9	-12	-28	-43
0.75	53	15	-15	-42	19.1	12	-9	-26	-41
0.8125	57	19	-12	-38	20.6	14	-7	-24	-39
0.875	61	23	-9	-36	22.2	16	-5	-23	-38
0.9375	65	27	-6	-33	23.8	18	-3	-21	-36
1.0	68	31	-3	-30	25.4	20	-1	-19	-35
1.0625	72	34	-1	-28	27.0	22	1	-18	-33
1.125	75	37	2	-26	28.6	24	3	-17	-32
1.1875	77	40	2	-23	30.2	25	4	-17	-31
1.25	80	43	6	-21	31.8	27	6	-14	-30
1.3125	82	45	8	-19	33.3	28	7	-13	-28
1.375	84	47	10	-18	34.9	29	8	-12	-28
1.4375	86	49	12	-16	36.5	30	9	-11	-27
1.5	88	51	14	-14	38.1	31	11	-10	-26
1.5625	90	53	16	-13	39.7	32	12	-9	-25
1.625	92	55	17	-11	41.3	33	13	-8	-24
1.6875	93	57	19	-10	42.9	34	14	-7	-23
1.75	94	58	20	-8	44.5	34	14	-7	-22
1.8125	96	59	22	-7	46.0	36	15	-6	-22
1.875	97	61	23	-6	47.6	36	16	-5	-21
1.9375	98	62	24	-5	49.2	37	17	-4	-21
2.0	99	63	26	-4	50.8	37	17	-3	-20
2.0625	100	64	27	-3	52.4	38	18	-3	-19
2.125	101	65	28	-2	54.0	38	18	-2	-19
2.1875	102	66	29	-1	55.6	39	19	-2	-18
2.25	102	67	30	0	57.2	39	19	-1	-18
2.3125	103	68	31	1	58.7	39	20	-1	-17
2.375	104	69	32	2	60.3	40	21	0	-17
2.4375	105	70	33	3	61.9	41	21	1	-16
2.5	105	71	34	4	63.5	41	22	1	-16
2.5625	106	71	35	5	65.1	41	22	2	-15
2.625	107	73	36	6	66.7	42	23	2	-14
2.6875	107	73	37	7	68.3	42	23	3	-14
2.75	108	74	38	8	69.9	42	23	3	-13
2.8125	108	75	39	8	71.4	42	24	4	-13
2.875	109	76	40	9	73.0	43	24	4	-13
2.9375	109	77	40	10	74.6	43	25	5	-12
3.0	110	77	41	11	76.2	43	26	5	-12
3.0625	111	78	42	12	77.8	44	26	6	-11
3.125	111	79	43	12	79.4	44	26	6	-11
3.1875	112	80	44	13	81.0	44	27	7	-11
3.25	112	80	44	14	82.6	44	27	7	-10
3.3125	113	81	45	15	84.1	45	27	7	-9
3.375	113	82	46	15	85.7	45	28	8	-9
3.4375	114	83	46	16	87.3	46	28	8	-9

TABLE 2.17 Tabular values for Fig. 2.10. (Code Table UCS-66). (*Courtesy of the American Society of Mechanical Engineers.*)

thickness of 0.280 in. and the MDMT of 18°F locate a point above curve B, so no impact testing is required. This is also exempt from impact testing, as stated in exemption 1. In this case, this material could have been exempt from impact testing utilizing exemption model 1.

Code Pars. UCS-66(c), (d), and (f) grant exemptions from impact testing for certain types of special items as discussed below.

Customary Units					SI Units				
Thick-ness, in.	Curve A, °F	Curve B, °F	Curve C, °F	Curve D, °F	Thick-ness, mm	Curve A, °C	Curve B, °C	Curve C, °C	Curve D, °C
3.5	114	83	47	17	88.9	46	28	8	-8
3.5625	114	84	48	17	90.5	46	29	9	-8
3.625	115	85	49	18	92.1	46	29	9	-7
3.6875	115	85	49	19	93.7	46	29	9	-7
3.75	116	86	50	20	95.3	47	30	10	-7
3.8125	116	87	51	21	96.8	47	31	11	-6
3.875	116	88	51	21	98.4	47	31	11	-6
3.9375	117	88	52	22	100.0	47	32	11	-6
4.0	117	89	52	23	101.6	47	32	11	-5
4.0625	117	90	53	23	103.0	47	32	12	-5
4.125	118	90	54	24	105.0	48	32	12	-4
4.1875	118	91	54	25	106.0	48	33	12	-4
4.25	118	91	55	25	108.0	48	33	12	-4
4.3125	118	92	55	26	110.0	48	33	12	-3
4.375	119	93	56	27	111.0	49	34	13	-3
4.4375	119	93	56	27	113.0	49	34	13	-3
4.5	119	94	57	28	114.0	49	34	13	-2
4.5625	119	94	57	29	115.0	49	34	13	-2
4.625	119	95	58	29	117.0	49	35	14	-2
4.6875	119	95	58	30	118.0	49	35	14	-1
4.75	119	96	59	30	119.0	49	35	14	-1
4.8125	119	96	59	31	120.0	49	35	14	-1
4.875	119	97	60	31	121.0	49	36	15	-1
4.9375	119	97	60	32	122.0	49	36	15	0
5	119	97	60	32	123.0	49	36	15	0
5.0625	119	98	61	33	124.0	49	36	15	0
5.125	119	98	61	33	125.0	49	36	15	0
5.1875	119	98	62	34	126.0	49	36	16	1
5.25	119	99	62	34	127.0	49	37	16	1
5.3125	119	99	62	35	128.0	49	37	16	1
5.375	119	100	63	35	129.0	49	37	16	1
5.4375	119	100	63	36	130.0	49	37	16	2
5.5	119	100	63	36	131.0	49	37	16	2
5.5625	119	101	64	36	132.0	49	38	17	2
5.625	119	101	64	37	133.0	49	38	17	2
5.6875	119	102	64	37	134.0	49	38	17	2
5.75	120	102	65	38	135.0	50	38	17	3
5.8125	120	103	65	38	136.0	50	39	17	3
5.875	120	103	66	38	137.0	50	39	18	3
5.9375	120	104	66	39	138.0	50	39	18	3
6.0	120	104	66	39	139.0	50	39	18	3

TABLE 2.17 *(Contd)*

Impact testing is not required for the following types of flanges when used at minimum design metal temperatures no colder than –20°F:

- ASME B16.5 flanges of ferritic steel;
- ASME B16.47 flanges of ferritic steel;
- Split loose flanges of SA-216 GR WCB with outside diameter and bolting dimensions per either ASME B16.5 Class 150 or Class 300 and with flange thicknesses that are not greater than either ASME B16.5 Class 150 or Class 300, respectively;
- Carbon and low-alloy steel, long weld neck flanges

No impact testing is required for Part UCS materials 0.10 in. in thickness and thinner and with design metal temperatures no colder than –55°F.

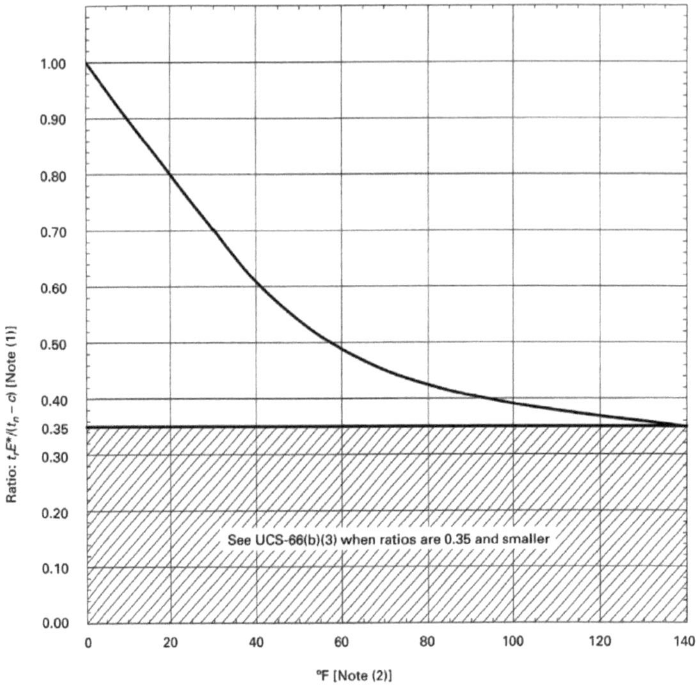

c = corrosion allowance, in.
E^* = as defined in Figure UCS-66.2, Note (3)
t_n = nominal thickness of the component under consideration before corrosion allowance is deducted, in.
t_r = required thickness of the component under consideration in the corroded condition for all applicable loadings [Figure UCS-66.2, Note (2)], based on the applicable joint efficiency E [Figure UCS-66.2, Note (3)], in.

NOTES:
(1) *Alternative Ratio* = $S^* E^*$ divided by the product of the maximum allowable stress value from Table UCS-23 times E, where S^* is the applied general primary membrane tensile stress and E and E^* are as defined in Figure UCS-66.2, Note (3)
(2) See UCS-66(b).

FIGURE 2.11 Reduction in MDMT without impact testing (Code Fig. UCS-66.1). *(Courtesy of the American Society of Mechanical Engineers.)*

One last important factor when considering impact testing is Form U-1, Manufacturer's Data Report. It should be noted in the Remarks section of the data report "...Impact test exempt per..." and then written which exemption was used. For example, UG-20(f), UCS-66(a), (b), or (c). In some cases, more than one exemption may be listed on the data report.

The important thing to remember is that vessels operating at any temperature must be evaluated for impact-test requirements and demand special attention in design and materials. Specific paragraphs in the *ASME Boiler and Pressure Vessel Code* prescribe the type of welded joints involved in low-temperature operations. These joints must meet the requirements of Code Par. UW-2, Service Restrictions.

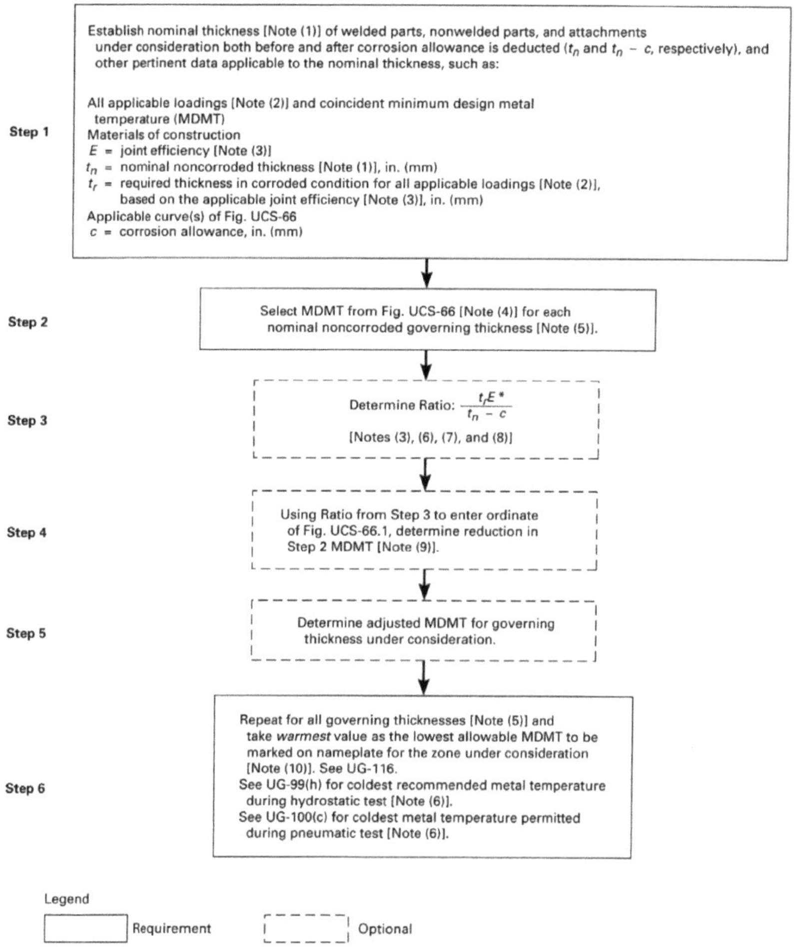

NOTES:

(1) For pipe where a mill undertolerance is allowed by the material specification, the thickness after mill undertolerance has been deducted shall be taken as the noncorroded nominal thickness t_n for determination of the MDMT to be stamped on the nameplate. Likewise, for formed heads, the minimum specified thickness after forming shall be used as t_n.

(2) Loadings, including those listed in UG-22, which result in general primary membrane tensile stress at the coincident MDMT.

FIGURE 2.12 Diagram of UCS-66 rules for determining lowest MDMT without impact testing (Code Fig. UCS-66.2). *(Courtesy of the American Society of Mechanical Engineers.)*

Code Par. UW-3, Welded Joint Category, determines the location of the joint. If the material is carbon steel, Code Pars. UG-20(f), UCS-66, and UCS-67 state whether impact tests are required as well as specify the exemptions for impact testing. If the tests are required, Code Par. UG-84 describes how they must be made. Code Par. UG-84 also gives instructions on the required size of test specimens and on the location of the areas in the specimens, taking impacts, test values,

NOTES (CONT'D):

(3) E is the joint efficiency (Table UW-12) used in the calculation of t_r; E^* has a value equal to E except that E^* shall not be less than 0.80. For castings, use quality factor or joint efficiency, E, whichever governs design.

(4) The construction of Figure UCS-66 is such that the MDMT so selected is considered to occur coincidentally with an applied general primary membrane tensile stress at the maximum allowable stress value in tension from Section II, Part D, Subpart 1, Table 1A. Tabular values for Figure UCS-66 are shown in Table UCS-66.

(5) See UCS-66(a)(1)(-a), (a)(1)(-b), and (a)(1)(-c) for definitions of governing thickness.

(6) If the basis for calculated test pressure is greater than the design pressure [UG-99(c) test], a Ratio based on the t_r determined from the basis for calculated test pressure and associated appropriate value of $t_n - c$ shall be used to determine the recommended coldest metal temperature during hydrostatic test and the coldest metal temperature permitted during the pneumatic test. See UG-99(h) and UG-100(c).

(7) Alternatively, a Ratio of $S^* E^*$ divided by the product of the maximum allowable stress value in tension from Section II, Part D, Subpart 1, Table 1A times E may be used, where S^* is the applied general primary membrane tensile stress and E and E^* are as defined in Note (3).

(8) For UCS-66(b)(1)(-b) and UCS-66(i)(2), a ratio of the maximum design pressure at the MDMT to the maximum allowable pressure (MAP) at the MDMT shall be used. The MAP is defined as the highest permissible pressure as determined by the design equations for a component using the nominal thickness less corrosion allowance and the maximum allowable stress value from the Section II, Part D, Subpart 1, Table 1A at the MDMT. For ferritic steel flanges defined in UCS-66(c), the flange rating at the warmer of the MDMT or 100°F (38°C) may be used as the MAP.

(9) For reductions in MDMT up to and including 40°F (22°C), the reduction can be determined by: reduction in MDMT = (1 − Ratio) 100°F (56°C).

(10) A colder MDMT may be obtained by selective use of impact tested materials as appropriate to the need (see UG-84). See also UCS-68(c).

FIGURE 2.12 *(Contd)*

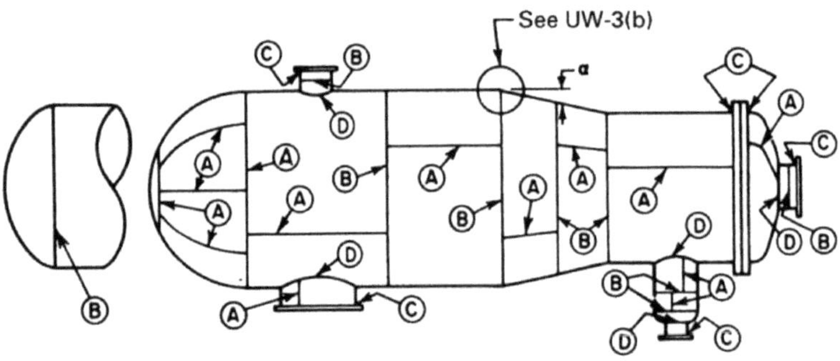

FIGURE 2.13 Illustration defining locations of welded joint categories (Code Fig. UW-3). *(Courtesy of the American Society of Mechanical Engineers.)*

and properties. However, impact testing is not a "must" for welds or materials if certain design requirements are met.

Other materials exempt from impact tests are austenitic stainless steels. [See Code Pars. UHA-51(d) thru (i) for details.]

Wrought aluminum alloys do not undergo a marked drop in impact resistance below –20°F. As indicated in Code Par. UNF-65, no impact tests are required down to a temperature of –452°F. Copper and copper alloys, nickel and nickel alloys, and cast aluminum alloys can be used up to –325°F without impact tests. Code Par. UNF-65 indicates the details for impact testing.

The major concern for metals at low temperature is brittle fracture, which can cause vessel failure. At normal temperatures, metals usually give warning by bulging, stretching, or leaking or show some other failure. Under low-temperature conditions, some normally ductile metals may fail at low-stress levels without warning of plastic deformation.

Four conditions must be controlled to avoid brittle fractures in low-temperature vessels:

1. The materials should not show a transition from ductile to brittle fracture.
2. The welding materials for low-temperature service must be selected carefully, and a proven welding procedure must be used.
3. Stress raisers from design or fabrication should not be permitted.
4. Localized yielding in the stress-raiser area must not be allowed.

Pressure vessel users should use a material that behaves in a ductile manner at all operating temperatures. This will allow for the possible stress raiser that may be overlooked in fabrication or design. Impact tests may assist in determining the correct materials and welding procedures for these vessels.

Service Restrictions and Welded Joint Category

Welded joints in vessels must meet the requirements in Code Par. UW-2, Service Restrictions, and Par. UW-3, Welded Joint Category, of the *ASME Boiler and Pressure Vessel Code,* Section VIII, Division 1.

The main joints fitting the definitions of Categories A thru D are as follows:

1. Longitudinal and spiral welded joints within the main shell, communicating chambers, transitions, or nozzles.

   Any welded joint within a sphere, within a formed or flat head, or within the side plates of a flat-sided vessel.

   Any butt-welded joint within a flat tubesheet.

   Circumferential welded joints connecting hemispherical heads to main shells, to transitions in diameters, to nozzles, or to communicating chambers.
2. Circumferential welded joints within the main shell, communicating chambers, nozzles, or transitions, including joints between the transition and a cylinder at either end. Circumferential welded joints connecting formed heads other than hemispherical to main shells, to transitions in diameter, to nozzles, or to communicating chambers. Circumferential welded joints are butt joints if the half-apex angle, α, is equal to or less than 30° and angle joints when α is greater than 30°.

3. Welded joints connecting flanges, Van Stone laps, tubesheets, or flat heads to main shell, to formed heads, to transitions in diameter, to nozzles, or to communicating chambers.

   Any welded joint connecting one side plate to another side plate of a flat-sided vessel.

4. Welded joints connecting communicating chambers or nozzles to main shells, to spheres, to transitions in diameter, to heads, or to flat-sided vessels, and those joints connecting nozzles to communicating chambers.

The paragraph on the welded joint category (see Fig. 2.13 and Code Fig. UW-3) was added to the Code to assist in clarification of Code Par. UW-2, Service Restriction. The requirements of Code Par. UW-2 have often been misinterpreted and at times not referred to until too late. For example, if a vessel designed for –140°F which would require impact tests was designed with flanges, the flanges would have to be welded joint category C and would require full-penetration welds extending through the entire section of the joint. Also, all longitudinal seams would have to be category A, without backing strips.

Other cautions are that vessels designed for cryogenic service requiring impact tests also require the design on the flanges to specify full-penetration welds, and the longitudinal seams cannot have metal backing strips which remain in place. It is important that the design engineers and pressure vessel manufacturers have a clear understanding of the paragraphs in the Code on service restrictions and welded joint category and that the vessels, where required, receive proper attention during design and in handling in the shop.

What are the vessels that demand special requirements listed in Code Par. UW-2, Service Restrictions? They are vessels intended for lethal service; vessels constructed of carbon and low-alloy steels which are required to be postweld heat treated; vessels designed to operate below certain temperatures designated by Code Part UCS (see Code Par. UCS-86); vessels of impact-tested materials, including weld metal, as required by Code Part UHA; unfired steam boilers with design pressures exceeding 50 psi; and pressure vessels subject to direct firing.

What must be remembered about the welded joint category? Only those joints to which special requirements apply are included in the categories. The term *category* only defines the location of the welded joint in the vessel and never implies the type of welded joint that is required. The *type* of joint is prescribed by other paragraphs in the Code suitable to the design, which may specify special requirements regarding joint type and the degree of inspection for certain welded joints.

The purpose of this review is to spell out the requirements of the Code with regard to service restrictions and welded joint category.

Designers and manufacturers should examine their practices to ensure that the service restrictions paragraph of the Code is receiving appropriate attention in the engineering department and the shop, for the designers and manufacturers must accept full responsibility for the design and construction of a Code pressure vessel when the vessel is completed and the data sheets are signed.

The ASME Code Form U-1, Manufacturer's Data Report, in the section for remarks requires a brief description of the purpose of the vessel and also requires that the contents of each vessel be indicated. It will be the Code Inspector's duty to determine whether this item is accurately completed and whether service restrictions were taken into consideration.

Therefore, in designing Code vessels, be sure you have taken the paragraphs on service restrictions and welded joint category into consideration. Also be sure the Authorized Inspector has reviewed the design before fabrication starts so that no difficulties will arise when the vessel is completed.

Noncircular-Cross-Section-Type Pressure Vessels

Code Appendix 13 lists the design, fabrication, and inspection requirements for vessels of noncircular cross section. These rules cover rectangular, stayed, and obround pressure vessels. Table 2.18 lists other code references along with helpful information in the design of such vessels.

Today many manufacturers are manufacturing jacketed vessels. Most jacketed vessels are designed for external pressure on the inner shell and internal pressure on the outer shell. These types may be of the braced and stayed type construction (see Code Appendix 9). Some configurations of such vessels are of the "dimpled" jacketed type (see Code Appendix 17). The outer plate, normally thin gage stainless steel, is formed by presses. Holes are punched in the material at approximately 3 in. centers and then a dimple is formed at each hole. This jacket is then welded to an inner shell by welding the hole in which the dimple is formed, thus creating the jacket. This type of vessel is frequently used in the pharmaceutical industry.

Special design considerations must be accounted for. Table 2.19 lists the Code references for such construction.

Pressure Vessels Stamped with the UM Symbol

Vessels stamped with the UM symbol are limited to size and design pressures in accordance with the ASME Code Section VIII, Division 1, Code Par. U-1(j). These vessels are exempt from inspection by Inspectors, as defined in UG-91, provided they comply with all other requirements of the Code. Table 2.20 lists the requirements for pressure vessels exempt from inspection.

Remarks	Code references
Scope a. Types of vessels, descriptions, and figures b. Materials	Par. 13-1 Par. 13-2 Par. 13-3
Design requirements and stress acceptance criterion	Par. 13-4
Thermal expansion reactions and loads due to attachments	Par. 13-4
Nomenclature and terms used in design formulas	Par. 13-5
End closures a. If ends are unstayed, flat plate design closure per UG-34 except a C factor of 0.20 shall be used b. All other end closures are special design per U-2(g), or they may be subject to proof test per UG-101 c. Braced and stayed surfaces may be designed per Code Par. UG-47	Par. 13-4 Pars. U-2(g), UG-101 Pars. UG-47, UG-48, UG-49, UW-19, 13-4
Special considerations for vessels of aspect ratio less than 4	Par. 13-4(h)
Vessels are designed for membrane and bending stress throughout the structure (for both the short-sided and long-sided plates). Membrane, bending, and total stresses are considered. Formulas for the stresses are given as follows: a. Vessels of rectangular cross section (nonreinforced) b. Reinforced vessels of rectangular cross section c. Stayed vessels of rectangular cross section d. Unreinforced vessels having an obround cross section e. Reinforced vessels of obround cross section f. Stayed vessels of obround cross section g. Vessels of circular cross section having a single diametral staying member h. Vessels of noncircular cross section subject to external pressure	Par. 13-4 Par. 13-7 Par. 13-8 Par. 13-9 Par. 13-10 Par. 13-11 Par. 13-12 Par. 13-13 Par. 13-14
Vessel fabrication	Par. 13-15
Examples illustrating application of formulas	Par. UG-16(f)
Corner joints	Par. UW-13
Special calculations Weld efficiencies Ligament efficiencies Aspect ratios between 1 and 2	Par. 13-18 Par. 13-18(a) Par. 13-18(b) Pars. 13-18(c), 13-18 (d)

TABLE 2.18 Vessels of Noncircular Cross Section

Remarks	Code references
Scope and minimum requirements a. Any combination of pressures and vacuum in vessel and jacket producing a total pressure over 15 psi on the inner wall is in the scope of Code Appendix 9 b. Half-pipe jackets are not included in the scope of Appendix 9	Par. 9-1 App. EE
Types of jacketed vessels a. Full jackets b. Partial jackets	Par. 9-2 Fig. 9-2 Par. 9-7 Fig. 9-7
Materials	Par. 9-3
Dimpled jackets	Par. UW-19 App. 17
Embossed assemblies	Par. UW-19 App. 17
Proof test of dimpled and embossed assemblies, and partial jackets	Pars. UG-101, 17-7
Design of jackets	Par. 9-4
Design of openings through jacket space	Par. 9-6, Fig. 9-6 Pars. UG-36 to UG-45
Braced and stayed surfaces	Pars. 9-4, UG-47, UW-19
Inspection openings need not be over 2 in.	Par. 9-4
Hydrostatic test pressure for glass-lined vessels with jacket	Par. 27-4

Table 2.19 Jacketed Vessels

Marking and Nameplate Requirements

The Code specifies that completed pressure vessels be marked and furnished with nameplates in order to positively identify the parameters associated with the vessel. These parameters are shown in Fig. 2.14 (Code Fig. UG-118). The stamping should be located in a conspicuous place, preferably near a manhole or handhole opening. For vessels having two or more pressure chambers, see Code Par. UG-116(j).

Remarks	Code references
Vessels not within the scope of the Code	
Even though exempted from the Code, any vessel meeting all the Code requirements may be stamped with the Code symbol	Par. U-1(c)(2)
Vessels containing water (or water cushioned with air) with a maximum design pressure of 300 psi and maximum design temperature of 210°F	Par. U-1(c)(2)(-f)
Vessels having an internal or external operating pressure of 15 psi or less with no limitation on size	Par. U-1(c)(2)(-h)
Vessels having inside diameter, width, height, or cross section diagonal of 6 in. or less with no vessel length or pressure limitation	Par. U-1(c)(2)(-i)
Pressure vessels for human occupancy	Par. U-1(c)(2)(-j)
See Par. U-1(c)(2) for complete list of exemptions	
Vessels within the scope of the Code	
Pressure vessels not required to be fully radiographed, not provided with quick-actuating or quick-opening closures, and not exceeding the following volume and pressure limits may be exempted from inspection by Inspectors: a. 5 ft^3 in volume and 250 psi design pressure, or b. 3 ft^3 in volume and 350 psi design pressure, or c. 1½ ft^3 in volume and 600 psi design pressure.	Par. U-1(j)
Straight line interpolation between the above volumes and pressures is permitted for intermediate values.	
Exempted vessels must comply with the Code requirements in all other respects	
Vessels fabricated in accordance with this rule shall be marked with the "UM" symbol	Par. UG-116 Fig. UG-116(b)

Table 2.20 Vessels Exempt from Inspection

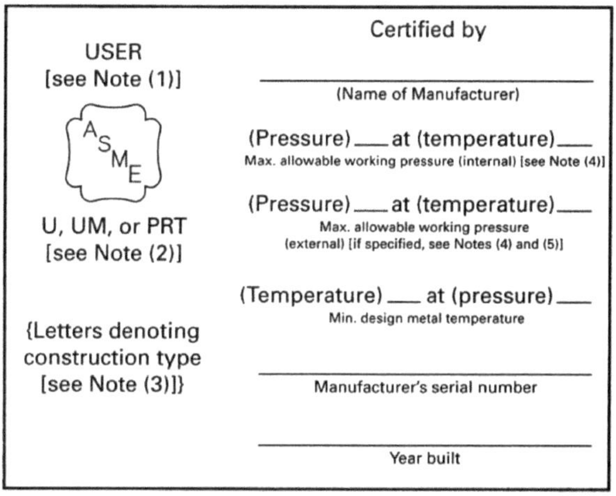

Remarks	Code references
Note (1): "USER" shall be included when the vessel is inspected by a user's Inspector as provided in UG-91	Par. UG-116(a)(1)(-a)
Note (2): • U Designator on vessels inspected in accordance with the requirements in UG-90 through UG-97 • UM Designator on vessels constructed in accordance with the provisions in 11-1(j) • PRT Designator on parts	 Par. UG-116(a)(1)(-a) Par. UG-116(a)(1)(-b) Par. UG-116(a)(1)(-c)
Note (3): Type of construction • W Arc or gas welded • P Pressure welded (except resistance) • B Brazed • RES Resistance welded • G Graphite	 Par. UG-116(b)(1)
Vessels intended for special service shall be marked as follows: • L Lethal service • UB Unfired steam boiler • DF Direct firing	Par. UG-116(c)
Vessels with radiographic or ultrasonic examination in accordance with UW-11 shall include the following marks:	Par. UG-116(e)
• RT 1 Full radiography requirements are satisfied	Par. UG-116(e)(1)

FIGURE 2.14 Stamping requirements and applicable form (Code Fig. UG-118). (*Courtesy of the American Society of Mechanical Engineers.*)

• RT 2 The complete vessel satisfies the requirements of UW-11(a)(5) and when the spot radiography requirements of UW-11(a)(5)(-b) have been applied	Par. UG-116(e)(2)
• RT 3 The complete vessel satisfies the spot radiography requirements of UW-11(b)	Par. UG-116(e)(3)
• RT 4 Only part of the complete vessel has satisfied the radiographic requirements of UW-11(a) or where none of the markings “RT 1,” “RT 2,” or “RT 3” are applicable	Par. UG-116(e)(4)
Heat treatment marking, if applicable	Par. UG-116(f)
• HT The complete vessel has been postweld heat treated as provided in UW-10	
• PHT Only part of the complete vessel has been postweld heat treated as provided in UW-10	
• Parts of vessels for which Partial Data Reports are required in UG-120(c) shall be marked with the U or PRT Designator above the word “PART”	Par. UG-116(h)(1)(-a)
For cases where the internal and external MAWP values have the same designated coincident temperature, the values may be combined on a single line as: Pint/FV (psi) at Temp (°F)	
The MAWP (external) is required only when specified as a design condition	

Figure 2.14 *(Contd)*

CHAPTER 3

Design for Safety

Basic Causes of Pressure Vessel Accidents

Two new vessels built for equal pressures and temperatures as well as the same type of service can fare very differently in different locations. In one plant, capable maintenance personnel can keep the vessel in top shape by means of a systematic schedule of inspection and maintenance; in another, an equal number of personnel, through ignorance, incompetence, or neglect, can bring about the failure of the vessel in a comparatively short time. Competent personnel are obviously a key factor in the longevity and safety of a pressure vessel.

Corrosion Failures

Most pressure vessels are subject to deterioration by either corrosion or erosion or both. The effect of corrosion may be pitting or grooving over either localized or large areas. Corrosion over large areas can bring about a general reduction in the plate thickness.

While some of the many causes of corrosion are inherent in the process being used, others are preventable. For example, vessels containing corrosive substances deteriorate more rapidly if the contents are kept in motion by the stirring paddles or by the circulation of steam or air. If the corrosion is local, it may not appear serious at first. However, as the metal thickness is reduced, the metal becomes more stressed—even under normal pressures. The more stressed the metal becomes, the faster it corrodes, at last reaching the point of sudden rupture. Corrosion in unstressed metal is less likely to cause such sudden failure.

Stress Failures

Many plate materials are not at all suitable for the service required of them. Other materials may be serviceable in an unstressed condition but will fail rapidly in a stressed condition. To give an example, a large chemical company once used stainless steel corrosion buttons to measure corrosion in a large fired cast-steel vessel. When the cast-steel vessel had to be replaced, the company ordered a new vessel made of the same stainless steel used for the buttons because of its proved resistance. After a few days of service, however, the operator noticed leakage through the 1 in. shell and immediately reduced the pressure and shut down the vessel. Examination showed many cracks in the stainless steel. The substance being processed was highly corrosive to the grain boundaries of the metal when under stress (see Code Appendix UHA-A). This fact was verified when the metal was tested in both stressed and unstressed conditions in the same solution. The stressed pieces failed within 100 hours.

When metals are subjected to repeated stresses exceeding their limits, they ultimately fail from *fatigue*. It may take thousands of stress applications to produce failure, but a vessel such as an air tank in which there are rapid pressure fluctuations may fail at the knuckle radius. In a large laundry facility, the authors once saw grooving that penetrated halfway into a ⅝ in. head in the knuckle radius of a hot water tank.

Excessive stresses can develop in vessels that are not free to expand and contract with temperature changes. If these stresses are allowed to recur too frequently, eventual failure from fatigue can be expected.

Other Failures

Vessels are often installed in positions that prevent entry through the inspection openings for the examination of interior surfaces. Any owner or user of a vessel who is interested in its actual condition will never let this happen.

If a vessel is supported on concrete saddles, the difference in the coefficients of expansion of the concrete and steel will sometimes cause the concrete to crack just beneath the tank, allowing accumulations of moisture and dust that will cause corrosion.

Safety valves or devices that become plugged or otherwise inoperative can lead to accidents. Safety valves may also be adjusted too high for the allowable pressure so that the vessel fails if overpressure occurs.

Defective material not visible at the time of fabrication, such as laminations in the plate or castings with internal defects, may bring on failure. Inferior technique at the time of fabrication, such as poor joint fit-up or bad welding that may appear satisfactory on the surface but suffers from a lack of penetration, slag, excessive porosity, or ductility, can also cause serious trouble.

It requires consistent effort on the part of plant personnel as well as a well-organized system of maintenance and inspection to safeguard pressure vessel equipment properly.

Design Precautions

The reliable functioning of a pressure vessel depends upon the foresight of the user and the information the user gives to the designer

Remarks	Code references
General requirements	Par. UG-25
Suggested good practice	Appendix E
For brazed vessels	Par. UB-13
For clad vessels	Pars. UCL-25, UG-26 Appendix F
Vessels subject to corrosion must have suitable drain openings	Par. UG-25(f)
Vessels subject to internal corrosion, erosion, or mechanical abrasion must have inspection openings	Par. UG-46

Table 3.1 Corrosion Allowances and Requirements

and fabricator. For example, if the process involved is corrosive, these data must include the chemicals to be used and, if possible, the expected rate of corrosion based on the history of vessels in similar service. (Corrosion resistance data are available for most materials and chemicals, but actual operating conditions can seldom be duplicated in the laboratory and must be determined by frequent, close examination after the vessel is placed in service.) Table 3.1 lists the requirements of the ASME Code Section VIII, Division 1 corrosion allowances and the applicable Code references.

Usually no one but the user is fully familiar with the operating conditions and the causes of defects and deterioration in the type of vessel being ordered. If the designer is presented with all the facts, however, the materials best suited for the job will be used and all openings will be placed in the most desirable locations.

Although the Code permits many types of construction, some of them may not be suitable for the service required. Therefore, merely to prescribe an ASME vessel is not always sufficient. For example, the Code permits small openings to be welded from one side only and also provides for the offset construction of girth seams and heads (see Code Par. UW-13; also Fig. UW-13.1). This type of construction, however, may not be suitable for some vessels because it leaves some inside seams and fittings not welded and thus subject to corrosion. The purchaser aware of such facts will be specific in ordering and will state the type of welded joints desired. The drawings will also be checked before fabrication to make sure that the openings are located where desired and that the inspection openings permit easy access to the interior.

The Code Inspector must also check the design of the vessel to be assured that the shell, head, and nozzle thicknesses are correct, the knuckle radius sufficient, the openings properly reinforced, the flange thickness sufficient, and the like, but there is no concern with the position of the outlets so long as they comply with the Code.

Code Par. UG-22 and Appendix G state that loadings must be considered when designing a vessel, but do not give clear information about how to take them into account. When vessels and piping systems operate out-of-doors, wind must be considered. Also to be taken into account are external loadings such as snow; the minimum ambient temperature; the reaction of piping systems; the restricted thermal expansion on the attachment of piping systems to the vessel and on its hangers or anchors; the weight of valves; and the pumps, blowers, and rotating machinery. It is imperative for the designer to see that all machinery, piping, and equipment are properly balanced and supported. Non-mandatory Code Appendixes D and G also give more useful information to be taken into consideration.

Internal corrosion usually develops on surfaces susceptible to moisture. For this reason, it is important to make sure that the vessel can be properly drained. In vertical vessels equipped with agitating devices, the greatest wear usually occurs on the bottom head. In horizontal vessels with agitators, the most wear will be between the six o'clock and nine o'clock positions for clockwise rotation and between the six o'clock and three o'clock positions for counterclockwise rotation. Horizontal vessels with an agitator should therefore be designed for turning, if possible, so that the thinned-out area can be rotated to the top half. The useful life of the vessel will thus be extended.

Vessels to be subjected to heavy corrosion, erosion, or mechanical abrasion require an extra corrosion allowance and telltale holes. Leakage from these holes will give a warning that the vessel has corroded to its minimum allowable thickness. For example, if the minimum thickness required is ⅜ in. but the plate thickness used (to increase the length of service) is ⅞ in., the corrosion allowance would be ½ in. Telltale holes placed on about 18 in. centers would then be drilled ⅜ in. deep from the surface not affected by corrosion.

Code Par. UG-25(e) states: "When telltale holes are drilled, they shall be 1/16 to 3/16 in. in diameter and shall be drilled to a depth not less than 80 percent of the thickness required for a seamless shell of like dimensions. These holes shall be drilled into the surface opposite to that at which deterioration is expected." For clad or lined vessels, see Code Par. UCL-25(b).

Table 3.2 lists the minimum thickness of plate which may be used when designing Code pressure vessels.

Inspection Openings

Accessibility to the inside of a vessel is essential. Carrying out regular internal inspections is of importance to the safety and extended life of a vessel. Code Par. UG-46 requires inspection openings in all vessels subject to internal corrosion or having internal surfaces subject to erosion or mechanical abrasion so that these surfaces may be examined for defects. Actually, almost all pressure vessels are subject to some

Material	Minimum thickness	Code references
	The minimum thickness permitted for shells and heads, after forming and regardless of the product form and material, shall be 1/16 in. exclusive of any corrosion allowance.	UG-16(b)
	The minimum thickness of shells and heads of unfired steam boilers shall be ¼ in. exclusive of any corrosion allowance.	UG-16(b)(3)
Table UCS-23 materials	The minimum thickness of shells and heads used in compressed air service, steam service, and water service shall be 3/32 in. exclusive of any corrosion allowance.	UG-16(b)(4)
Clad vessels	The minimum thickness shall be the total thickness for clad material with corrosion resistant integral or weld metal overlay cladding and the base-material thickness for applied-lining construction.	UCL-20(b)
5%, 8%, and 9% nickel steels	The minimum thickness after forming of a section subject to pressure shall be 3/16 in. (Maximum thickness: 2 in.)	ULT-16(b)
Cast iron dual metal cylinders	The minimum wall thickness shall be 5 in. (Maximum OD: 36 in.)	UCI-29
Layered vessels	The minimum thickness of any layer is ⅛ in.	ULW-16(d)

The minimum thickness of plate after forming and without allowances for corrosion shall not be less than the minimum thickness allowed for the type of material being used (see Code Pars. UG-16 and UG-25).

Table 3.2 Minimum Thickness of Plate

kind of interior corrosion, and its seriousness can be determined only by examining the inside of the vessel. The alternative is possible failure and consequent injury to, or even loss of, human life.

Minimum sizes for the various types of openings have been established by the Code. An elliptical or obround manhole opening must not be less than 12 in. × 16 in. and a circular manhole's inside diameter not less than 16 in. Handholes must not be less than 2 in. × 3 in.

On small vessels, it is sometimes difficult to provide extra openings. The Code states that vessels more than 12 in. but less than 18 in. in inside diameter must have not less than two handholes or two threaded plugged openings of at least 1½ in. pipe size. Vessels 18 in. to 36 in., inclusive, in inside diameter must have one manhole or two handholes or two threaded plugs of not less than 2 in. pipe size. Vessels with inside diameter more than 36 in. must have a manhole. Those vessels of a shape or use that make a manhole impracticable must have at least two 4 in. × 6 in. handholes or two openings of equal area.

Removable heads or cover plates may be used in place of inspection openings, provided they are large enough to permit an equivalent view of the interior.

If vessels with inside diameter not less than 12 in. or more than 16 in. are so installed that they must be disconnected from an assembly to permit inspection, removable pipe connections of not less than 1½ in. pipe size can be used for inspection openings.

Only vessels 12 in. or less in inside diameter are not required to have openings if there are at least two removable pipe connections of at least ¾ in. pipe size.

Inspection openings may be omitted in vessels covered in Code Par. UG-46(b). When openings are not provided, the manufacturer's data report shall include the notations required under remarks.

When a plant orders a vessel, it should be sure that the design provides for inspection openings that will be accessible after the vessel has been installed. The plant should also have the vessel examined periodically by competent inspectors to make sure that the internal surfaces remain free from deterioration.

Quick-Opening Closures

Code Par. UG-35 cautions users about removable closures on the ends of pressure vessels, utility access manways, and similar openings that are not of the multi-bolt type (such as charge openings, dump openings, quick-opening doors, and cover plates). It states that these openings must be so arranged that the failure of one holding device will not release all the others.

Because of the serious failures incurred by such closures, fabricators and inspectors must give them special attention and see that the closures are equipped with a protective device that will ensure their being completely locked before any pressure can enter the vessel and that will prevent their being opened until all pressure has been released.

As there are many types of quick-opening closures, it is not possible to give detailed requirements for protective devices. One example will have to suffice. The most common of these closures is the breech-lock door, which is supported on a hinge or davit and is locked or unlocked by being revolved a few degrees. When locked, the lugs in the door are in a position behind the lugs in the door frame or locking ring. An interlocking device can be easily installed to make sure that these lugs are properly engaged before steam is fed to the vessel. If a positive locking device is not possible, then some type of warning device, such as a locking pin, should be installed to prevent the door or locking device from turning until an exhaust valve opens. The steam escaping from the opened valve will warn the operator to release all the pressure in the vessel before the door is opened. Likewise, the valve cannot be closed until the door is in the proper closed position. All quick-opening closure vessels must have a pressure gage visible from the operating area.

Figures 3.1 and 3.2 show the result from the loosening of a set screw in the collar on a valve stem which allowed the valve in a safety locking device to be closed when the door was not fully locked. Vibrations from the vessel being under pressure caused the locking device to become fully disengaged, allowing the door to be blown away. The 8 ft, 8 in. diameter door traveled about 500 ft through the air before shearing off the roof of a car on a highway. The door continued down the highway about 400 ft, then veered, tore down a

FIGURE 3.1 Drying cylinder explodes when an almost foolproof safety device failed. (*Courtesy of The Hartford Steam Boiler Inspection and Insurance Company.*)

FIGURE 3.2 Door of cylinder. (*Courtesy of The Hartford Steam Boiler Inspection and Insurance Company.*)

chain-link fence, and cut down a tree of 10 in. diameter. In reaction, the 129 ft long cylinder vaulted from the 2 ft deep pit in which it was located, traveled 200 ft before ramming through a building, and then struck, overturned, and demolished a railroad freight car half loaded with cement. The total force acting on the head of a quick-opening closure when it is under maximum pressure is very large.

Safety interlocking devices will help prevent many failures, but they are not infallible. The condition of the operating parts of a door should be carefully examined, and the door itself should be observed to be in its intended position before pressure is applied. Precautions should be taken to ensure that the closed door cannot be moved while the pressure is on and that the pressure will be entirely released before the door is opened. The importance of the maintenance of doors, door frames, and locking rings—especially the lubrication of locking surfaces and the wear on locking wedges of quick-opening closures—must not be overlooked. A precise examination schedule should be carried out.

Pressure Relief Devices

The importance of pressure relief devices in the safe operation of pressure vessels cannot be overemphasized. Improper functioning can result in disaster, both to life and to equipment. The *ASME Boiler and Pressure Vessel Code,* in Par. UG-125, states that all pressure vessels must be provided with a means of overpressure protection. It is the manufacturer's responsibility to see that the vessels are fabricated according to the ASME Code and that they meet the minimum design requirements. Simply because the customer did not indicate the size and location of the safety valve opening on the purchase order or drawing does not relieve the vessel manufacturer of this responsibility.

The safety devices need not be provided by the vessel manufacturer, but overpressure protection must be installed before the vessel is placed in service or tested for service; a statement to this effect should be made in the remarks section of the manufacturer's data report.

The user of a pressure vessel should ensure that the requirements of the pressure relief devices—be they pressure relief valves (Code Par. UG-126), non-reclosing pressure relief devices such as rupture disk devices (Code Par. UG-127), or liquid relief valves (Code Par. UG-128)—are designed in accordance with the ASME Code requirements and so stamped in accordance with Code Par. UG-129.

Rupture disk devices shall be marked by the manufacturer as required in Code Par. UG-129 in such a way that the marking will not be obliterated in service.

Safety valves must be installed on the pressure vessel or in the system without intervening valves between the vessel and protective device or between devices and the point of discharge. (See also Code Appendix M.)

To prevent excessive pressure from building up in the pressure vessel, a safety valve is set at or below the maximum allowable working pressure for the vessel it protects.

Safety valve construction is important in the selection of safety relief valves. Proper fabrication may be ensured by specifying that the construction must conform to ASME requirements.

Before a manufacturer of relief devices is allowed to use the Code stamp on a relief device, the manufacturer must apply to the Boiler and Pressure Vessel Committee in writing for an authorization to use the stamp. A manufacturer or assembler of safety valves must demonstrate to the satisfaction of a designated representative of the National Board of Boiler and Pressure Vessel Inspectors that the manufacturing, production, testing facilities, and quality control procedures will ensure close agreement between the performance of random production samples and the capacity performance of valves submitted to the National Board for capacity certification. (For capacity conversion formulas, see Code Appendix 11.)

The escape pipe discharge should be so located as to prevent injury to anyone where it is discharging. It is important to have the discharge pipe of the relief device at least equal to the size of the relief device. It must be adequately supported, and provisions must be made for the expansion of the piping system.

No pressure vessel unit can be considered completed until it is provided with a properly designed safety valve or relief valve of adequate capacity and in good operating condition.

All safety devices should be regularly tested by a person responsible for the safe operation of the pressure vessel.

CHAPTER 4

Guide to Quality Control Systems for ASME Code Vessels

A Practical Guide to Writing and Implementing a Quality Control Manual

To assure that a manufacturer has the ability and integrity to build vessels according to the Code, the *ASME Boiler and Pressure Vessel Code* requires each manufacturer of Code power boilers, heating boilers, and pressure vessels to have and to demonstrate a quality control system. (For pressure vessels, see Code Pars. U-2 and UG-90 and Code Appendix 10.) This system must include a written description or checklist explaining the quality-controlled manufacturing process in detail. The written description—which for clarification we will call a manual—must explain the company's organized and systematic way of specifying procedures involving each level of design, material, fabrication, testing, and inspection. An effective quality program will find errors before fabrication and will reduce or prevent defects during fabrication. To be effective, this system must have managerial direction and technical support. Proper administration is essential in order to provide the quality control and quality assurance necessary for the fabrication of ASME Code pressure vessels.

Before the issuance or renewal of an ASME Certificate of Authorization to manufacture Code vessels or piping is granted, the manufacturer's facilities and organization are subject to a joint review by an inspection agency and the legal jurisdiction concerned. In areas where a jurisdiction does not review a manufacturer's facility, it may be carried out by a representative of the National Board of Boiler and Pressure Vessel Inspectors.

The quality control manual will, therefore, form the basis for understanding the manufacturer's controlled manufacturing system. It will be reviewed by the inspection agency, the legal jurisdiction, or a representative of the National Board. The manual will serve as the basis for the continuous auditing of the system by the Authorized Inspectors and later for review when the manufacturer's Certificate of Authorization is to be renewed.

The Code emphasizes that the manual must reflect the actual process by which the manufacturer produces a vessel. The Code states just a few specific requirements for the manual. For example, Code Appendix 10 sets out the various details to be described; however, it does not mandate a certain way to do this. Code Par. 10-4 includes the following statement: "The Code does not intend to encroach on the Manufacturer's right to establish, and from time to time to alter, whatever form of organization the Manufacturer considers appropriate for its Code work." The suggestions and requirements implied in this book reflect a system that has worked and been accepted by the Code and, therefore, can be considered guidelines for the manufacturer intending to develop a quality control system.

The scope and details of the system will depend upon the size and type of vessel being manufactured and the work to be performed. All welded pressure vessels require certain basic provisions which must be included in the manual. An outline of the procedures to be used by the manufacturer in describing a quality control system should include the following items:

1. **Statement of authority and responsibility.** Management's statement of authority—signed by the president of the company or someone else in top management—must specify that the quality control manager shall have the authority and responsibility to establish and maintain a quality assurance program and the organizational freedom to recognize quality control and quality assurance problems and to provide solutions to these problems.
2. **Organization.** Manufacturer's organizational chart giving titles and showing the relationship between management, quality control, purchasing, engineering, fabrication, testing, and inspection must be presented.
3. **Drawings, design calculations, and specification control.** The quality control program shall include and address those

responsible and their responsibilities regarding the design of pressure vessels which shall be produced under the *ASME Boiler and Pressure Vessel Code.* These procedures shall ensure that the latest applicable drawings, design calculations, specifications, and instructions required by the Code, as well as authorized changes, are used for manufacture, examination, inspection, and testing. In an approved Code shop, an Authorized Inspector's first step in the shop inspection of Code vessels is to review the design. It is the manufacturer's responsibility to submit the drawings and calculations; therefore, in the quality control manual, the following items should be listed under design specifications:

a. The manufacturer or fabricator must verify that the specifications have been reviewed and are available to the Authorized Inspector for his or her review and acceptance. Additionally, a representative of the manufacturer should be responsible to review, accept, and approve any customer-supplied calculations, specifications, and drawings to assure Code compliance prior to the ordering of materials or releasing the job to the shop floor for fabrication (see Code Pars. U-2 and UG-90).

b. For those design calculations, drawings, and specifications prepared by the manufacturer, a responsible person shall be assigned to review, accept, and approve these documents. Design drawings, calculations, and specifications for the pressure vessel must be signed by a responsible engineering department person and reviewed by the Authorized Inspector prior to the start of work so that the Authorized Inspector may designate desired hold or inspection points. Figure 4.1 details a typical calculation sheet for ASME Code pressure vessel design. Computerized design programs must be verified for both input and output data. Controls must be described in the quality program regarding the security of the system to preclude tampering.

c. The method of handling specification and drawing changes, so that any changes in design specifications and calculations can be verified with the Authorized Inspector, must be stated.

d. The table of revisions, revision number, and the date of revision should be given.

An important factor in designing a vessel is the accessibility of welding and the position of the joint to be welded. For example, if it is difficult to weld a joint because the welder cannot reach it, see it, and be comfortable at the same time, it is impossible to expect and get good quality work as a result. This is especially important in field erection. Start quality control by reviewing the drawings, for it is here that difficult-to-weld joints may be discovered and possibly corrected before fabrication is begun.

In reviewing a drawing, the Inspector will be looking for the following design information: dimensions, plate material and thickness, weld details, corrosion allowance, design pressure and temperatures, hydrostatic testing pressure, calculations for the reinforcement of openings, and the size and type of flanges. It is especially important to indicate whether the vessel is to be postweld heat-treated and whether it is to be radiographed or spot-radiographed because the Code allows 100 percent efficiency for double-welded butt joints that are fully radiographed, 85 percent for joints that are spot-radiographed, and 70 percent efficiency for joints not radiographed.

Any comments by the Inspector on the design should be brought to the attention of the manufacturer's engineering department immediately. It is easy to make corrections before construction. To make them afterward is costly and sometimes impossible.

Another important function that the engineer performs in the control of quality is the selection of materials. The method of material procurement control must be shown in the manufacturer's quality control manual.

1. **Material control.** Material control is the quality function which ensures that only acceptable materials are utilized in the fabrication or construction of Section VIII, Division 1 pressure vessels. The quality control program set forth at the shop must address how the program will control the ordering, purchasing, receiving, storage, and material identification during the receipt inspection of materials.

   The quality control manual shall address the following:

   a. Engineering should have control of all materials which will be ordered for an ASME Code pressure vessel. The purchase requisition, bill of materials, or any other form used to document purchasing requirements shall clearly state the appropriate SA, SB, SFA, or other Code-allowed material specification for the material ordered. A job order number or another method of identifying a particular job order shall be issued.

   b. The system of material control used in purchasing, to assure the identification of all materials used in Code vessels and to check compliance with the Code material specification, should be explained (see Code Par. UG-93).

   c. The engineering department must supply all the quality assurance requirements on its purchase requisitions, bills of materials, or other requisitions before these documents are given to the purchasing department for procurement.

Calculation Sheet

CUSTOMER'S NAME								ORDER NO.		
Design pressure		*Material specification*		*Stress*	*Radiography*		*Postweld heat treat*		*Joint efficiency*	
Design temperature		Shell			None		Yes		Percent	
Corrosion allowance		Heads			Spot		No		Percent	
Inside diameter		Bolts			Complete					
Head radius		Nuts								

Shell thickness

$$t = \frac{PR}{SE - 0.6P} = \underline{\qquad\qquad} = \underline{\qquad}$$

\+ ______ Corrosion allowance

______ Thickness required

______ Thickness used

SE	
$0.6P$	
$SE - 0.6P$	

Code Par. UG-27

Torispherical head thickness

$$t = \frac{0.885PL}{SE - 0.1P} = \underline{\qquad\qquad} = \underline{\qquad}$$

\+ ______ Corrosion allowance

______ Thickness required after forming

______ Thickness used

SE	
$0.1P$	
$SE - 0.1P$	

Code Par. UG-32

Ellipsoidal head thickness

$$t = \frac{PD}{2SE - 0.2P} = \underline{\qquad\qquad} = \underline{\qquad}$$

\+ ______ Corrosion allowance

______ Thickness required after forming

______ Thickness used

$2SE$	
$0.2P$	
$2SE - 0.2P$	

Code Par. UG-32

Flat head thickness

$$t = d\sqrt{\frac{CP}{S}} = \sqrt{\frac{\qquad}{\qquad}} = \underline{\qquad}$$

\+ ______ Corrosion allowance

______ Thickness required

______ Thickness used

Code Par. UG-34

Maximum Allowable Working Pressure for New and Cold Vessels (see Code Por. UG-99)

Shell

R	
$0.6t$	
$R + 0.6t$	

$$P = \frac{SEt}{R + 0.6t} = \underline{\qquad\qquad} = \underline{\qquad}$$ psi

Torispherical head

$0.885L$	
$0.1t$	
$0.885L + 0.1t$	

$$P = \frac{SEt}{0.885L + 0.1t} = \underline{\qquad\qquad} = \underline{\qquad}$$ psi

Ellipsoidal head

D	
$0.2t$	
$D + 0.2t$	

$$P = \frac{2SEt}{D + 0.2t} = \underline{\qquad\qquad} = \underline{\qquad}$$ psi

Flat head

$$p = \frac{St^2}{d^2c} = \underline{\qquad\qquad} = \underline{\qquad}$$ psi

Maximum allowable working pressure = ____________ psi (limited by ____________)

Hydrostatic test pressure = ____________ X ____________ psi = ____________ psi

Figure 4.1 Calculation sheet for an ASME pressure vessel.

d. In the cold-forming of shell sections and heads or other pressure parts, the purchase order should state that the requirements of Code Pars. UCS-79 and/or UHT-79 must be satisfied.

Suppliers can sometimes cause problems. For example, while examining the material and fit-up of a flange on a carbon steel vessel, the Inspector noticed that the flange was 2.25 percent chromium, 1 percent molybdenum material. The vendor had substituted 2.25 percent chromium material because there was no carbon steel flange this size in stock. The vendor thought that a flange of better material was being shipped. But the 2.25 percent chromium, 1 percent molybdenum material required heat treatment and the carbon steel did not, in addition to the fact that this shop did not have welder qualifications for chromium–molybdenum welding or the special welding procedure specification for the welding of that material.

2. **Examination and inspection program.** The examination and inspection program is at the heart of the entire quality control system and is generally not described in enough detail. This section of the quality control manual should describe all the functions of inspection, tests, and examinations from the time the material is delivered to the shop until the Code vessel is stamped, certified, and shipped from the manufacturing facility to the customer.

In order to stress the importance of performing the correct material-receiving inspection at the time of delivery, it was noted that while performing an inspection at a Code manufacturing plant, a welder was attempting to weld a nozzle to a stainless steel type 321 vessel, but the weld metal was not flowing as it should. The welder notified the Inspector, who always carried a small magnet as part of the testing equipment. As the Inspector touched the magnet to the nozzle, there was a magnetic attraction. This easy test proved that the nozzle material was not type 321 stainless steel, since all 300 series stainless steels are nonmagnetic. Somehow a type 400 series stainless steel material—which is magnetic—was shipped and not properly checked when received. This is an example of why material-receiving inspection is so important. In addition to material-receiving inspections, a description of all inspections relative to the completed vessel shall be prepared as follows:

a. The quality control department must verify the mill test reports with the material received by checking the heat numbers, material specifications, physical properties, and tensile strength (see Code Pars. UG-77 and UG-85).

b. Procedures must be established for receiving the inspection and examination of plates, forgings, casting, pipe, nozzles, etc. This inspection shall require looking for laminations, scars, and surface defects to assure that all materials meet the quality assurance requirements of the Code. Formed heads and shells shall be checked for dimensions such as ovality, flange length, and curvature. Table 4.1 lists the Code-required markings and the

Remarks	Code references
General recommendations	Par. UG-4
Material must conform to a material specification given in Code Section II	Pars. UG-5 thru UG-15
One set of original identification marks should be visible: name of manufacturer, heat and slab numbers, quality, and minimum of the range of tensile strength. If unavoidably cut out, the manufacturer must maintain traceability by one of the methods described in Code par. UG-77(a). Inspector need not witness stamping but must check the accuracy of transfer	Pars. UG-77(a), UG-94
Where the service conditions prohibit die-stamping for material identification, and when so specified by the user, the materials manufacturer shall mark the required data on the plates in a manner which will allow positive identification upon delivery	Par. UG-77(b)
Plates given full heat treatment by the producing mill shall be stamped with the letters MT	SA-20 Section II, Part A
When the required heat treatments of material are not performed by the mill, they must be made under the control of the fabricator who shall place the letter T following the letter G in the mill plate marking. To show that the heat treatments have been completed, the fabricator must show a supplement to the Material Test Report	Par. UG-85
Vessel manufacturer shall obtain Material Test Report or Certificate of Compliance for plate material	Par. UG-93
Inspector shall examine Material Test Report and material markings	Pars. UG-93, UG-94
When stamping ferritic steels enhanced by heat treatment, use low-stress stamps	Par. UHT-86
Other markings in lieu of stamping	Pars. UG-77, UG-93
For some product forms, marking of container or bundle is required	Par. UG-93
Material must conform to a material specification given in Code Section II except as otherwise permitted	Pars. UG-10, UG-11, UG-15

TABLE 4.1 Material and Plate Identification and Inspection

RECEIVING REPORT

PURCHASE ORDER NUMBER: ____________ Hold Tag: YES NO
JOB NUMBER: ____________ Circle One
INSPECTED BY: ____________ DATE: ______ Released: YES NO

Plate

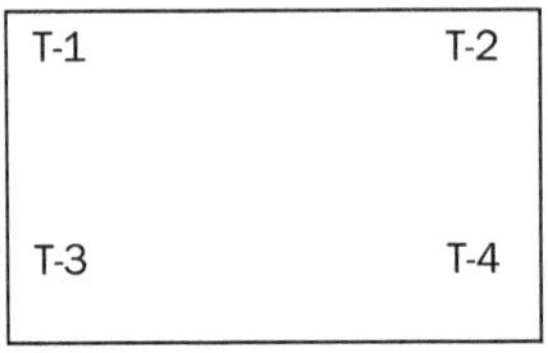

Specification: ______ Grade: ______ Class: ______

Heat Number: ______ Number of sheets: ______

T-1 ______ Size: L ______ by W ______
T-2 ______
T-3 ______ Normalized: YES NO NA
T-4 ______ Circle One

Pipe

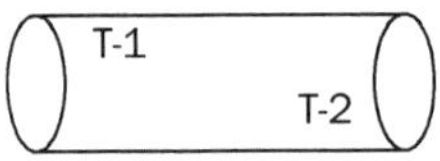

Specification: ______ Grade: ______ Class: ______

Heat Number: ______ Quantity: ______
Length: ______ Diameter: ______
T-1: ______ A.S.M.E. Stamp: Yes No NA
T-2: ______ NB or Serial No.: ______
¶ UCS 79 Requirements are met? Yes No NA
Out of Roundness: Yes No
Circle one each

Head

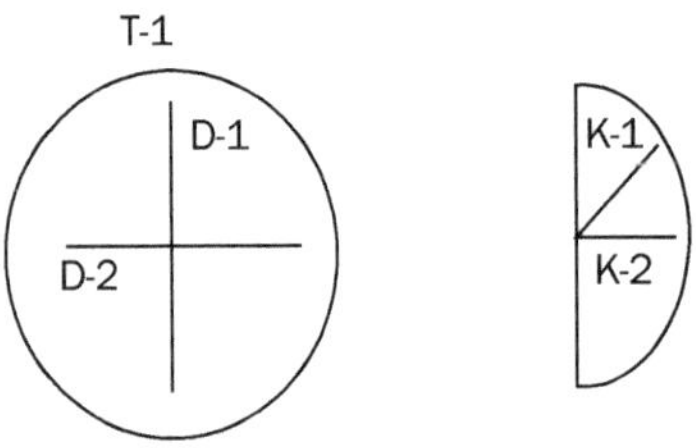

Specification: ______ Grade: ______ Class: ______
Hot Formed: ______ or Cold Formed: ______
D-1: ______ Quantity: ______
D-2: ______ Diameter: ______
K-1: ______ Heat Number if applicable:
K-2: ______ ______
T-1: ______
Head Type: 2:1 Elips. Hemi Toris Semi Ellips
¶ UCS 79 Requirements are met? Yes No NA
A.S.M.E. Stamp: Yes No NA
NB or Serial No. if Codestamped: ______ NA
Circle one each

MISCELLANEOUS:
Couplings: ______
Flanges: ______
Fittings: ______
Welding Rods ______ Type ______

NCR NUMBER: ______
SPECIFICATION: ______
SPECIFICATION: ______
SPECIFICATION: ______

Figure 4.2 Receiving report.

required identification of materials during receiving inspection and during fabrication. Inspection procedures should be developed to address these areas of inspection.

The thickness of the plates can be gaged at this time to see if they meet Code tolerances; a thorough examination can be made for defects such as laminations, pit marks, and surface scars. These dimensions and inspections shall be documented on a Receiving Inspection Report (see Fig. 4.2; see Code Par. UG-93).

The Inspector's first check on the job itself is making sure that the plate complies with the mill test reports and Code material specifications. All material should be identifiable and should carry the proper stamping or markings required by the specification and/or Section VIII, Division 1.

c. Nonconforming materials, components, and parts that do not meet the Code requirements or purchase requirements shall be corrected in accordance with the Code requirements or shall not be used for the construction of Code vessels.

d. Partial data reports when required by the Code must be requested for welded parts that require Code inspection and that are furnished by another shop (see Code Par. UG-120).

The ASME and National Board have always advocated better management procedures and require an orderly pervasive system of quality assurance in all the phases of Code vessel engineering and construction. A quality control manual must include the manufacturer's system of process control.

e. The scope of this section is to detail the system established to control quality in the manufacturer's operational sequence. This documented sequence shall be in the form of detailed process flowcharts, inspection records, or traveler sheets.

f. Detailed process flowcharts, inspection records, or traveler sheets should be prepared so that each operation can be examined and signed off by both the manufacturer's quality control representatives and the Authorized Inspector. The Authorized Inspector shall be given the opportunity to establish inspection points on the inspection records prior to the beginning of the work.

g. The chief engineer must take the responsibility for correct design instructions given to the shop, specifying on the drawings, written specifications, or any special Code requirements which would affect the fabrication or quality control operations.

h. The quality control manager must prepare and maintain procedures for controlling fabrication and testing processes—such as welding, nondestructive examinations, heat-treating, hydrostatic or pneumatic testing, and inspection—in addition to providing an operations process sheet with sign-off space for use by the manufacturer's representatives and the Authorized Inspector, since even the best check-off list may not indicate the hold points an Authorized Inspector will want to see. The list should include a remarks space for the Inspector to indicate the hold points of parts that are to be examined (see Fig. 4.3; see also Code Pars. UG-90 to UG-99).

The person (by title) responsible for conducting the final pressure test shall also be stated. Before the hydrostatic test is performed, the vessel should be blocked properly to permit the examination of all the parts during the test. Large, thin-walled vessels may require extra blocking to guard against undue strains caused by the water load. Adequate venting must be provided to ensure that all the parts of the vessel can be filled with water.

The water temperature should be maintained at least 30°F above the minimum design metal temperature to minimize the risk of brittle fracture, but in no case should it exceed 120°F. If the temperature of the test medium is too low, moisture in the air may condense on the surfaces, making proper testing difficult. Two gages should be used to provide a double check on the pressure. If only one gage is used and it is incorrect, the vessel can be damaged by overpressure.

The hydrostatic test must be performed per a test pressure derived from one of the methods specified in UG-99. Following the application of the hydrostatic test pressure, an inspection of all joints and connections shall be made. This inspection shall be made at a pressure not less than the test pressure divided by 1.3. After the vessel has been drained, a final examination should be made of the internal and external surfaces, the alignment, and the circularity (see Chap. 5, the section titled Hydrostatic Testing, Case 3).

i. The procedures for covering processes such as preheat, postweld heat treatment, hot-forming, cold-forming, etc. should be explained.

j. Material identification should be recorded on an as-built sketch, tabulation of materials, or the inspection record or traveler sheet. If the plates are to be sheared or cut, the shop is required to transfer the original mill stamping to each section (see Code Par. UG-77). The Authorized Inspector need not witness the stamping but may wish to check the accuracy of the transfer. Where the service conditions prohibit die-stamping for material identification, the manufacturer shall mark the required data on the plates

Inspection Checklist

Job Number: ____________________ Drawing Number: ______________
Part Number: ____________________ Prepared by: ____________
Q.C.Manager: ____________________ Date: __________
A.I. Review: ____________________ Date: __________

Attribute	WPS/NDE NCR's Etc.	Q.C.M. Inspection Accept and Date		A.I. Inspection Accept and Date	
1. Certificate of Authorization Number.					
2. Drawing and/or Design Calculation Number.					
3. Material Test Reports/C of C Review.		H		H	
4. Fit-up inspection:					
Lower Head to shell					
Upper Head to Shell					
Nozzle # ___ to Shell					
Nozzle # ___ to Shell					
Nozzle # ___ to Shell					
Nozzle # ___ to Shell					
Nozzle # ___ to Shell					
Nozzle # ___ to Shell					
Long. Seam					
Course #1 to Course #2					
Course #2 to Course #3					
5. Verify Material Identification is maintained					
6. Concurrence of A.I. Prior to Weld Repair.		H		H	
7. Root Pass Inspection:					
Lower Head to shell					
Upper Head to Shell					
Nozzle # ___ to Shell					
Nozzle # ___ to Shell					
Nozzle # ___ to Shell					
Nozzle # ___ to Shell					
Nozzle # ___ to Shell					
Nozzle # ___ to Shell					
Course #1 to Course #2					
Course #2 to Course #3					

H = Hold Point (Stop Work Notify Q.C.M.)
W = Witness Point (Notify Q.C.M.)

Page 1 of ____

FIGURE 4.3 Inspection checklist.

Inspection Checklist

Job Number: ______________________ Drawing Number: ______________
Part Number: ______________________ Prepared by: ____________
Q.C.Manager: ______________________ Date: __________
A.I. Review: ______________________ Date: __________

Attribute	WPS/NDE NCR's Etc.	Q.C.M. Inspection Accept and Date		A.I. Inspection Accept and Date	
8. Heat Treatment Procedure Review.					
9. Acceptance of Repair upon completion.		H		H	
10. PT of Root Pass.					
11. Final Weld Inspection:					
Lower Head to shell					
Upper Head to Shell					
Nozzle # ___ to Shell					
Nozzle # ___ to Shell					
Nozzle # ___ to Shell					
Nozzle # ___ to Shell					
Nozzle # ___ to Shell					
Nozzle # ___ to Shell					
Long. Seam					
Course #1 to Course #2					
Course #2 to Course #3					
12. NDE Examination: RT UT MT PT VT Circle one.					
13. RT Film Review		H		H	
14. UT, MT, PT, VT Reports					
15. Hydrostatic Test: Test Pressure __________ PSI at ________ F		H		H	
16. Check of welder I.D. Markings.					
17. Affix Required ASME Code Data Plate.					
18. Certify Manf. Data Report(s).		H		H	
19. Affix Code Symbol Stamp.		H		H	

H = Hold Point (Stop Work Notify Q.C.M.)
W = Witness Point (Notify Q.C.M.)

Page 2 of ___

Figure 4.3 (*Continued*)

in a manner that will allow positive identification. The method of transfer must be acceptable to the Inspector (see Code Par. UG77).

k. Include a section to address the control of the ASME Code symbol stamp—who has the custody of the stamp, who applies it, and when and how it is to be applied to the ASME Code data plate which will be attached to the completed vessel.

l. A description of the person responsible for the preparation of the ASME Code Data Report and the designated representative from the manufacturer who will certify the applicable data sheet form is needed. Additionally, if the vessel is required by the jurisdiction and/or customer to be registered with the National Board, a description of the person responsible for forwarding the forms to the National Board must be given. It should be noted that vessels manufactured outside the United States or in a Canadian province are required to be registered with the National Board of Boiler and Pressure Vessel Inspectors in Columbus, Ohio.

m. If the ASME Certificates of Authorization are extended to cover field construction or fabrication, this section should include controls relative to and functions necessary for controlling ASME-field activities, which must include the Authorized Inspector.

Many Code vessels, especially boilers, are fabricated partly in the shop and partly in the field. When the construction of a Code vessel is completed in the field, some organization must take the responsibility for the entire vessel, including the issuing of data sheets on the parts fabricated in shop and field and the application of Code stamping. Possible arrangements include the following:

1. The vessel manufacturer assumes full responsibility, supplying the personnel to complete the job.
2. The vessel manufacturer assumes full responsibility, completing the job with personally supervised local labor.
3. The vessel manufacturer assumes full responsibility but engages an outside erection company to assemble the vessel by Code procedures specified by the manufacturer. With this arrangement, the manufacturer must make inspections to be sure these procedures are being followed.
4. The vessel may be assembled by an assembler with the required ASME certification and Code symbol stamp, which indicates that the organization is familiar with the Code and that the welding procedures and operators are

Code-qualified. Such an assembler, if maintaining an approved quality control system, may assume full responsibility for the complete assembly.

Other arrangements are also possible. If more than one organization is involved, the Authorized Inspector should find out on the first visit which one will be responsible for the completed vessel. If a company other than the shop fabricator or certified assembler is called onto the job, the extent of association must be determined. This is especially important if the vessel is a power boiler and an outside contractor is hired to install the external piping. This piping, often entirely overlooked, lies within the jurisdiction of the Code. It must be fabricated in accordance with the applicable rules of the ASME B31.1 Code for pressure piping by a certified manufacturer or contractor and must be inspected by the Inspector who conducts the final tests before the master data sheets are signed.

Once the responsible company has been identified, the Inspector must contact the person directly responsible for the erection to agree upon a schedule that will not interfere with the erection but will still permit a proper examination of the important fit-ups, chip-outs, and the like. The first inspection will include a review of the company's quality control manual, welding procedures, and test results to make sure they cover all the positions and materials used on the job. Performance qualification records of all the welders working on the parts of the vessel under the jurisdiction of the Code must also be examined.

It is the responsibility of the manufacturer, contractor, or assembler to prepare drawings, calculations, and all pertinent design and test data so that the Inspector may make comments on the design or inspection procedures before the field work is started and thereby expedite the inspection schedule. New material received in the field for field fabrication must be acceptable Code material and must carry proper identification markings for checking against the original manufacturer's certified mill test reports. The material may be examined at this time for thickness, tolerances, and possible defects. The manufacturer who completes the vessel or parts of the vessel has the responsibility for the structural integrity of the whole or the part fabricated. Included in the responsibilities of the manufacturer is the documentation of the quality assurance program to the extent specified in the Code. The type of documentation needed varies in the different sections of the Code.

6. **Correction of nonconformities.** A definition of nonconformities must be given in the quality manual. Generally, there are two types of nonconformities—those found in material and/or parts during receiving inspection and those found during fabrication and final testing. Both must be explained as follows:

a. A nonconformity is any condition in material, manufacturing or fabrication, or quality of the finished product that does not comply with all of the applicable Code rules. Nonconforming materials, components, and parts that do not meet the Code requirements shall be corrected prior to the completion of the vessel in accordance with the quality control system and/or ASME Code requirements or they shall not be used for Code vessels.

b. When a nonconforming situation is found, a hold tag or other sticker should be placed on the vessel, vessel part, or vessel material by the company Inspector describing the nature of the nonconformity and also giving all information possible on the tag and on the nonconformity report (see Figs. 4.4 and 4.5).

c. In order to determine a proper disposition of the nonconformity, it should be brought to the attention of the appropriate company person or nonconformity review

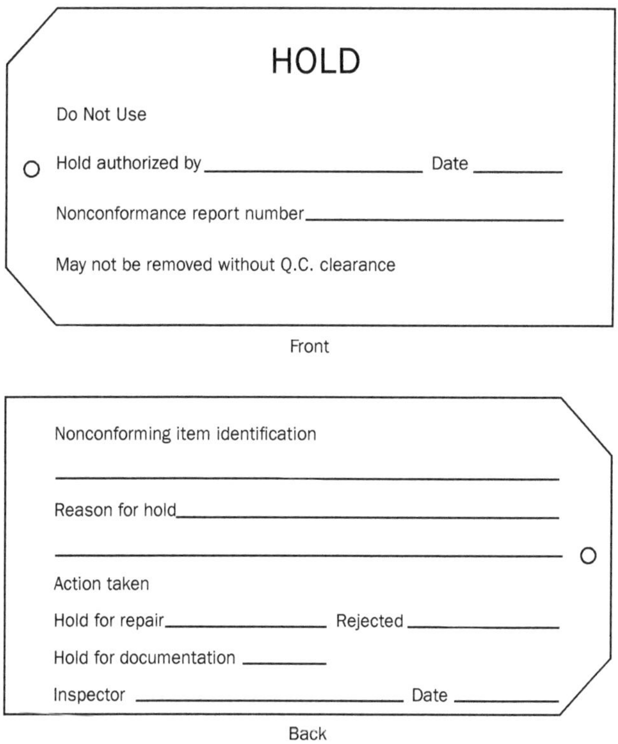

FIGURE 4.4 Hold tag used for identifying nonconforming items.

Nonconformance Report

Nonconformance report number: ______________________

Date: ______________________

Identification and description: ______________________

____

____

____

____

Corrective action: ______________________

____

____

____

____

Corrective action completed date: ______________________

Hold tags removed by: ______________ Date: __________

Quality control manager: ______________ Date: __________

Authorized inspector: ______________ Date: __________

Figure 4.5 Nonconformance report.

board as well as the Authorized Inspector. The method of repair or corrective action should be documented on the nonconformity report and hold tag.

d. When the corrective action has been completed, the report should be signed and dated by the company representatives—normally the quality control manager and any other required company personnel stated in the quality control system, such as engineering or purchasing in conjunction with the Authorized Inspector.

To assure the correct use of a nonconformity report, some companies print instructions on the backside of the report giving information on how the report should be routed to the responsible department or person whose job is to verify the disposition of the nonconformity and the correct action to take.

7. **Welding.** Welded pressure vessels must be fabricated by qualified welders and who are utilizing qualified welding procedures in accordance with ASME Code Section IX. Welding procedures and qualified welders shall be tested, certified, and qualified prior to welding the production joints. The following items must be addressed in the quality control program in order to assure that the Code requirements are attained:

    a. A sound weld is almost impossible without good fit-up. Therefore, the importance of proper edge preparation and fitting of edges on longitudinal seams, girth seams, heads, nozzles, and pipe should be clearly emphasized in the manual. Tack welds, if left in place, shall be feathered at the stops and the starts of the weld. If the tack welds are not prepared in such a manner, they shall be removed (see Code Par. UW-31).

    b. The procedure for fit-up and inspection before welding using either welding procedures or other written instructions should be listed. All tack welds shall be made utilizing qualified welding procedures and qualified welders (see Code Par. UG-90).

    c. Sign-off spaces should be initialed by the Inspectors for fit-up inspection before welding. The Inspectors must assure themselves that Code tolerances for the alignment of the abutting plates of the shell courses, heads, and openings are met. A multiple-course vessel may also require the staggering of the longitudinal joints. To avoid a flat section along the longitudinal seams, the edges of the rolled plates should be prepared with the proper curvature. The out-of-roundness measurement must be within the amount specified by the Code for the diameter of the vessel and the type of pressure (either internal and/or external) used (see Code Par. UG-96).

Following the fit-up of a part for welding, the quality of the weld is controlled by utilizing qualified welding procedures and qualified welders. Therefore, the quality control manual must show the system for welding quality assurance (see Code Pars. UW-26, UW-28, UW-29, and UW-31).

d. The quality control program must include the manufacturer's system for assuring that all welding procedures and welders or welding operators are properly qualified in accordance with the ASME Code Section IX, Welding Qualifications. A description of who within the company is responsible for certifying and dating the procedure qualification record and welder performance qualification records must be addressed.

e. The engineering department, with the concurrence of the manager of quality control, must determine which welds require new procedure qualification and must provide the methods for qualifying the welding procedure, welding operators, and welders. A system for revisions to welding procedures and procedure qualification records must be described.

f. The method of surveillance of the welding during fabrication to assure that only qualified procedures and welders are being used should be given.

g. The method for the qualification of personnel and the assigning of the welders' identification symbols (see Code Pars. UW-28 and UW-29) must be written.

h. The method of identifying the person making welds on a Code vessel, such as stamping the welds every 3 ft or utilizing a weld map which identifies each weld joint and who welded that joint (see Code Par. UW-37), should be described.

i. The method of welding electrode control, that is electrode ordering to an SFA specification, inspection when received, storage of electrodes, how they are issued, and the in-process protection of the electrode (see Code Section II, Part C, Table A-1), should be explained.

j. The title of the person responsible for seeing that item (i) above is carried out must be listed.

k. The method used on the drawing to tell what type and size of weld and electrode are to be used, that is whether the drawing shows weld detail in full cross section or uses welding symbols, should be provided.

l. The method of tack welding must be explained. Do the tack welds become part of the weld or are they removed? (See Code Par. UW-31.) The control of tack welding must be a part of the welding quality control system.

m. The continuity of welders' qualification records should be maintained, and a log of all welders' activities should be kept to indicate the welders' continuity in specific welding processes. If there is a gap of time in excess of six months in a specific process, the welder shall be requalified. Figure 4.6 gives an example of a welder's qualifications and performance record log.

WELDERS QUALIFICATION & PERFORMANCE RECORD LOG

Welder's name	Welder's symbol	Welding process qualified	Jan.	Feb.	Mar.	Apr.	May	June	July	Aug.	Sept.	Oct.	Nov.	Dec.

FIGURE 4.6 Welder's qualification and performance record log.

Tack welds to secure alignment are often under high stress when fit-up is made. Also, the rapid cooling of the tack weld to the parent metal can form brittle crack-sensitive microstructure and can cause cracks in the base material as well as in the tack weld. All tack welds must be made with a qualified welding procedure specification and by qualified welders.

One of the most important controls in welding is actual surveillance on the job. This cannot be minimized. The welding of the vessel must be done in accordance with a qualified procedure. The back-chip of the welded seams must be inspected to see that all defects have been removed and that the chipped or ground groove is smooth and has a proper bevel to allow welding without entrapping slag. Each piece of welding must be cleaned of all slag before the next bead is applied; many defects can be caused by improper cleaning.

When a backing strip is used, it is important to see that the strip is closely fitted and that the materials to be welded have sufficient clearance to permit good penetration on the first pass. The first pass is more susceptible to cracking because of the large amount of parent metal involved and the fact that the plate may be cold. As the root pass solidifies quickly, it tends to trap slag or set up stresses that can induce cracking if the fit is not good. The importance of good fit-ups to a sound weld cannot be overemphasized.

The finished butt welds should be free of undercuts and valleys and also should not have too high a crown or too heavy an overlay. Fillet welds should be free of undercuts and overlap. The throat and leg of the welds should be the size specified by the Code.

8. **Nondestructive examination.** A weld may look perfect on the outside and yet reveal cracks, lack of penetration, heavy porosity, or undercutting at the backing strip when the weld is given a nondestructive examination (NDE) by radiography or ultrasonic tests. As with welding, nondestructive examinations may require a procedure, and like the welding procedure, NDE must satisfy the requirements of the Code and be acceptable to the Authorized Inspector. Various sections of the ASME Code refer to Section V, Nondestructive Examinations, for nondestructive examinations. These examinations should be made to follow a detailed written procedure according to Section V and the referencing Code sections (i.e., Section I, Section VIII, Division 1, etc.). The manufacturer's quality control manual must include the following:

   a. A description of the manufacturer's system for assuring that all nondestructive examination personnel are qualified in accordance with the applicable code is needed. The quality control program must define what NDE will be utilized and whether it will be performed in house or subcontracted.

b. Written procedures to assure that nondestructive examination requirements are maintained in accordance with the Code shall be addressed.

c. If an NDE is subcontracted, a statement of who is responsible for checking the subcontractor's site, equipment, and qualification of procedures to assure that they are in conformance with the applicable Code must be given. In an NDE certification program, the manufacturer has the responsibility for the qualification and certification of its NDE personnel. To obtain these goals, personnel may be trained and certified by the manufacturer, or an independent laboratory may be hired. If an independent laboratory is hired, the manufacturer must state in the quality control manual the department and person (by title) responsible for assuring that the necessary quality control requirements are met, because the manufacturer's responsibility is the same as if the parent company had done the work.

d. If the manufacturer performs radiography testing (RT) or ultrasonic testing (UT), NDE personnel shall be qualified in accordance with the recommended practice known as SNT-TC-1A, utilizing the current Code-accepted edition.

For the pressure vessel fabricator, nondestructive testing provides a means of spotting faulty welds, plates, castings, and forgings before, during, and after fabrication. For the user, it offers a maintenance technique that more or less assures the safety of a vessel until the next scheduled inspection period.

9. **Heat treatment.** Because of certain materials utilized in construction and fabrication—due to material thickness limitations or product form—heat treatment may not be required or desirable. If a company does not have heat-treatment facilities, a paragraph should address this fact and simply state that if heat treatment is required, it will be subcontracted. The following should be addressed regarding heat treatment when such treatment is required:

   a. For an explanation of when preheat is necessary, the welding preheat and interpass temperature control methods should be provided.

   b. For postweld heat treatment, a description of how the furnace is loaded, how and where the thermocouples are placed, the time–temperature recording system, marking on charts, and how these details are recorded for each thermocouple (see Code Pars. UW-40 and UW-49) should be given.

c. The methods of heating and cooling, metal temperature, metal temperature uniformity, and temperature control must be explained.

d. The methods of vessel or parts support during postweld heat treatment should be listed.

10. **Calibration of measurement and test equipment.** A system shall be established documenting the method of calibration of measurement and test equipment as follows:

a. The system used for assuring that the gage measuring and testing devices used are calibrated and controlled should be described.

b. An explanation of how records and identification are kept to assure proper calibration must be given. These records shall be available to the Authorized Inspectors.

c. All gages should be calibrated with a calibrated master gage or a standard deadweight tester whenever there is a reason to believe that they are in error. It is highly recommended that all gages be tested and if necessary calibrated at least every six months. The method for testing gages and instruments with references to the standards of the National Institute of Standards Testing (NIST) or other recognized national standards should be detailed. In the case of foreign Code manufacturers, for example, the Republic of Korea, records traceable to the Korean Standards (KS) are acceptable. Having accurate and calibrated gages will assure the correct testing pressure when the vessel is undergoing hydrostatic testing or pneumatic testing.

Many fabricators take the testing equipment for granted and seldom check the accuracy of gages and measuring equipment. We have often had gages checked that proved to be inaccurate. For example, in testing a vessel 12 ft in diameter with a quick-opening door, the pressure gage was found to be 25 lb off on the low side. On a head 12 ft in diameter, its total area is 16,350 in.^{2}. For each pound over the test pressure, the head would carry 16,350 lb. If this gage had been used, the extra load on the head would be 25 × 16,350 lb, or 410,000 lb above the test requirement. This extra pressure would have been enough to cause considerable damage to this type of opening. The quality control manual must state how the measuring and test equipment is controlled.

11. **Record retention.** The ASME Code Section VIII, Division 1 has limited retention requirements relative to records. This topic must be addressed as follows:

a. The manufacturer must have a system for the maintenance of manufacturer's data reports, radiographs, impact-test results, and the like.

b. The manager of quality control shall assemble all the manufacturing and inspection documents required by the referencing Code section. These records shall be filed until the data report is signed by the Authorized Inspector and shall include, but not be limited to, the following: design specifications and drawings, design calculations, certified material test reports, nondestructive examination reports and radiographic films, heat-treatment records, hydrostatic and pneumatic test records, impact-test results, copies of signed ASME data reports, nonconformity records, inspection check-off list, and sign-offs. When the data report is signed by the Authorized Inspector, the manufacturer may dispose of these records. If the manufacturer has not registered the vessel with the National Board of Boiler and Pressure Vessel Inspectors, the data reports must be kept in the manufacturer's files for a minimum of five years. UT reports shall also be kept for five years.

c. All the records shall be kept in a safe environment.

The maintenance of records and revisions in accordance with the Code should be the quality control manager's responsibility. A system should be maintained for controlling the issuance of the latest quality control manual revisions. Revisions generally are controlled through the issuance of the revised section, which is documented on a revisions control sheet or table of contents placed in the front of the manual. Notify the Authorized Inspector of any contemplated revisions for the quality control manual. Verify any revisions with the Authorized Inspector. Revisions and revision record forms can be listed as an appendix of the manual. No revisions are to be issued without prior review and acceptance by the Authorized Inspector.

12. **Sample forms.** The forms used by the company which are generated and which affect and control the quality of the vessel shall be shown by exhibit in the quality control manual as follows:

    a. The type of check-off form used, including those for welding procedures and welding performance qualification, traveler sheets, welder's log records, and other charts used in the completion of records for material purchase, identification, examinations, and tests that were taken before, during, and on the completion of the vessel, should be listed.

    b. The examples of all forms should be given and their use, such as the check-off list and data reports which will be presented to the Authorized Inspector for review, should be described. The Inspector will carefully check these documents to make sure that they are properly listed in the data reports, that the vessel complies with the Code, and that the data reports are signed by the manufacturer before being signed by the Inspector. Documentation requires good procedure and organizational review. Experience has proven that with good documentation a manufacturer is able to assess the operation accurately and improve the efficiency of production.

13. **Authorized Inspector** (inspection of vessel and vessel parts).
    a. The manual should state that the Authorized Inspector is the Code Inspector and it is the obligation of the company to assist in the performance of these duties.
    b. The Inspector can indicate the hold points and discuss with the quality control manager the arrangements for additional inspections and audits to establish the company's compliance with the ASME Code.
    c. The manual should state that the checklist of records, drawings, calculations, process sheets, and any other quality control records shall be available for the Authorized Inspector's review.
    d. The final hydrostatic or pneumatic tests required by the Code shall be witnessed by the Authorized Inspector, and any examinations deemed necessary shall be made before and after these tests.
    e. The manufacturer's data reports shall be carefully reviewed by a responsible representative of the manufacturer or assembler. If this representative finds them properly completed, they shall be signed for the company. The Authorized Inspector, after gaining assurance that the requirements of the Code have been met and the data sheets properly completed, may sign these data sheets.

The manufacturer should be carefully acquainted with the provisions in the Code that establish the duties of the Authorized Inspector. This third party in the manufacturer's plant, by virtue of authorization by the state to do Code inspections, is the legal representative. Having this uninterested third party at the shop of fabrication permits the manufacturer to fabricate under state laws. Under the ASME Code rules, the Authorized Inspector has certain duties as specified in the Code.

It is the obligation of the manufacturer to see that the Inspector has the opportunity to perform as required by law. The details of an authorized inspection at the plant or site are to be worked out between the Inspector and the manufacturer. The manufacturer's quality control manual should state provisions for assisting the Authorized Inspector in performing duties (see Code Par. UG-92).

The Authorized Inspector, after witnessing the final hydrostatic tests or pneumatic tests and carrying out the examinations that are necessary, will check the vessel data report sheets to make sure they are properly filled out and signed by the manufacturer and that the material and quality of the work in the vessel are accurately identified. The Authorized Inspector will then allow the Code symbol to be applied in accordance with the Code and any local or state requirements. If everything is in order, the data sheets are signed by the Inspector.

14. **Audits.** Audits are not required by all the sections of the Code, but in order to make certain that the quality control system is being properly used during manufacture, a comprehensive system of planned and periodic audits should be carried out by the manufacturer to assure compliance with all the aspects of the quality control program. The manufacturer's quality control manual should describe the system of self-auditing.
    a. The manufacturer's system of planned periodic audits to determine the effectiveness of the quality assurance program must be explained.
    b. Audits should be performed in accordance with the manufacturer's procedure, using the checklist developed for each audit, to ensure that a consistent approach is used by the audit personnel.
    c. The reports of audit results should be given to top management. Positive corrective action to be taken, if necessary, should also be indicated.
    d. Corrective action must be taken by the quality control manager to resolve the deficiencies, if any, revealed by the audits.
    e. A re-audit should be made within 30 days to confirm the results when corrective action is taken.
    f. The results of the audit must be made available to the Authorized Inspector.
    g. The name and title of the person responsible for the audit system should be specified. Good management will have a self-auditing system that regularly reviews the adequacy of its quality control program and provides for immediate and effective corrective measures when required.
15. **Appendix.** This section of the quality control manual is not mandatory; however, this section may contain the necessary forms, exhibits, and other items or such items may be placed in the applicable sections of the quality control manual. The following are ideas that may be placed here:

a. The table of revisions, the revision number, and the date of revision can be presented.

b. All forms can be shown, and the form usage can be described.

In our comments, we have given the reader an in-depth outline for the inspection and quality control systems that the average-sized manufacturer of boilers and pressure vessels might use in establishing a quality control program and/or quality manual as required by the ASME Code Section I, Power Boilers, Section IV, Heating Boilers, and Section VIII, Divisions 1 and 2, Pressure Vessels. A quality assurance program for nuclear vessels would require a more comprehensive program and manual (see Chap. 10). We have not included Section X, Fiberglass Reinforced Plastic Pressure Vessels. For quality control systems required by these Code vessels, the applicable Code sections must be followed.

It is important to remember that before the issuance or renewal of an ASME Certificate of Authorization to manufacture Code vessels is granted, the manufacturer's facilities and organization will be reviewed by an inspection agency and/or legal jurisdiction, or a representative of the National Board of Boilers and Pressure Vessel Inspectors.

In reviewing the various operations listed in the manufacturer's quality control manual, their objective will be to see the controlled manufacturing system described by the manufacturer's quality control manual in operation. This survey must be jointly made before the survey team can write a report of the manufacturer's system for the American Society of Mechanical Engineers.

The responsibility for establishing the required quality control system is that of the manufacturer. In writing a quality control manual, effective means of planning during all the phases of design, fabrication, and inspection should be considered in order to assure a properly constructed Code vessel. Rear Admiral Charles Curtze, in a report to the congressional committee investigating the loss of the USS Thresher, talked about the need for a higher IQ (in this case, integrity and quality) in industry. If your plant has both the ability to build Code vessels and a good IQ, you should have no trouble acquiring and maintaining the ASME Code stamp.

CHAPTER 5

Inspection and Quality Control of ASME Code Vessels

Shop Inspection

An Authorized Inspector's first step in the shop inspection of Code vessels is to review the design. It is the manufacturer's responsibility to submit drawings, calculations, and the following design information: dimensions, plate material specifications and thickness, weld detail drawings, calculated corrosion allowance requirements, design pressure and temperature calculations (see Code Pars. UG-20 and UCS-66), impact-test exemptions if allowed (see Code Pars. UG-20(f) and UCS-66), hydrostatic testing pressure and the part of the vessel to which the maximum pressure is limited, calculations for the reinforcement of openings, and the design requirements for the size and type of flanges which may be used on the vessel. It is especially important to indicate whether the vessel is to be postweld heat-treated and what its degree of radiography is because the Code generally allows 100 percent efficiency for double-welded butt joints that are fully radiographed, 85 percent efficiency for joints that are spot-radiographed, and 70 percent efficiency for joints not radiographed. Table 5.1 lists some of the various inspections, tests, and examinations which are established within the Code. This list is given to assist the reader in determining when certain tests, examinations, and inspections must be made.

The Inspector's first check on the job itself is to make sure that the materials of construction comply with the purchase order requirements, mill test reports supplied with the material, and ASME Section II, Material Specifications. All materials should be identifiable and should carry the proper stamping. The thickness of the plates can be gaged at this time to see if they meet Code tolerances, and

Remarks	Code references
Hydrostatic tests	
The test must be at least 1.3 times the maximum allowable working pressure multiplied by the lowest ratio of the stress value for the test temperature to that for the design temperature	Par. UG-99, UG-98
If the allowable stress at design temperature is less than the allowable stress at test temperature, the hydrostatic test pressure must be increased proportionally. Thus	
$\text{Test pressure} = 1.3 \dfrac{\text{allowable stress at test temperature}}{\text{allowable stress at design temperature}} \text{design pressure}$	
Combination units must be so tested that hydrostatic pressure is on one chamber, without pressure in the other parts	Par. UG-99(e)
Include corrosion allowance in calculating test pressure	
Test cast ductile iron at 2 times the maximum allowable working pressure	Par. UCD-99
Test cast iron vessels at 2 times the design working pressure for MAWP greater than 30 psi; test at 2½ times the design pressure for MAWP under 30 psi but not to exceed 60 psi	Par. UCI-99
Low-temperature vessels per Part ULT	Par. ULT-99
Pneumatic tests	
A pneumatic test may be used instead of a hydrostatic test when: 1. The vessels are so designed and supported that they cannot be filled with water safely 2. Vessels are for a service in which traces of testing liquid cannot be tolerated and are not easily dried	Par. UG-100
The test pressure must not be less than 1.1 times the maximum allowable pressure and not more than 1.1 times the basis for the calculated test pressure as defined in Code Par. 3-2	Par. UG-100(b)
Special precautions should be taken when using air or gas for testing	Par. UG-100
Magnetic particle and liquid penetrant tests	
When a vessel is to be pneumatically tested, the welds around the openings and the attachment welds of throat thickness greater than ¼ in. must be given a magnetic particle or liquid penetrant examination for ferromagnetic materials (liquid penetrant for nonmagnetic materials)	Par. UW-50
All austenitic chromium–nickel alloy steel and austenitic/ferritic duplex steel welds which exceed a nominal size of 3/4 in. shall be given a liquid penetrant examination for the detection of cracks. If the material is heat-treated, the examination shall be made following the heat treatment	Par. UHA-34

(Continued)

Vessels with heat-treated enhanced materials; all welds must be given a magnetic particle examination after hydrostatic testing. Liquid penetrant is an acceptable alternative	Par. UHT-57
Pressure part welded to a plate thicker than ½ in. to form a corner joint must be examined by magnetic particle or liquid penetrant method	Pars. UG-93(d)(3), UW-13, Fig. UW-13.2
Layered vessels	Par. ULW-56
Low-temperature vessel construction; all attachment welds and joints not radiographically examined shall be given a liquid penetrant examination	Par. ULT-57
Qualification of personnel:	
Magnetic particle	App. 6
Liquid penetrant	App. 8

Table 5.1 ASME Code: Required Tests, Inspections, and Nondestructive Examinations

a thorough examination can be made for defects such as laminations, pit marks, and surface scars.

If the plates are to be sheared or cut, the shop is required to transfer the original mill stamping to each section. The Inspector need not witness the stamping but must check the accuracy of the transfer.

Code tolerances for the alignment of the abutting plates of the shell courses, heads, and openings must be met (see Code Par. UW-33 and Table UW-33). Table 5.2 instructs the reader on how to inspect these tolerances for strict compliance to the ASME Code. Such inspections verify the out-of-roundness measurement of the shell of the vessel. Table 5.2 lists such tolerances and the location of the acceptance criteria in the Code. A multiple-course vessel may also require the staggering of the longitudinal joints. To avoid a flat section along the longitudinal seams, the edges of the rolled plates should be prepared with the proper curvature. The out-of-roundness measure must be within the amount specified by the Code for the diameter of the vessel and the type of pressure (internal or external) used (see Code Pars. UG-80 and UG-81 and Code Figs. UG-80.1 and UG-80.2).

The welding of the vessel must be done according to a qualified welding procedure (see Chap. 6). The back-chip of the welded main seams must be inspected to see that all defects have been removed and that the chipped or ground groove is smooth and has a proper bevel to allow welding without entrapping slag. Each bead of welding must be cleaned of all slag before the next bead is applied; many defects can be caused by improper cleaning.

When a backing strip is used, it is important to see that the ring is closely fitted and that the plates have sufficient clearance to permit

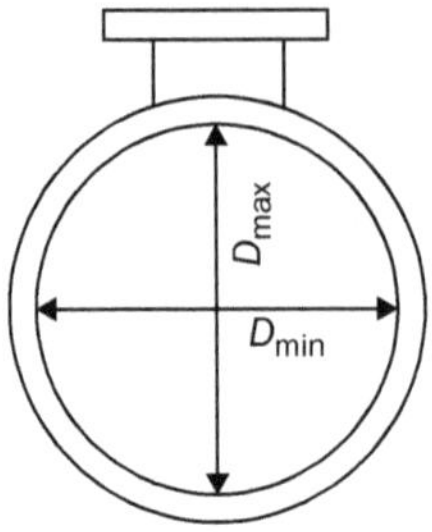

Tolerance	Remarks	Code references
Out-of-roundness measure in cylindrical shells under internal pressure	$D_{max} - D_{min}$ must not exceed 1% of nominal diameter	Par. UG-80(a) Fig. UG-80.2
	At nozzle openings, increase above tolerance by 2% of inside diameter of opening	Par. UG-80(a)
	Tolerance must be met at all points on the shell	
Out-of-roundness measure in cylindrical shells under external pressure	Minimum of tolerance for internal pressure, above, or Code Par. UG-80(b)	Par. UG-80(b) Fig. UG-80.1
	Take measurements on the unwelded surface of plate	
	For shells with lap joint, increase tolerance by value *t*	Par. UG-80(a)
	Do not include corrosion allowance in value *t*	
Permissible variation in plate thickness	No plate shall have an under-tolerance exceeding 0.01 in. or 6% of the ordered thickness, whichever is less	Par. UG-16(c)
Formed heads	Heads must not deviate outside of the specified shape by more than 1–1/4% of *D* or inside the specified shape by more than 5/8% of *D*, where *D* is the nominal inside diameter of the head skirt	Pars. UG-81, UW-13
	Hemispherical heads or any spherical portion of a head designed for external pressure shall also meet the tolerances specified for spheres in UG-80(b)	Par. UG-81(b)

Tolerance	Plate thickness	Offset limits	Code references
Longitudinal joints (category A)	$t \le 1/2$ in. $1/2$ in. $< t \le 3/4$ in. $3/4$ in. $< t \le 1\frac{1}{2}$ in. $1\frac{1}{2}$ in. $< t \le 2$ in. $t < 2$ in.	$1/4t$ 1/8 in. 1/8 in. 1/8 in. Lesser of $1/16t$ and 3/8 in.	Par. UW-33 Table UW-33
Circumferential joints (categories B, C, and D)	$t \le 3/4$ in. $3/4$ in. $< t \le 1\frac{1}{2}$ in. $1\frac{1}{2}$ in. $< t \le 2$ in. $t < 2$ in.	$1/4t$ 3/16 in. $1/8t$ Lesser of $1/8t$ and 3/4 in.	Par. UW-33 Table UW-33
Offset type joint	5/8 in. maximum; 3/8 in. maximum if connecting hemi head to shell; longitudinal weld within offset area shall be ground flush with the parent metal prior to offset and examined by the magnetic particle method after offset		Par. UW-13(c) Fig. UW-13.1(i)
Welding of plates of unequal thickness			Pars. UW-9, UW-42 Fig. UW-9
Spherical vessels and hemispherical heads	Offset must meet longitudinal joint requirements		Par. UW-33
	Offset must be faired at a 3-to-1 taper over the width of the finished weld		Pars. UW-42, UW-9 Fig. UW-9
	Additional weld metal may be added to the edge of weld if examined by the liquid penetrant or the magnetic particle method. If the weld buildup is in the area to be radiographed, the buildup shall also be radiographed		Pars. UW-42, UW-9

Table 5.2 Code Tolerance Requirements

good penetration on the first pass. The first pass is more susceptible to cracking because of the large amount of parent metal involved and the fact that the plate may be cold. As the root pass solidifies quickly, it tends to trap slag or set up stresses that can induce cracking if the fit is not good. The importance of good fit-up to a sound weld cannot be overemphasized.

The finished butt welds should be free of undercuts and valleys and also should not have too high a crown or too heavy an overlay. Fillet welds should be free of undercuts and overlap. The throat and leg of the weld should be of the size specified by the drawing and in compliance with the Code. Automatic welding properly performed is superior to hand welding, but it requires rigid supervision and inspection. The best check is provided by spot-radiographing. If radiography is unavailable, a test plate can be welded and sectioned (see the section titled Sectioning in this chapter).

Tests

A weld may look perfect on the outside and yet reveal deep cracks, lack of penetration, heavy porosity, or undercutting at the backing strip when it is radiographed.

If the vessel is to be radiographed, all irregularities on the weld plate surfaces must be removed so as to not interfere with the interpretation of subsurface defects. If a vessel is to only have a pneumatic test, all the attachment welds with a throat thickness greater than 1⁄4 in. must be given a magnetic particle examination or liquid penetrant examination before the pneumatic test.

If a vessel is constructed of austenitic chromium–nickel alloy steels over 3⁄4 in. thick, all welds must be given a liquid penetrant examination (after heat treatment if such treatment is performed).

If the vessel is to be fully postweld heat-treated, the furnace should be checked to make sure that the heat will be applied uniformly and without flame impingement on the vessel.

The vessels with thin walls or large diameters should be protected against deformation by internal bracing. The vessel should be evenly blocked in the furnace to prevent deformation at the blocks and sagging during the stress-relieving operation. Furnace charts should be reviewed to determine if the rates of heating, holding time, and cooling time are correct.

All the work should be completed before a vessel is presented for final testing, especially if it is to be postweld heat-treated. A minor repair can assume major proportions because of lack of time to make the repair and repeat postweld heat treatment and testing prior to shipping.

A vessel undergoing a hydrostatic test should be blocked properly to permit the examination of all the parts during the test. Large, thin-walled vessels may require extra blocking to guard against undue strains caused by the water load. Adequate venting must be provided to ensure that all the parts of the vessel can be filled with water. The water should be approximately at room temperature. If the temperature is lower, moisture in the air may condense on the surfaces, making proper testing difficult. Two gages should be used to provide a double-check on the pressure. If only one gage is used and it is incorrect, the vessel can be damaged by overpressure. The hydrostatic test must be performed at the agreed-upon pressure in accordance with Code Par. UG-99. After the vessel has been drained, a final examination should be made of the internal and external surfaces, the alignment, and the circularity. If everything is satisfactory, the vessel may be stamped in accordance with the Code and any local or state or jurisdictional requirements.

The data report sheets should then be checked to make sure that they are properly filled out, signed, and certified by the manufacturer and that the material and workmanship of the vessel are accurately stated. If everything is in order, the data sheets are signed by the Inspector.

ASME Manufacturer's Data Reports

The importance of having the ASME Manufacturer's Data Reports fully and correctly completed as well as the record retention of complete vessel material and fabrication records must not be underestimated.

When a pressure vessel accident happens, the cause can be poor operating procedures, insufficient vessel maintenance, inadequate training and indoctrination of the operating personnel, or incorrect sizing and type of relief devices. All of the above have been seen by the authors while doing forensic work involving product liability.

When an employee is injured in an accident, sometimes it appears that there are no laws under which to sue the employer. The employer only pays for worker's compensation. The employee and the hired attorney then go after the manufacturer of the vessel for product liability and try to prove that the vessel was improperly designed and fabricated. In the event of litigation, the ASME data reports may be subpoenaed as legal documents. It is here that the vessel manufacturer's properly completed ASME data report and preservation of records even beyond the time required by the ASME Code will be most helpful. (To assist the manufacturer in completing the data report correctly, refer to Appendix W of Section VIII, Division 1 for guidance.)

Records should include the customer's purchase order, the user's and vessel manufacturer's design reports, certified mill test reports of materials, the records of qualified welding procedures, a list of all qualified welding operators who welded on the vessel, quality control and/or quality assurance checklists, and certified ASME data reports signed by the vessel manufacturer.

The authors have seen situations where a data report was not filled out completely. For example, the data report item on safety relief outlets was left blank. The vessel manufacturer should have tried to find out from the vessel user the type and the size and setting of the relief device. Code Par. UG-125 states that all pressure vessels must be provided with a means of overpressure protection. The number, size, and location of relief devices are to be listed on the Manufacturer's Data Report.

There are times when the vessels are designed for a system and properly protected against overpressure by one safety relief device. If this is the case, the vessel manufacturer may add the statement under the pressure relief devices item as *provided by others* or *provided by owner.*

The pressure relief device item should never be left blank, nor should it contain a statement that no overpressure protection is provided.

It is the manufacturer's responsibility to see that the vessel's fabricated items conform to the ASME Code and meet the minimum design requirements. Just because the customer did not indicate the size, pressure, and location of the safety relief device on the purchase order or drawing does not relieve the manufacturer of this responsibility.

The safety devices need not be provided by the vessel manufacturer, but overpressure protection must be installed before the vessel is placed in service or tested for service; a statement to this effect should be made in the Remarks section on the Manufacturer's Data Report.

No pressure vessel unit can be considered completed until it is provided with a properly designed safety relief device of adequate capacity and in good operating condition.

Field Assembly Inspection

Many code vessels—especially boilers—are fabricated partly in the shop and partly in the field. Although this book deals primarily with pressure vessels, many plants will want to know about the field assembly of boilers as well.

Responsibility for Field Assembly

When the construction of a Code vessel is completed in the field, some organization must take the responsibility for the entire vessel, including the issuing of data sheets on the parts fabricated in shop and field and the application of Code stamping. The possible arrangements include the following:

1. The vessel manufacturer assumes full responsibility, supplying the personnel to complete the job.

2. The vessel manufacturer assumes full responsibility, completing the job with local labor under the manufacturer's supervision.
3. The vessel manufacturer assumes full responsibility but engages an outside erection company to assemble the vessel by Code procedures specified by the manufacturer. With this arrangement, the manufacturer must make inspections to be sure that these procedures are being followed.
4. The vessel may be assembled in the field by an assembler with the required ASME Certificate of Authorization whose scope allows field construction. This certificate along with the applicable ASME Code stamps indicates that the assembler's organization is familiar with the Code and that the welding procedures and operators are Code qualified. Such an assembler may assume full responsibility for the complete assembly.

Other arrangements are also possible. If more than one organization is involved, the Authorized Inspector should find out on the first visit which one will be responsible for the completed vessel. If a company representative other than the shop fabricator or an authorized assembler is called onto the job, the extent of association must be determined. This is especially important if the vessel is a power boiler and an outside contractor is hired to install the external piping. This piping, often entirely overlooked, lies within the jurisdiction of the Code. It must be fabricated in accordance with the applicable rules of the ASME B31.1 Code for pressure piping by a certified manufacturer or contractor and must be inspected by the inspector who conducts the final tests before signing the master data sheets.

Inspection Procedure

Once the responsible company has been identified, the inspector must contact the person directly responsible for the erection to agree upon a schedule that will not interfere with the erection but will still permit a proper examination of the important fit-ups, chip-outs, and the like. The first inspection may include a review of the welding procedures (and test results) to make sure that they cover all the positions and materials used on the job. Performance qualification records of all the welders working on the parts of the vessel under the jurisdiction of the Code must also be examined.

It is the responsibility of the manufacturer, contractor, or assembler to submit drawings, calculations, and all pertinent design and test data (see the section titled Shop Inspection in this chapter) so that the Inspector may make comments on the design or inspection procedures before the field work is started and thereby expedite the inspection schedule.

New material received in the field for field fabrication must be acceptable Code material and must carry proper identification markings for checking against the mill test reports. It may be examined at this time for thicknesses, tolerances, and defects such as laminations, pit marks, and surface scars.

Field examinations of the vessel for fit-up and finished welds, and during and after hydrostatic or pneumatic testing, should be conducted in the manner already prescribed for shop inspections.

If all is satisfactory, the Inspector will check the data sheets to see that they are properly filled out and signed by the manufacturers and shop inspectors and that the portion of the vessel that was field-constructed is accurately described and the material and quality of the work are accurately stated. The Inspector then signs the certificate of field assembly.

Radiography

Radiography is a tool permitting the examination of the invisible parts of a welded joint. Radiographic equipment requires skilled personnel for safe and efficient handling. Special provisions must be made for their protection, and unauthorized persons must be kept from entering the surrounding area while the equipment is in use.

Radiography is used extensively in the welded pressure vessel industry, especially because of the higher weld joint efficiencies allowed by the Code for vessels that are radiographically examined. Such vessels can be built for the same pressure and diameter with thinner shells and heads and consequently with less weld metal and welding time.

Conditions Requiring Radiography

The ASME Code requires the radiographing of pressure vessels for certain design and service conditions—with the type of radiography varying with materials, thicknesses, joint efficiency, type of construction, and the like. The conditions for which full or spot radiography is specified are as follows:

1. To increase the basic joint efficiency, either full or spot radiography may be required (see Code Pars. UW-11 and UW-12; Code Table UW-12).
2. The vessels using lethal substances require full radiography of all longitudinal and circumferential joints [see Code Pars. UW-2(a) and UW-11(a)(1)].
3. If the plate or vessel wall thickness at the weld joint exceeds the thicknesses given in Code Pars. UW-11, UCS-57, Table UCS-57, Pars. UNF-57, UHA-33, UCL-35 or UCL-36, and UHT-57, then full radiography is required (also see Table 5.3).

4. The cast ductile iron repair of defects requires radiography [see Code Par. UCD-78(a)(6)].
5. The vessels conforming to type 405 materials that are butt welded with straight chromium electrodes and to types 410, 429, and 430 base materials welded with any electrode require radiography for all thicknesses.
6. Clad and lined vessels may require either full or spot radiography (see Code Pars. UCL-35 and UCL-36).
7. The welded repairs of clad vessels damaged by seepage in the applied lining may require radiography if the Inspector so requests (see Code Par. UCL-51).
8. If the longitudinal joints of adjacent courses are not staggered or separated by a distance at least 5 times the plate thickness of the thicker plate, each side of the welded intersection must be radiographed for a distance of 4 in. [see Code Par. UW-9(d)].
9. Unreinforced openings located in a circumferential seam must be radiographed for a distance 3 times greater than the diameter of the opening (see Code Par. UW-14).
10. The castings for which a higher quality factor is desired require radiography at all critical sections (see Code Par. UG-24).
11. Imperfections that permit the leakage of cast iron vessels to be repaired by plugging require the radiography of metal around the defect (see Code Par. UCI-78).
12. Forgings repaired by welding may require radiographing (see Code Par. UF-37).
13. Unfired steam boilers with a design pressure exceeding 50 psi must be fully radiographed [see Code Par. UW-11(a)].
14. Nozzle butt welds joining the flange or saddle to a vessel requiring radiography must be fully radiographed (see Code Par. UW-11 and Code Fig. UW-16.1, sketches f-1 thru f-4).
15. When weld buildup of the base material is utilized for the purpose of restoring the material thickness to its original thickness or for modifying the configuration of welded joints to provide a uniform transition, if radiography is required of the weld joint, the buildup area is also required to be radiographed (see Code Par. UW-42).
16. Full or spot radiography must be used in order to increase the allowable stress and joint efficiency in the calculations for shell and head thicknesses (see Code Pars. UW-11 and UW-12).

Thickness Above Which Full Radiographic Examination of Butt-Welded Joints is Mandatory	
P-No. and Group No. Classification of Material	Nominal Thickness Above Which Butt-Welded Joints Shall Be Fully Radiographed, in. (mm)
1 Gr. 1, 2, 3	1¼ (32)
3 Gr. 1, 2, 3	¾ (19)
4 Gr. 1, 2	⅝ (16)
5A Gr. 1, 2	0 (0)
5B Gr. 1	0 (0)
5C Gr. 1	0 (0)
15E, Gr. 1	0 (0)
9A Gr. 1	⅝ (16)
9B Gr. 1	⅝ (16)
10A Gr. 1	¾ (19)
10B Gr. 1	⅝ (16)
10C Gr. 1	⅝ (16)

Table 5.3 Full Radiographic Thickness Requirements—Carbon and Low-Alloy Steels (Code Table UCS-57) (*Courtesy of the American Society of Mechanical Engineers.*)

Interpretation of Welding Radiographs

Some people falsely assume that because a weld had been radiographed, no defect which has not been revealed by the radiographing process can exist in the weld. On the other hand, some people interpret each visible mark on the film as a defect in the weld and request chip-out and rewelding. In both these cases, the radiograph is subjected to false or poor interpretation.

In order for the viewer to interpret the film correctly to find all possible defects in the weld, a proper understanding of the nature and appearance of weld defects, of common processing defects, of the degree of contrast and detail possible in radiography, of the film and film sensitivity, and of the technique of making a radiograph is necessary.

Radiography uses either x-rays or gamma rays. X-ray machines operate through a range of 10,000 to 2 million volts. Gamma rays, measured in curies, are produced from radioactive materials of various strengths. (The most common sources of gamma rays in the welding industry are cobalt-60, cesium-136, iridium-192, and radium.)

Both x-rays and gamma rays are a form of electromagnetic waves of short wavelength, which allows them to penetrate metals impervious to ordinary light. Some of the radiation is absorbed after penetration, the amount depending on the wavelength and the thickness and density of the material. If a defect such as a crack, slag inclusion, or porosity exists in a material of uniform thickness and density, the thickness and density are less at the point of defect and thus more radiation will pass through. When this radiation is recorded on a sensitive film, the defect will show up as an x-ray shadow. The shadow picture is called a *radiograph*.

X-ray radiographs show the defects more sharply than gamma-ray radiographs—a fact that must be taken into account in the interpretation of the film. When persons experienced with gamma rays are shifted to x-ray work, they tend to see any defect as a cause for rejection, whereas those experienced with an x-ray film may tend to underestimate the seriousness of a defect seen on a gamma-ray film.

Penetrameters are used as a check on the radiographic technique. The thickness of the penetrameters is given in the ASME Boiler and Pressure Vessel Code, Section V, Nondestructive Examination, Article 2, Table T-233.1. Penetrameters shall not be more than 2 percent of the thickness of the plate being radiographed, but for thin plates, it need not be less than 0.005 in. As measured by the penetrameter, sensitivity is understood to be the smallest fractional difference in the thickness of the material that can be detected visually on a radiograph. The appearance of the penetrameter image on the film defines a vertical sensitivity of 1 to 2 percent. The required appearance of the 2 T-hole or slit in the penetrameter proves a lateral sensitivity of 2 to 4 percent of the thickness being radiographed.

If the weld reinforcement or backing strip of a welded seam is not removed, a shim of the same material as the backing strip must be placed under the penetrameter so that the thickness being radiographed under the penetrameter is the same as the total thickness of the weld, including the reinforcement and backing strips.

Each penetrameter must have three holes with diameters that are usually 1, 2, and 4 times the penetrameter thickness, one of which must be of diameter equal to 2 times the penetrameter thickness but not less than 1⁄16 in. However, smaller holes are permitted. The smallest hole must be distinguishable on the film.

The penetrameters should be placed on the side nearest to the source of the radiation. Where it is not possible to do this, they may be placed on the film side of the joint provided that the radiographic technique can be proved adequate, and a lead letter F with a dimension as high as the penetrameter identification numbers is placed adjacent to the penetrameter or placed directly on the penetrameter, provided it does not mask the essential holes; at least one penetrameter must be used for each exposure. The penetrameter should be placed parallel and adjacent to the welded seam except when the weld metal is not similar to the base material, in which case it may be placed over the weld metal with the plane of the penetrameter normal (perpendicular) to the radiation beam. This provides for penetrameter image sharpness and minimum image distortion and allows the proper evaluation of radiographic quality.

To assure getting good contrast and seeing a weld defect, the ASME Code gives specific requirements for the film density, which are based on the Hurter and Driffield (H&D) curve. The minimum density through the radiographic image of the body of the penetrameter shall be 1.8 for single film viewing for radiographs made with an x-ray source and 2.0 for radiographs made with a gamma-ray source. Each radiograph of a composite set shall have a minimum density of 1.3. The maximum density shall be 4.0 for either case. A tolerance of 0.05 in density is allowed for variations between density readings. In the upper limits of density, the viewing equipment is important for the correct interpretation of the film. It is essential to have a strong source of light, available with the flexibility of light intensity and area both.

In recent years, the Code has allowed the inception of alternate methods of determining the acceptable densities of radiographs. One alternative method of measuring the film density is the use of wire penetrameters (see Fig. 5.1). Wire penetrameters are just as their name implies. Six wires whose composition is of material similar to the material being radiographed are encapsulated or "sandwiched" between two sheets of 0.030 in. clear vinyl plastic. Wire penetrameters are classified by sizes which range from 0.0032 in. to 0.32 in. These wires are categorized in sets known as sets A, B, C, and D.

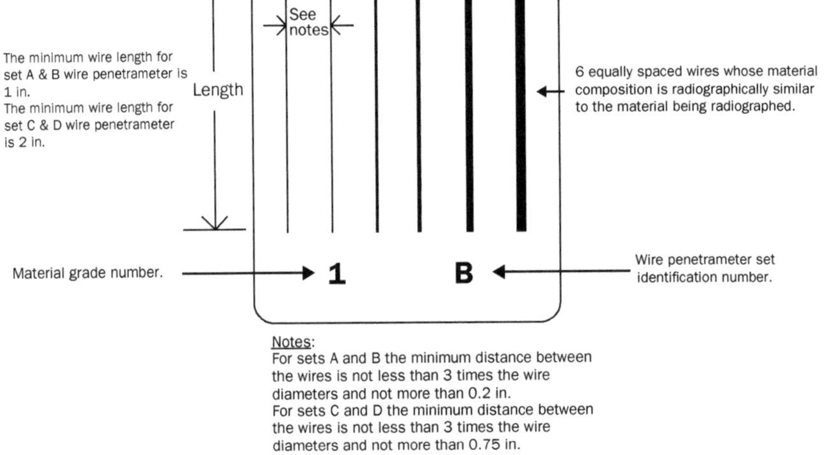

FIGURE 5.1 Wire penetrameter.

For set A and B wire penetrameters, the minimum distance between the wires is 3 times the wire diameter and not more than 0.2 in. The length of the wires for sets A and B is a minimum of 1 in. For sets C and D, the wire distance separations are to be not less than 3 times the wire diameter and not more than 0.75 in. The length of set C and D wire penetrameters is a minimum of 2 in. Each plastic case shall have a lead material grade-identifying number in the lower left-hand corner of the casing as well as the ASTM set designation.

Much like hole penetrameters, at least one wire penetrameter shall be placed abutting the weld to be radiographed. The radiographic density measured adjacent to the penetrameter must not vary by more than +30 to –15 percent. When the density does vary more than that, two sets shall be used.

If the source of radiation is placed at the axis of the joint and the complete circumference is radiographed with a single exposure, three uniformly spaced penetrameters are to be used.

Many people in the welding industry who know how to interpret a radiographic film sometimes err in their judgment. Moreover, radiographs often require retakes because of improper techniques or differences of opinion in interpretation. Only a retake can prove whether the technique or the interpretation of the original film was wrong. Basically, the important decisions in a radiographic inspection are the following:

1. Has the radiograph been properly taken?
2. Has the complete penetration of the weld been obtained?
3. Are the revealed defects acceptable or should they be repaired?

Proper radiography will show any change in the metal thickness greater than 2 percent, which is the contrast sensitivity required by the Code. Any defect surpassing the allowable minimum—whether slag entrapment, porosity, lack of penetration, undercutting, or cracks—will be recorded on the film as a dark area compared to the normal density of the sound metal. Weld reinforcements, backing strips, splatter, and icicles will show on the film as light areas because they represent the metal sections thicker than the average plate. Burn-through shows up as a dark, irregular area, but the drip-through surrounding the burn-through presents a lighter image. In a highly absorbent material such as the burn-off from an electrode used in heliarc welding, the burn-off shows up as a light area that is usually spherical in shape.

In heavy wall welds, knowing how deep a defect is from each side is helpful. We have seen chip-outs 2 in. deep made in a 3 in. weld that could have been only 1 in. deep from the opposite side if the stereoscopic principles of radiography had been used to make proper depth interpretation. (See Fig. 5.2 for an example.) When available

D = Shift of defect image

M = Shift of marker image

† = Thickness of weld

L = Depth of defect from film side

$$L = \frac{D\,\dagger}{M}$$

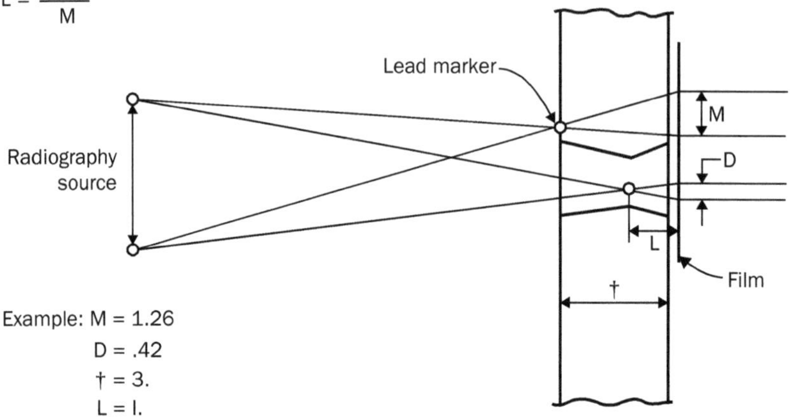

Example: M = 1.26

D = .42

† = 3.

L = l.

$$1'' = \frac{.42 \times 3}{1.26}$$

Figure 5.2 Stereoscopic principle in radiography.

ultrasonic techniques can be and often are utilized to determine the depth of many types of defects. Not every discontinuity in the film is objectionable if the vessels are fully radiographed (see Code Par. UW-51 and Appendix 4). See Code Appendix 4 for the details of acceptable and unacceptable defects. With full radiography, any type of crack or lack of penetration is not acceptable and must be repaired. Porosity is judged by comparison with the standards given in the Code.

If only spot radiography is required, the standards are less demanding (see Code Par. UW-52). The length of the film need be only 6 in. Porosity, moreover, ceases to be a factor in the acceptability of welds. There is a difference in the other allowed defects as well. Any weld that shows cavities or elongated slag inclusions is unacceptable if these have a length greater than two-thirds of the thickness of the thinner plate of the welded joint.

Any group of imperfections in a line with a total length less than the plate thickness in a length of 6 times the plate thickness is acceptable only if the longest defects are separated by acceptable weld metal equal to 3 times the length of the longest defect. (As the minimum length required for a spot radiograph is 6 in., the total length of imperfections in a line shall be proportional for radiographs shorter than 6 times the plate thickness.) Any defect shorter than 1⁄4 in. is acceptable in spot radiography, with the exception of any type of crack or lack of penetration, which is never acceptable and must be repaired. Any defect longer than 3⁄4 in. is not acceptable.

If a spot radiograph shows welding that violates Code requirements, two additional spot radiographs must be taken in the same seam at locations chosen by the Inspector away from the original spot—preferably one on each side of the defective section. If these additional spots do not meet the acceptable standard, the rejected welding must be removed and the joint rewelded, or the fabricator may choose to reradiograph the entire unit of weld represented, and the defective welding need only be repaired. The rewelded areas must then be spot-radiographed to prove that they meet Code requirements.

For the proper interpretation of films, the view box should be lighted with sufficient intensity to illuminate the darkest area of the film. The view box should be located in a room that can be darkened. Film handling and processing will at times cause marks that give the impression of defects. A fine scratch on the film surface can look like a crack; water marks, stains, and streaks can resemble other defects. If the film surface is viewed in reflected light, these defects can be properly detected. Other apparent but nonexistent flaws in the weld can be caused by a weld overlay that contrasts sharply at the weld edge, by scratches on intensifying screens, or by lines of weld penetration into the backing ring if a backing strip is used. Slag under the backing strip and tack welds on the strip will produce a lighter image and should not be interpreted as defects.

Such conditions may be obvious to an experienced technician, if not to a novice. However, there are often real defects in a weld that are difficult to determine, and it is these one must watch for. Some critical materials are more prone to crack than others, and their film images must be evaluated with extreme care.

A crack that runs perpendicular to the plate surface can be seen easily, but cracks along the fusion line are often at an angle, and the defect image may not be sharp. Defects that are parallel to the plate also may not show up clearly. For example, if a laminated plate is radiographed with the radiographic beam perpendicular to the plate surface, the lamination most likely will not be seen. However, if the plate is cut and radiographed with the beam parallel to the plate as rolled, these laminations will show clearly. (See Figs. 7.1 and 7.2 for examples.)

Most cracks show up longitudinally to the weld, in the weld itself, in the fusion line of the weld, at craters, and occasionally in the plate adjacent to the weld. Transverse cracks are not seen as often but do appear in the weld and plate edges. Slag inclusions are usually sharp and irregular in shape. As stated before, certain small slag inclusions are acceptable to the Code, but an elongated slag and too many inclusions in a line are not permitted.

Incomplete penetration—which is not acceptable—is usually sharply defined. Undercuts are usually easily seen at the edge of a weld. Porosity shows up as small, dark spots. These spots may be defined as either *fine* or *coarse porosity*; if full radiography is required, their acceptability is measured by the standards of the Code.

In viewing radiographs, it is important to be sure that the penetrameters are properly placed so that the penetrameter holes show and the H&D density requirements are given in the negative. This will assure better contrast to show any possible defects. The appearance of the essential hole in the penetrameter is the indication that the defects in the weld will be seen in the film (also see Table 7.1).

Sectioning

The Code permits the examination of welded joints by sectioning if this method is agreed to by both the user and the manufacturer, but the sectioning of a welded joint is not considered a substitute for spot-radiographic examination and permits no increase in joint efficiency; therefore, it is seldom used (see Code Par. UW-41 and Code Appendix K).

Sectioning may be accomplished with a trepanning saw, which cuts a round specimen, or with a probe saw, which cuts a boat-shaped specimen. The specimen removed must be large enough to provide a full cross section of the welded joint.

A flame-cut section must be sawed across the weld. A saw-cut surface must show the full width of the weld. The specimens must be ground or filed smooth and then etched in a solution that will reveal the defects. Care must be taken to avoid deep etching.

Suggested etching solutions are given in Code Par. K-1. The etching process should be continued just long enough to reveal all defects and clearly define the cross-sectional surfaces of the weld.

Many shops prefer to use ammonium persulfate because the method of its application allows a better observation of the specimen and prevents a deep etch. The solution, one part ammonium persulfate to nine parts water, is used at room temperature. A piece of cotton saturated with the solution is rubbed hard on the surface to be etched. The etching process is continued until a clear definition of the cross-sectional surfaces of the weld and any defects present emerges.

Other solutions , such as iodine, potassium iodide, and nitric acid, are used to etch metal specimens, but hydrochloric acid and ammonium persulfate are the most common ones. To be acceptable, the specimens should not show any type of crack or lack of penetration.

The plugs or segments should be suitably marked or tagged after removal with a record of their original location, the job number, and the welder or welding operator who did the welding. After etching, they should be oiled and kept in an identifiable container. A record should be made on a drawing of the vessel of all these specimens (with identification marks) after the vessel has been completed and stamped. The plugs may be retained by the purchaser, or they may be discarded.

The holes left by the removal of the trepanned plugs may be closed by any welding procedure approved by the Inspector. Code Par. K-2 offers some suggested methods for closing the holes. The best method is to cut a plug from the same plate material and then: (1) bevel the plate around the hole and the plug, (2) insert the plug, (3) weld from one side, (4) back-chip the other side, and (5) reweld. When gas welding is used, preheat the area surrounding the plug. If it is possible to weld from only one side, the bottom of the plug should be fitted to permit full penetration of the weld closure. Each layer of the weld should be peened to reduce the stresses set up by welding.

If the plate thickness is less than one-third of the diameter of the hole, a backing strip may be placed inside the shell over the hole and the hole filled completely with the weld metal. If the plate thickness is not less than one-third or greater than two-thirds of the diameter of the hole, the hole can be completely filled with weld metal from one side and then chipped clean on the other side. Other methods exist, but the ones described here are the most common.

Hydrostatic Testing

Most fabricating shops take pride in their work and do everything possible to turn out a good job. All this effort can be spoiled by carelessness in preparing for or conducting a hydrostatic test. Several actual incidents may serve as illustrations.

Case 1. A large fabricating shop—one of the best—put all its care and know-how into a critically needed vessel. However, during hydrostatic testing, a 3 in. shell plate bulged because of overpressure. This near-accident resulted in a costly delay in shipping that should never have occurred. The company had been cautioned by an Authorized Inspector on the first day in its shop, and several times thereafter, that the water pressure on the line to the test floor was 2200 lb and that a reducing valve or some other means of protection was required. The shop's chief inspector replied that dependable people worked on the test floor and that the two shut-off valves provided for each vessel were sufficient.

The Authorized Inspector told the shop's inspector that, as a precautionary measure, the presence of an Authorized Inspector would be necessary at all the hydrostatic tests of the Code vessels. The general manager was then informed of the possibility of a vessel being damaged during these tests. The general manager said that the matter would be taken up with the chief inspector, but the precautionary safety devices recommended were never installed. A few months later, an urgently needed non-Code vessel was badly damaged.

The Code does not specify an upper limit for the hydrostatic test pressure. However, if this pressure exceeds the required test pressure to the degree that a vessel is visibly distorted, the Inspector shall reserve the right to reject the vessel.

Case 2. A thin-shelled vessel was hydrostatically tested at 30 lb. It was filled late in the afternoon and left overnight with the pressure on and all valves closed. During the night, the water in the vessel warmed up, expanded, and bulged out a flat stayed section. This accident occurred despite the cautionary footnote to Code Par. UG-99(h), which states: "A small liquid relief valve set to $1\frac{1}{3}$ times the test pressure is recommended for the pressure test system, in case a vessel, while under test, is likely to be warmed up materially with personnel absent." (For a graph showing the increase in water pressure caused by the increase of water temperature, see Fig. 5.3.)

Case 3. A vessel 6 ft in diameter and 15 ft long was damaged while being drained. A tester opened a large opening at the bottom of the tank without first opening a valve or other inlet on top of the vessel to permit the air to enter.

The large opening at the bottom drained the water rapidly, creating a partial vacuum in the tank. As the vessel was not designed for external pressure, this vacuum caused the shell to crack between two nozzle outlets.

All three of these mishaps occurred in shops whose work is far above average; this proves that similar accidents are possible in any shop if care is not taken.

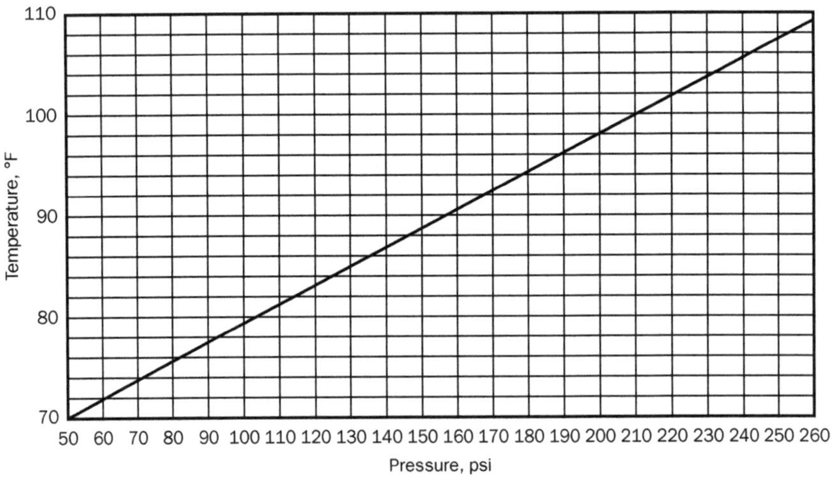

FIGURE 5.3 Water pressure and temperature graph.

Test Gage Requirements

Code Par. UG-102 states: "An indicating gage shall be connected directly to the vessel. If it is not readily visible to the operator controlling the pressure applied, an additional gage shall be provided where it will be visible throughout the duration of the test." It also recommends that a recording gage be used in addition to indicating gages on larger vessels. Not all shops, especially small ones, have recording gages, but good practice indicates the use of two gages on all tests because they are sensitive instruments and can be damaged easily.

The Code requires vents at all the high points of a vessel (in the position in which it is being tested) to purge air pockets while the vessel is filling. During a recent hydrostatic test of a large vessel, the frame of a large quick-opening door failed and the door dropped to the floor, thereby flooding the shop with several inches of water. Had there been any air in the vessel, serious damage to the building and possible injury to personnel might have occurred.

The Code requires the test equipment to be inspected and all low-pressure filling lines and other appurtenances that should not be subjected to test pressure to be disconnected before test pressure is applied. The reason can be seen in the recent test of a vessel for 1500 lb pressure with a 125 lb valve in the line. As this valve was leaking at the bonnet, the Inspector immediately had the pressure reduced and the valve replaced. The Inspector had seen the damage in another shop when a bonnet, valve, and stem had been blown through the roof, leaving a hole a foot wide (fortunately, no one was over the valve when it failed).

In another instance, a tester was seen placing a cast iron plug in a vessel that was to be tested to 1400 lb pressure. The tester was stopped by the Inspector, who requested that a steel plug be used. Replying, "Is that why the plug blew out?" the tester then showed the Inspector a dent in the concrete wall.

During a 2500 lb hydrostatic test of a large vessel, a leak was noted at a flanged connection. A worker who was responsible for the hydrostatic test picked up a sledgehammer and tapped on the flange cover trying to stop the leak. The cover then exploded off the vessel, rocketed to the far end of the shop, and struck a fellow worker who was walking across the end of the shop. His leg was severed at the knee. The ignorance of correct procedures can be fatal in testing vessels.

Seven Simplified Steps to Efficient Weld Inspection

A manufacturer's preparation of pressure vessels for examination by a Code Inspector or a customer can be either routine and easy or complicated and time-consuming. The speediest and most efficient inspection takes place in the shop that prepares for the inspection in an orderly manner. Such preparation will save not only time but very possibly money as well.

A job should be advanced to the point at which the specifications call for an inspection before one is undertaken; if there is no such specification, the job should have advanced to the point requested by the Inspector. Any vessel failing to meet minor requirements should be corrected before being shown to the Inspector. Major defects must be shown to an Authorized Inspector before the repairs are made. The Inspector makes an inspection on the basis of a written code or customer specification only. It is the responsibility of the manufacturer, and not of the Inspector, to see that these standards are met.

A good Inspector tries to find all the defects before the final or hydrostatic tests, especially if the vessel is to be postweld heat-treated. The Inspector knows—and certainly the manufacturer knows—that a minor repair can assume major proportions because of a lack of time to make the repair and to repeat postweld heat treatment and testing operations before the scheduled date of delivery.

A careful observation of the following items will help expedite the inspection:

1. The Inspectors should be provided with drawings that clearly show the weld details and dimension, types of material, corrosion allowances, and design pressures and temperatures. If the vessel is to be fully radiographed or spot-radiographed, this fact should be clearly stated because the Code allows different efficiencies for fully radiographed, spot-radiographed, and nonradiographed joints. Hydrostatic test pressure should be based on the maximum working pressure allowed for a new and cold vessel, and the part of the vessel to which the maximum pressure is limited should be defined.
2. A Code Inspector must check and verify all the calculations. If they are neatly arranged and include all pertinent design information, the job will be much easier.
3. If the vessel is a Code vessel, the data sheet should be checked and all corrections made before it is printed. Data sheets prepared at the start of a job after the drawings are made keep the engineering department and the Inspector from having to review the design and calculations a second time.
4. The mill test reports of material should be filed with the Inspector's copies of drawings to save the manufacturer's personnel from being interrupted while performing normal tasks to produce these reports.
5. The stamping information, including the location of stamps, should be readily available. The corrections should be made beforehand.

6. A written copy of the welding procedure (or a note stating that both the procedure and the operator are qualified) and of the test results should be available on request. These items should be clipped or filed together.
7. A place for the Inspector to sit and prepare the written part of the inspection should be provided to avoid the Inspector feeling like an intruder, which could happen if someone else's desk has to be used. This could even cause the Inspector to do the rest of the work at home or in a hotel room. This inconvenience could conceivably cause important data to be overlooked that might not have been missed had the report been written in the shop.

CHAPTER 6

Welding, Welding Procedure, and Operator Qualification

Examination of Procedure and Operator

One of the most important functions of an Authorized Inspector is examining welding procedures and operators for their compliance with the intent of Code requirements. The Code does not dictate the methods to the manufacturer, but it does require proof that the welding is sound (Code Par. QW-201) and that operators are qualified in accordance with the ASME Code Section IX, Welding and Brazing Qualifications (see Code Par. QW-301). *To avoid confusion, it should be noted that all Code references in this chapter, unlike the rest of this book, refer to ASME Code Section IX, Welding and Brazing Qualifications.*

An Inspector must be familiar with the physical properties of materials and electrodes and should have experience in the evaluation of the written procedures and test plates required by Section IX. An Inspector also must know how to determine the cause of test plate failures. The manufacturer should actually be more interested than the Inspector in finding the cause, for it is far less expensive to prevent failures than to make expensive repairs. This is especially true if the procedure is to be used on a production line where individual items may miss inspection. For example, in a plant where liquified propane gas tanks of a high-tensile-strength material were being constructed, the manufacturer had difficulty in performing satisfactory bend tests on an offset joint for the head and girth seam. After various changes in welding procedures, a test finally satisfied the Code requirements. However, because of the history the plant had had with this material and the type of welded joint, the Inspector realized that one successful test might not be sufficient. Knowing that any repairs would be costly and failure could be disastrous, additional bend tests before approving the procedure were requested.

A qualified welding procedure is the most important factor in sound welding. If one has been properly established, welders need only demonstrate their ability to make a weld in the position to be welded as well as other required variables, so that they will pass root and face bend or side bend tests in order to gain qualification. It should be remembered that weld excellence depends not only on good welding electrodes but also on the way they are used. Moreover, an Inspector should be satisfied by not only that the manufacturer's procedure is acceptable and the operators are qualified but also that the welding procedure is used and followed during actual fabrication.

Record of Qualifications

Once a welding procedure has been approved, it should be filed for future review by Authorized Inspectors or customers' representatives. Some manufacturers are not as careful as they should be in keeping these records. In one instance, an Inspector asked to see the written welding procedure of a certain shop and also the test results for material similar to that used on the job that was to be inspected. The written procedure could not be located for several days. After the Inspector explained the importance of proper filing and the unnecessary expense of having to requalify procedures, the manufacturer had records brought up-to-date and sets filed with both the shop superintendent and the engineering office. When the representative of a large refining company made a survey of the plant a few months later, the general manager felt very pleased at being able to show the procedures promptly upon request and felt that this helped the company's business position.

Another company with many approved welding procedures lost the file containing them and was forced into an expenditure of several thousand dollars to replace them. Needless to say, this firm now keeps duplicate copies of all welding qualifications and test results.

Currently, there are several commercial programs available that help the manufacturer keep the required records. Besides utilizing the strong database capabilities of computers, they greatly reduce the search time for specific types of welding procedure specifications and associated procedure qualification records and keep the welder qualifications up-to-date. The larger the number of welders or procedures, the greater the efficiency these programs bring to the manufacturer.

Welding Procedures and Qualification Simplified

Welding procedure specifications (WPS) and *procedure qualification records* (PQR) are the most essential factors contributing to sound welding.

A welding procedure specification records the ranges of variables that can be used and gives direction to the welder using the procedure. The procedure qualification record states exactly the variables used in qualifying the WPS, shows test results, and furnishes proof of the weldability of the variables described in the WPS. If these have been properly established, welders need only demonstrate their ability to make a weld that will pass root and face or side bend tests in the position to be welded in order to gain qualification. Other variables do apply, such as thickness limits, diameter, and the type of electrode used.

In building welded ASME Code vessels, one requirement is to have written welding procedure specifications for the materials to be welded (see Code Par. QW-201), properly completed records of qualification for this procedure (Code Par. QW-200.1), and the records of qualification for the welders working on the vessel (Code Par. QW-301.4). These procedures should be used in production by the fabricator. If narrative welding procedures are utilized, the wording of the procedure should be such that the meaning is clear and easily understood by all concerned. If standard ASME Code welding procedure documentation is utilized, then the fabricator need only fill in the welding procedures in writing on the recommended WPS, Code Form QW-482, and make test plates, the results of which are recorded on the PQR, Code Form QW-483. The purpose of the welding procedure is to provide welders with direction for making Code welds, which in turn will assure Code compliance. Once a welding procedure has been qualified, it should be filed for future review by Authorized Inspectors, customers' representatives, and any other interested parties. These records are one of the first things an Authorized Inspector of ASME vessels will examine. Code Par. QW-201 requires each manufacturer or contractor to have a qualified welding procedure specification. Code Par. QW-301 requires each welder or welding operator doing Code welding to be qualified in accordance with a welding procedure specification utilizing a process he or she will be using during actual fabrication. Proven sound welding procedures are a prerequisite to the fabrication and construction of ASME Code vessels. Section IX of the *ASME Boiler and Pressure Vessel Code* currently consists of two parts. Part QW covers the requirements for welding, and Part QB contains the requirements for brazing; the previous distinction between ferrous and nonferrous metals has been eliminated. For a specific material with a special requirement, such requirements are listed by the appropriate variables. By far the greatest percentage of fabrication involves welding. For this reason, most examples given here apply to Part QW, Welding. For brazing procedures, see Section IX, Part QB, Brazing.

Article V, allowing prequalified *standard welding procedure specifications* (SWPSs), has been added to Section IX. There are several steps a manufacturer must follow prior to using these SWPSs (see Code Par. QW-510); Code Form QW-485 is suggested for documenting the steps. This and other forms and records showing compliance and use shall be available to the Authorized Inspector. A listing of the permitted SWPSs is in mandatory Appendix E.

In the case of face, root, and some side bends for nonferrous groupings of base metals for those P numbers listed in Code Table QW-422, coupons may be ⅛ in. thick instead of ⅜ in. as required by most ferrous materials (see Code Section IX, Table and Code Figs. QW-462.2 and QW 462.3; see also Fig. 6.5). It is therefore clear that each Code welding procedure must be carefully read and separately evaluated. Also note, when welding clad or lined material, the procedure shall be qualified in accordance with Section IX of the Code, with modified provisions as required in the referencing section, Part UCL for clad vessels, in Section VIII, Division 1, of the *ASME Boiler and Pressure Vessel Code.* As stated previously, before a manufacturer can build or repair ASME Code vessels, proof must first be given that welding procedures and welders or welding operators qualify under the provisions of welding qualifications in Section IX (Code Par. QW-201). WPS Code Form QW-482 for recording welding procedure specifications and PQR Code Form QW-483 for procedure qualification records make it easy to fill in the required information. These forms may be purchased from the ASME and the National Board of Boiler and Pressure Vessel Inspectors. The following suggestions should be observed in preparing a WPS Code Form QW-482.

Welding procedure specification forms

The welding procedure should be titled and dated and should have an identifying WPS number and supporting PQR number.

Welding variables (Code Pars. QW-400 and QW-401). The procedure should state the welding processes and other variables as well as the type of equipment—automatic, semiautomatic, or manual. If inert-gas metal-arc welding is used, it should be specified whether a consumable or nonconsumable electrode is used as well as the material of which the electrode is made. Essential, supplementary, and nonessential variables for procedure qualification and performance qualification records must be addressed.

Joints (Code Par. QW-402). Preparation of base material should include group design, that is double butt, single butt with or without a backing strip, or any type of backing material used (e.g., steel, copper, argon, flux, etc.), showing a sketch of groove design on the PQR. The groove design should include the number of passes for each thickness of material and whether the welds are strings or weaving beads. The sketches should be comprehensive and should cover the full range of base metal thickness to be welded, the details of the welding groove, the spacing position of pieces to be welded giving maximum fit-up gap, the welding wire or electrode size, the number of weld passes, and the electrical characteristics; if automatic welding is used, the rate of travel and the rate of wire feed should be recorded.

Base metals (Code Par. QW-403). The type of material to be welded, giving the ASME specification number, the P number, group number under which it is listed, and the thickness range should be specified on the WPS. On the PQR, show the ASME materials specification, type, and grade number; P numbers to be welded together; and the thickness and diameter of pipes or tubes. If this procedure is for an unlisted material, either the ASTM or other Code designations or the chemical analysis and physical properties should be specified. This should be done only when you intend to work with a special material permitted by a Code case.

Filler metal (Code Par. QW-404). The SFA specification number and the AWS classification when the type of ASME filler metal is given with the filler metal group F number (Code Table QW-432) and the weld metal analysis A number (QW-404) should be stated. Any manufacturer's product bearing these classifications can be used; if there is no A or F number, the electrode manufacturer's trade name should be listed. In submerged arc welding, the nominal composition of the flux must be stated. For inert-gas arc welding, the shielding gas must be described. The size of the electrode or wire, amperage, voltage, number of passes, rate of travel for automatic welding, and rate of shielding gas flow for inert-gas arc welding must be listed. If a consumable insert is used on single butt welds, it should be stated.

Position (Code Par. QW-405). The position in which the welding will be done should be stated, and if it will be done in the vertical position, it should be listed whether uphill or downhill in the progression specified for any pass of a vertical weld. The inclusion of more than one welding position is permitted as long as the intent for each position is clear and shown by attaching sketches to the procedure.

Preheat (Code Par. QW-406). Preheating and temperature control should be stated only if they are to be used, and any preheating and control of interpass temperature during welding that will be done should be described. For example, materials in the P-4 and P-5 groups must be preheated. Carbon steels for pressure vessels in the P-1 group if preheated to 200°F can be welded up to 1½ in. thickness before postweld heat treatment is required, whereas if not preheated, the limit is 1¼ in. thickness [see Section VIII, Division 1, Table UCS-56-1, General Note (b)(2)].

Postweld heat treatment (Code Par. QW-407) (see also Section VIII, Division 1, Code Table UCS-56). Any postweld heat treatment given, the temperature, time range, and any other treatment should be described.

Gas (Code Par. QW-408). The shielding gas type (e.g., Ar, CO2, etc.), the composition of gas mixture used (e.g., 75 percent CO_2 and 25 percent Ar), and the shielding gas flow rate range should be stated. It should be listed whether gas backing, gas trailing, or shielding gas is used; the gas backing or trailing shielding gas flow rate should also be listed.

Electrical characteristics (Code Par. QW-409). When describing electrical characteristics, it should be stated whether the current is ac or dc, and if dc current is used, it should be specified whether the base material will be on the negative or positive side of the line. The mean voltage and current for each pass should be given. The mean should be the average of the specified upper or lower limits within 15 percent of the mean; the travel speed should also be given.

Technique (Code Par. QW-410). The joint welding procedure may be shown on the same sketches with the preparation of base material or on a separate sketch. The sketch should show the number of passes for each thickness of material, whether the welds are strings or weaving beads, and single or multiple electrodes. The orifice or gas cup size range, the oscillation parameter ranges covered by WPS (i.e., width, frequency, etc.), and the contact tube to work distance range should be given. The initial and interpass cleaning, including the method and extent of any imperfection removal, should be described. The method of back gouging should be explained. It is important that the method of defect removal is stated clearly. For example, flame gouging is detrimental to some materials. If this is so, make sure that another method of defect removal is used. Peening, if mentioned at all, should be used only when the desired result cannot be obtained by postweld heat treatment. If peening is used, all the details of the peening operation such as not peening first and last passes should be stated clearly (see Section VIII, Division 1, Code Par. UW-39). The treatment of the underside of the welding groove should be clearly indicated, for example "chipped to clean metal and welding applied as shown in sketch number...."

Variables [Code Pars. QW-400, QW-250, and QW-350 and Tables QW-415 (WPS) and QW-416 (WPQ)]

Any variable which would affect the quality or characteristics of the finished weld may require a new welding procedure. If any such variables exist, they should be clearly indicated. Change of essential variables will require the requalification of the procedure.

Additional requirements in the Code are known as supplementary essential variables. These variables are to be included on the WPS only when notch toughness is required by the referencing Code section and are in addition to the essential variables already stated. Changes in nonessential variables as well as editorial changes may be made without setting up new procedures if the changes are recorded and documented as such. These nonessential variables can be changed without requalification, but the changes must be indicated.

When writing a welding procedure specification, it is important to remember that qualification in any position for procedure qualification will qualify for all positions except when notch toughness or stud welding is a requirement (see Code Pars. QW-203 and QW-250). When notch toughness is a factor, qualification in the vertical, uphill position will qualify for all positions (Code Par. QW-405). If the weld must also meet notch toughness requirements as required by another section of the Code, the supplementary essential variable must be included (see Code Tables QW-252 to QW-262 and Code Pars. QW-280 to QW-283 and QW-400). Test plates are required for all procedure qualifications. If the completed WPS Code Form QW-482 and PQR Form QW-483 are clear and the test plates are approved and recorded, the manufacturer will be qualified to weld pressure vessels of the same or similar type of material used in making the test plate and also any other material in the P-number grouping. Starting with the 1974 Code, the P-number listing for base metals has been divided into groups. The purpose of the groupings is to cover the qualifications in which notch toughness is a requirement. When notch toughness is required, each group must be separately qualified. Many shops use a simple jig with a hydraulic jack similar to the one shown in Fig. 6.1 to make their own root, face, and side bends on the assumption that if the bend tests satisfy the requirements of Section IX, Code Par. QW-163 with no defect exceeding ⅛ in. (or $\frac{1}{16}$ in. for corrosion resistance weld overlay cladding), then the tensile test should present no difficulty. The advantage of conducting your own bend tests is that it keeps a shop from having to go to the expense of sending tensile bars to be machined and pulled if the root, face, or side bends should fail. Moreover, this jig can always be used to qualify

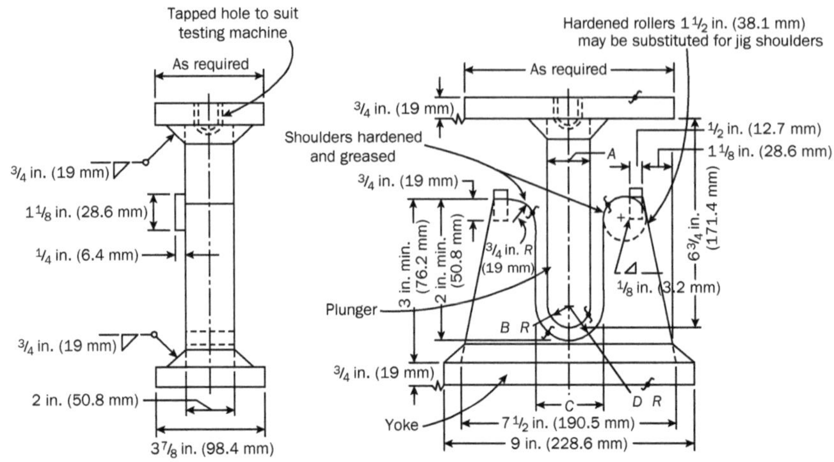

Customary Units					
Material	Thickness of Specimen, in.	A, in.	B, in.	C, in.	D, in.
P-No. 23 to P-No. 21 through P-No 25; P-No. 21 through -No. 25 with F-No. 23; P-No. 35; any P-No. metal with F-No. 33, 36, or 37	$\frac{1}{8}$ t = $\frac{1}{8}$ or less	$2\frac{1}{16}$ $16\frac{1}{2}t$	$1\frac{1}{32}$ $8\frac{1}{4}t$	$2\frac{3}{8}$ $18\frac{1}{2}t + \frac{1}{16}$	$1\frac{3}{16}$ $9\frac{1}{4}t + \frac{1}{32}$
P-No. 11; P-No. 25 to P-No. 21 or P-No. 22 or P-No. 25	$\frac{3}{8}$ t = $\frac{3}{8}$ or less	$2\frac{1}{2}$ $6\frac{2}{3}t$	$1\frac{1}{4}$ $3\frac{1}{3}t$	$3\frac{3}{8}$ $8\frac{2}{3}t + \frac{1}{8}$	$1\frac{11}{16}$ $4\frac{1}{3}t + \frac{1}{16}$
P-No. 51	$\frac{3}{8}$ t = $\frac{3}{8}$ or less	3 $8t$	$1\frac{1}{2}$ $4t$	$3\frac{7}{8}$ $10t + \frac{1}{8}$	$1\frac{15}{16}$ $5t + \frac{1}{16}$
P-No. 52, P-No. 53, P-No. 61, P-No. 62	$\frac{3}{8}$ t = $\frac{3}{8}$ or less	$3\frac{3}{4}$ $10t$	$1\frac{7}{8}$ $5t$	$4\frac{5}{8}$ $12t + \frac{1}{8}$	$2\frac{5}{16}$ $6t + \frac{1}{16}$
All others with greater than or equal to 20% elongation	$\frac{3}{8}$ t = $\frac{3}{8}$ or less	$1\frac{1}{2}$ $4t$	$\frac{3}{4}$ $2t$	$2\frac{3}{8}$ $6t + \frac{1}{8}$	$1\frac{3}{16}$ $3t + \frac{1}{16}$
All others with less than 20% elongation	t = (see Note b)	$32\frac{7}{8}t$, max.	$16\frac{7}{16}t$, max.	$34\frac{7}{8}t + \frac{1}{16}$, max.	$17\frac{7}{16}t + \frac{1}{32}$, max.

(continued)

FIGURE 6.1 Test jig dimensions (Code Fig. QW-466.1).

welders. The qualification of a procedure is affected by the type of electrode used. For example, if a manufacturer has been welding Code vessels of SA-285 grade C materials using an E-6020 electrode classed as type F-1 (Section IX, Code Table QW-432, F numbers) but wants to change to an E-6015 or E-6016 electrode classed as type F-4, the manufacturer must have this procedure qualified separately (see Section IX, Code Par. QW-404). The F-number grouping of electrodes and welding rods is based on their usability qualities, which essentially determine the ability of welders to make dependable welds with a given filler metal.

Procedure qualification for both groove welds and fillet welds is made on groove welds (Code Par. QW-202.2).

Material	Thickness of Specimen, mm	A	B	C	D
	SI Units				
P-No. 23 to P-No. 21 through P-No 25; P-No. 21 through -No. 25 with F-No. 23; P-No. 35; any P-No. metal with F-No. 33, 36, or 37	3.2 t = 3.2 or less	52.4 $16\frac{1}{2}t$	26.2 $8\frac{1}{4}t$	60.4 $18\frac{1}{2}t + 1.6$	30.2 $9\frac{1}{4}t + 0.8$
P-No. 11; P-No. 25 to P-No. 21 or P-No. 22 or P-No. 25	9.5 t = 9.5 or less	63.3 $6\frac{2}{3}t$	31.8 $3\frac{1}{3}t$	85.8 $8\frac{2}{3}t + 3.2$	42.9 $4\frac{1}{3}t + 1.6$
P-No. 51	9.5 t = 9.5 or less	76.2 $8t$	38.1 $4t$	98.4 $10t + 3.2$	49.2 $5t + 1.6$
P-No. 52, P-No. 53, P-No. 61, P-No. 62	9.5 t = 9.5 or less	93.2 $10t$	47.6 $5t$	117.5 $12t + 3.2$	58.7 $6t + 1.6$
All others with greater than or equal to 20% elongation	9.5 t = 9.5 or less	38.1 $4t$	19.0 $2t$	60.4 $6t + 3.2$	30.2 $3t + 1.6$
All others with less than 20% elongation	t = (see Note b)	$32\frac{7}{8}t$, max.	$16\frac{7}{16}t$, max.	$34\frac{7}{8}t + 1.6$ max.	$17\frac{7}{16}t + 0.8$ max.

GENERAL NOTES:

(a) For P-Numbers, see QW-422; for F-Numbers, see QW-432.

(b) The dimensions of the test jig shall be such as to give the bend test specimen a calculated percent outer fiber elongation equal to at least that of the base material with the lower minimum elongation as specified in the base material specification.

$$\text{percent outer fiber elongation} = \frac{100t}{A + t}$$

The following formula is provided for convenience in calculating the bend specimen thickness:

$$\text{thickness of specimen } (t) = \frac{A \times \text{percent elongation}}{[100 - (\text{percent elongation})]}$$

(c) For guided-bend jig configuration, see QW-466.2, QW-466.3, and QW-466.4.

(d) The weld and heat-affected zone, in the case of a transverse weld bend specimen, shall be completely within the bend portion of the specimen after testing.

Figure 6.1 *(Contd)*

Many vessels used for cryogenic service require an impact test. Pressure vessels built of austenitic chromium–nickel steels have shown a reliable service record plus consistent good quality as demonstrated by the record of the impact tests required by the Code for vessels in low-temperature service. Therefore, Section VIII, Division 1, Code Par. UG-84 does not require the impact tests of some vessels fabricated of these steels if these are conducted as part of the welding procedure qualification tests. (See Code Par. UHA-51 and Fig. 6.2 on zone locations for the removal of impact-test specimens.)

To summarize, we have listed several suggestions to be followed in the preparation of welding procedures.

1. A WPS Code Form QW-482 should be written by a person with the knowledge of production-shop needs and engineering-performance requirements.
2. The PQR Code Form QW-483 should be filed with the WPS Code Form QW-482.

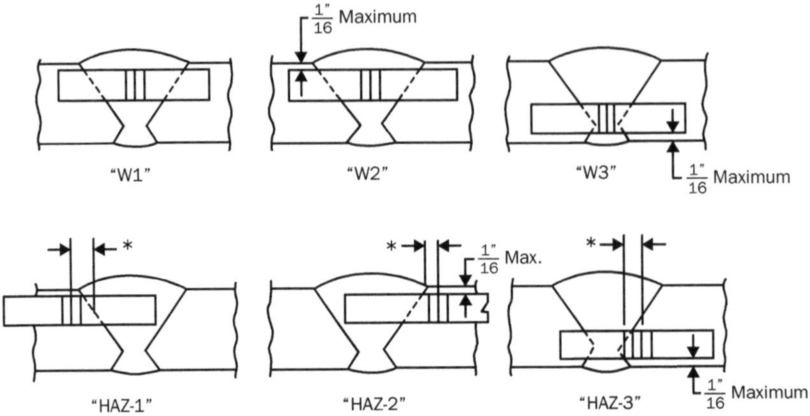

* Locate after etching so that maximum amount of heat-affected zone material is at the notch.

For welding procedure the heat-affected zone need not be impact tested when base material is exempted from impact tests. (CODE PAR. UG 84)

FIGURE 6.2 Weld metal and heat-affected zone locations for impact-test specimens.

3. Sketches and tables should be shown.
4. The existing Code should be reliably interpreted and the Code or specification translated to shop terminology.
5. Written procedures should be checked by reliable shop personnel who are to use these procedures. Writing a welding procedure specification requires the contributions of knowledge and experience from the shop personnel as well as the engineering department. If this is considered, a successful procedure can be used by the designing engineer to assist in designing a better weld at a lower cost.

Welder Qualifications

Unlike the welding procedure specification, which is primarily used to determine that the weldment is capable of providing the required mechanical properties of the welding to be performed, the welder qualification test and subsequent welder qualification record are used to determine the ability of a welder to deposit sound weld metal. Thus, the welder qualification does not require a tension test (see Table 6.1).

A welder is automatically qualified for a particular process if the test plates prepared for the welding procedure qualification pass the requirements (see Section IX, Code Pars. QW-301.2 and QW-303). A differentiation must be made here between a welder and a welding operator (Section IX, Code Par. QW-492 contains the definitions). A welder is

Remarks	Code references
Manufacturer must assign an identifying number, letter, or symbol to permit the identification of welds made by a welder or welding operator	Par. UW-29
Manufacturer must maintain a record of welders and welding operators, showing date, test results, and identification mark assigned to each	
Records must be certified by manufacturer and made accessible to Inspector	
Welder must stamp symbol on plate at intervals of 3 ft or less on work done on steel plate ¼ in. thick or more; for steel plate less than ¼ in. thick or nonferrous plate less than ½ in. thick, suitable stencil or other surface marking must be used	Par. UW-37
Records of all welders and welding operators working on a vessel and the welds made by each must be kept by manufacturer so that these data will be available for the Inspector	Par. UW-48
No production work can be done until both the welding procedure and the welders or welding operators have been qualified	Pars. UW-5, UW-26(c), UW-28, UW-29, UW-47, Section IX
Procedure for clad vessels	Pars. UCL-31, UCL-40, UCL-42, UCL-46
Lowest permissible temperature for welding	Par. UW-30
No welding should be done when temperature of base metal is lower than 0°F	
Heat the surface area to above 60°F where a weld is to be started when the temperature is between 0 and 32°F	

Table 6.1 Welding Procedure Qualification and Welders' Qualification Records, Welder Identification Markings, and Welding Records

one who has the ability to perform a manual or semiautomatic welding job. A welding operator is one who operates machine (automatic) welding equipment.

Welders passing the groove weld test need not take the fillet weld test. Welders passing the test for fillet welds are qualified to make fillet welds only [Section IX, Code Par. QW-303.1, Code Figure QW-462.4(b), and Code Par. QW-180]. The welder who is qualified to use type F-4 electrodes (E-6015 and E-6016), however, is also qualified to use type F-1 (E-6020 and E-6030), type F-2 (E-6012, E-6013, and E-6014), and type F-3 electrodes (E-6010 and E-6011) (see Section IX, Table QW-432 and Code Par. QW-404).

The welding operator may be qualified with a single test plate or pipe from which bend tests are removed. Although most shop welding is accomplished in the flat position with the weld metal deposited from above, this procedure is often not possible. Therefore, welds must be made in the position used in actual fabrication in order to qualify the welder (Section IX, Code Par. QW-303 and Code Fig. QW-461). Some fabricators set up their procedures (Section IX, Code Pars. QW-303 and QW-461) for position 6G, that is, pipe with fixed position and with the axis inclined to 45° to horizontal. A qualification in the inclined fixed position (6G) will also qualify for all the positions.

For performance qualification only, all position qualifications for pipe may also be done by welding in the horizontal position (2G), the horizontal fixed position (5G), or the 6G position (see Fig. 6.3). Qualification in these positions will also qualify for the flat, vertical, and overhead positions (Section IX, Code Par. QW-303). The six positions are shown in Fig. 6.3.

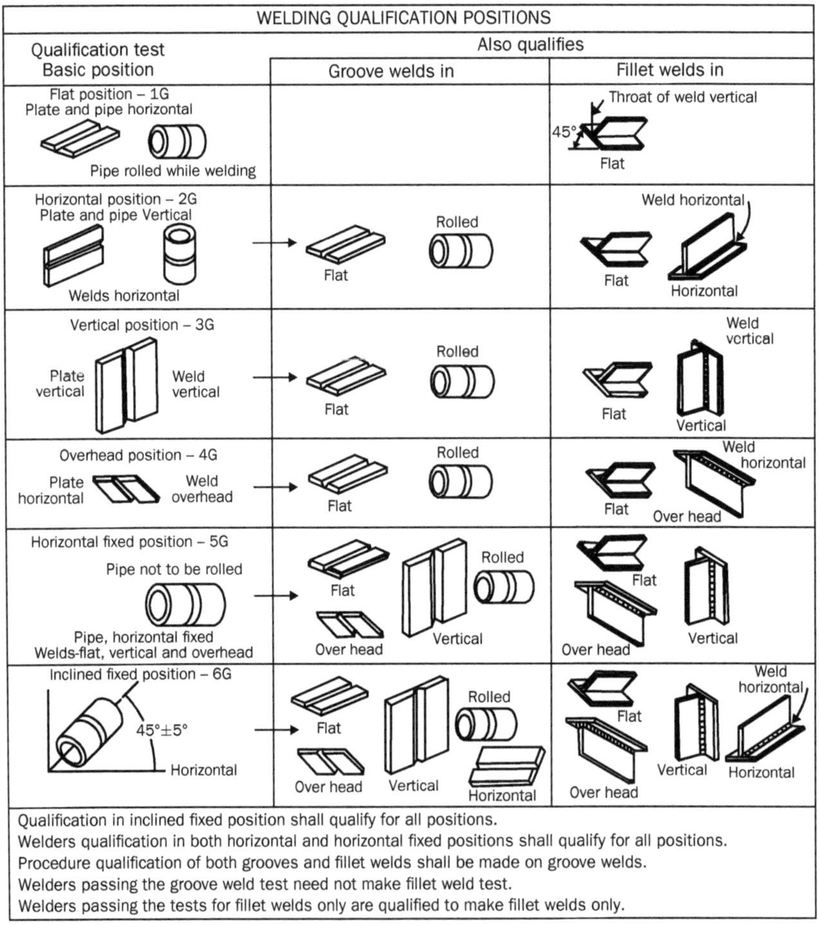

FIGURE 6.3 Welding qualification positions.

In addition to position, Section IX also places diameter restrictions under which a welder may weld in production. For example, if a welder qualifies on a 6 in. diameter piece of pipe and wishes to weld 2 in. diameter pipe by means of a butt joint, he or she must requalify for the small diameter pipe (see Code Par. QW-452.3). Further restrictions concern the thickness of materials to be welded and P number and F number.

The order of removal and the number and types of the test specimen from the welded test plates are shown in Fig. 6.4, transverse, and Fig. 6.5, longitudinal test specimens. The preparation of the transverse test specimens is shown in Figs. 6.6 and 6.7.

As an alternative to bend tests, radiographic examinations may be employed to prove the performance qualification of welders to make a sound weld, using the welding of groove welds with the manual

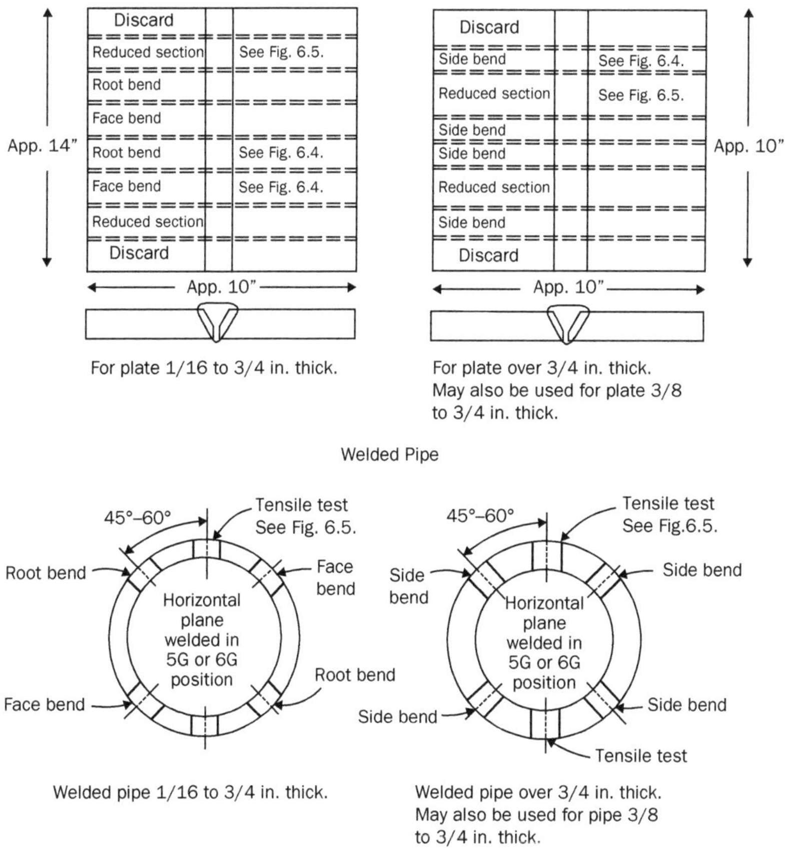

FIGURE 6.4 Order of removal of welded transverse test specimens.

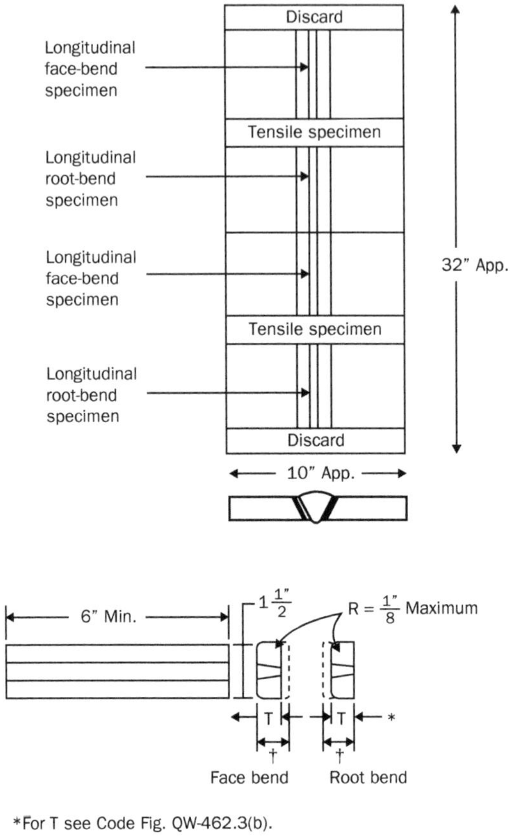

Figure 6.5 Longitudinal test specimens for plate materials.

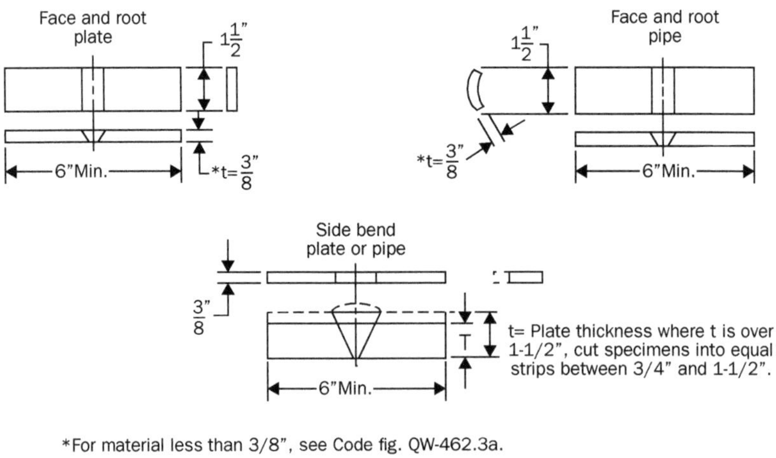

Figure 6.6 Preparation of face, root, and side bend specimens.

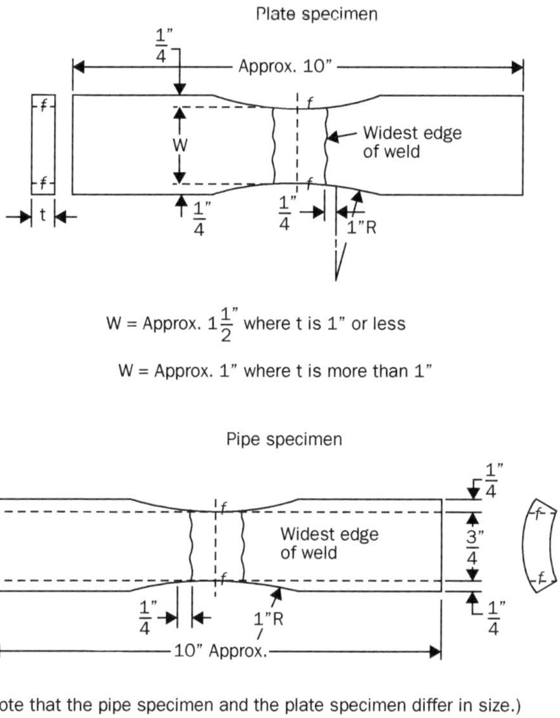

FIGURE 6.7 Preparation of reduced section tension test specimens.

shielded metal-arc welded process with covered electrodes and the gas tungsten-arc welding (Code Pars. QW-304 and QW-305). A record of the welders' and welding operators' test results should be kept on forms similar to the recommended Code Form QW-484, Section IX. The record of the tests shall be certified by the manufacturer and shall be made accessible to the Authorized Inspector.

It is to the fabricator's advantage to have one welder qualify for position 2G. Should this welder pass the test, this position also brings qualification for welding in the flat position (1G). Another welder could then qualify for the horizontal fixed position (5G). Should the tests be passed, this position also qualifies the welder for the flat, vertical, and overhead positions. It is also better to qualify for a position other than flat because the inclined fixed position (6G), horizontal (2G), vertical (3G), overhead (4G), and horizontal fixed position (5G) also automatically qualify the welder for the flat (1G) position, which, in the long run, is more economical.

A continuity of welders' qualification records should be maintained and a log kept of all welders' activities to indicate their continuity in specific welding procedures. If there is a gap of time in excess of six months in a specific process, the welder shall be requalified. See Fig. 4.6 for an example of a welders' qualifications and performance record log. If a manufacturer must qualify a welder on hard-to-obtain or costly material, such as stainless steel or alloy steel, the qualification can be obtained by using type F-5 austenitic electrodes on carbon steel plate.

The Code allows single-welded butt joints without backing strips for some circumferential joints in positions 1G and 2G. If a manufacturer intends to use such joints on shells or pipes, the welders (Section IX, Code Par. QW-303) must be qualified. For this reason, some manufacturers have their welders qualified in metal-arc welding without a backing strip (Section IX, Code Fig. QW-469.2) so that they will also be qualified to metal-arc weld with a backing strip.

Welding Details and Symbols

Welding details and symbols are an essential part of every shop drawing used in the design of welded pressure vessels; they guide the welder in the type of weld required, the proper size of weld metal to be deposited, and the size of bevel. Despite the rapid advances in welding proficiency in the past 25 years, many fabricating shops surprisingly still do not use either the details or the symbols.

There is a tendency in some pressure vessel shops toward overwelding, which not only increases the production costs but can also result in the distortion of the part welded, especially on stainless steel and aluminum vessels around nozzles, flanges, and accessory fittings. Other shops, especially those with an incentive rate or piecework rate, have a tendency to underweld. When an Authorized Inspector discovers underwelding on a Code vessel, the proper size of weld will be insisted upon. The extra handling and welding this entails will add to the cost of the vessel. Underwelding that goes undetected, on the other hand, will result in a weakened vessel and possible failure.

Without realizing the possible consequences, many vessel manufacturers leave the amount of weld to be deposited to the discretion of the shop supervisor. As a result, weld sizes do not always meet Code requirements.

Some welders or supervisors given the responsibility to produce a safe weld are inclined to make the weld "heavy enough." Usually, this is too heavy. Others, especially those on an incentive rate, take advantage of the fact and underweld. In the long run, underwelding may be more costly because of the possibility of failure and the loss of future business.

The responsibility for correct design instructions to the shop must be taken by the engineering department. The designer knows the exact thicknesses required for the shell, heads, nozzles, and flanges and the stress loads to be carried by the welds. The type and size of the welds required must be stated and clearly shown on the shop drawings either as a separate detail or by the use of the appropriate symbol. Both underwelding and overwelding may often be attributed to inadequate welding information on shop drawings.

The Code gives fillet weld sizes for nozzles and flanges in terms of throat dimensions (to facilitate load calculations). *The effective stress-carrying area is the throat dimension times the length of the weld.* The American Welding Society's *Standard Welding Symbols* gives fillet weld sizes in terms of leg dimensions, and it is this measurement that should be given to the shop rather than the throat dimensions. Calculations for leg sizes are illustrated and explained in Fig. 6.8. For example, if the Code-required throat dimension of a nozzle is 0.75 in., the leg size would be 0.75 divided by 0.707, or 1.06 in., and it is this last figure that should appear on the drawing. If this is not done when the designer calls for a throat measurement of 0.75 in., the welder may assume a leg size of 0.75 in. The person in the shop certainly cannot take the time to calculate the leg sizes of welds and should not be asked to do so.

It can easily be seen why all dimensions given to the shop should be prepared in advance. Some Code welds naturally do not meet Code requirements. Correct weld design will be assured only if weld sizes are given to the shop as leg dimensions (shown either in the weld detail or in weld symbol). The American Welding Society's

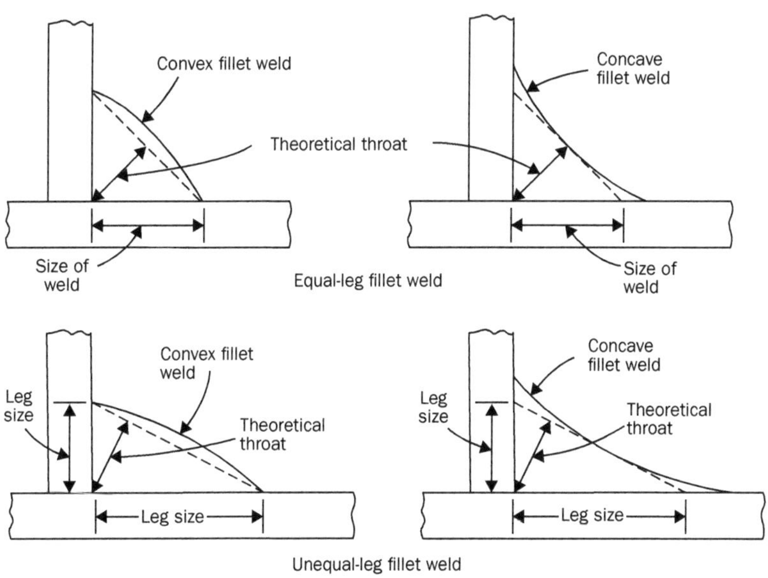

FIGURE 6.8 Fillet weld size.

Standard Welding Symbols are standard for almost all Codes and companies, although some companies make slight variations to meet their particular needs. A chart illustrating their use (see Fig. 6.9) can be obtained from the American Welding Society, P. O. Box 351040, Miami FL 33125.

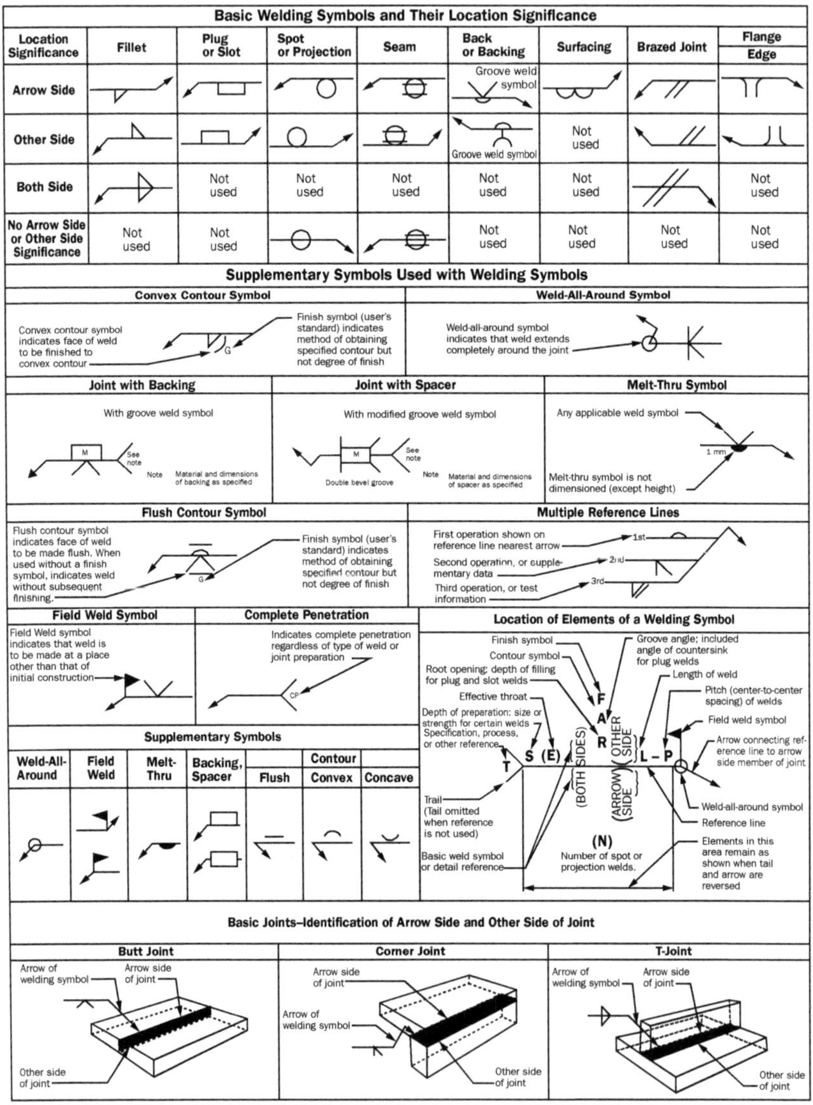

FIGURE 6.9 American Welding Society's standard welding symbols. (Courtesy of the American Welding Society.)

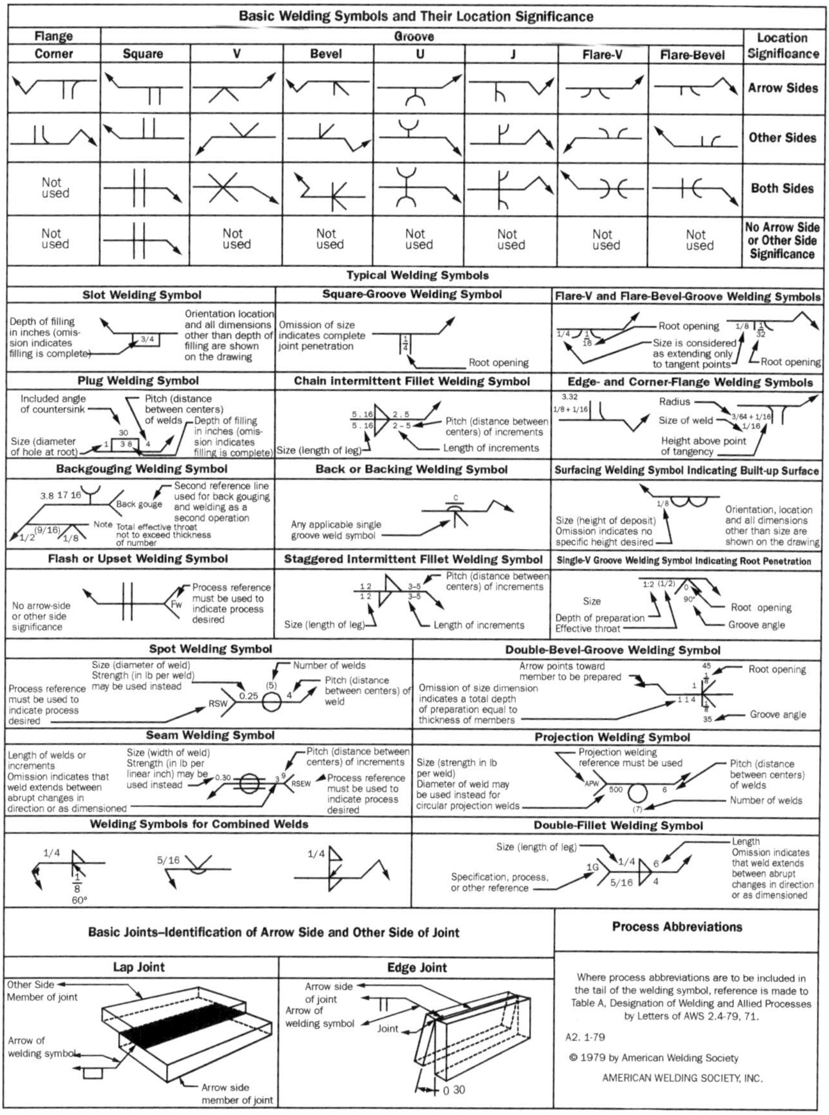

FIGURE 6.9 *(Contd)*

Many of these symbols, which are intended to cover all types of welding, are not used in the fabrication of pressure vessels. In fact, most pressure vessel manufacturers require only a few of them. Figures 6.10 to 6.12—which make no attempt to explain the function of all the AWS symbols—include the details of the welded joints used most often in pressure vessel construction and the proper symbols to indicate those joints.

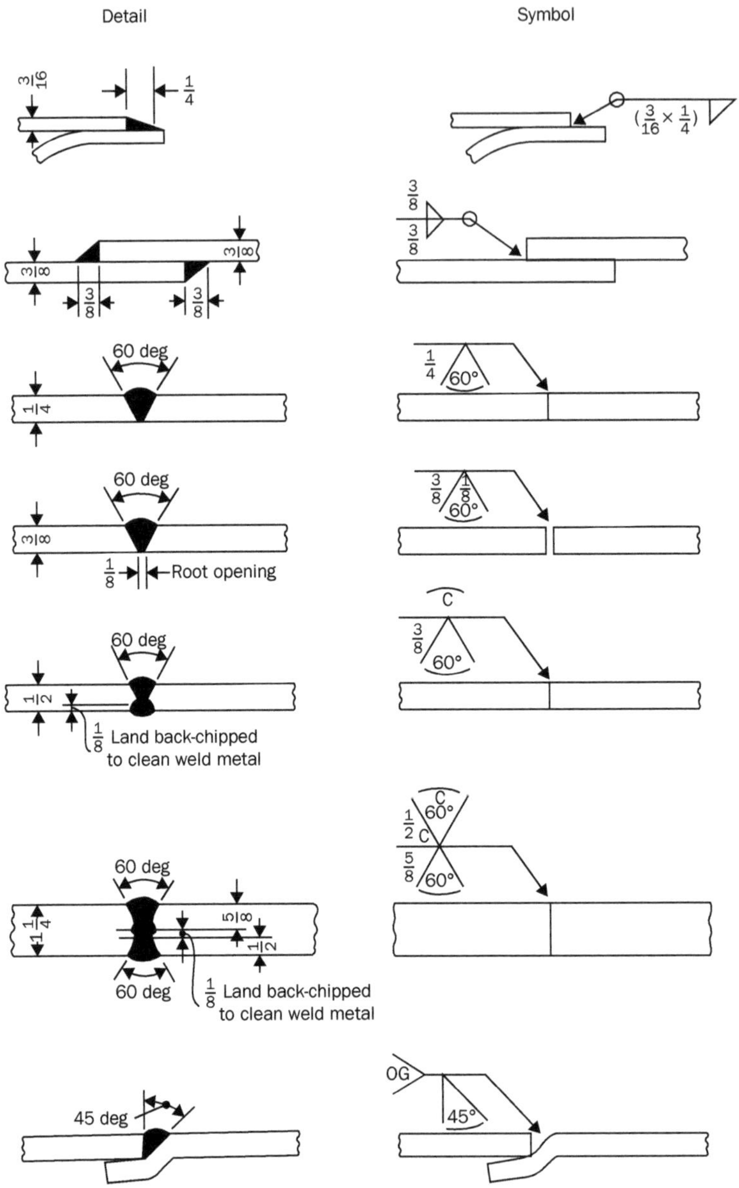

FIGURE 6.10 Welding details and symbols for shell plates and heads.

Joint Preparation and Fit-Up

A sound weld is almost impossible without good fit-up. The importance of proper joint preparation of welding edges on longitudinal seams, girth seams, heads, nozzles, and pipe cannot be

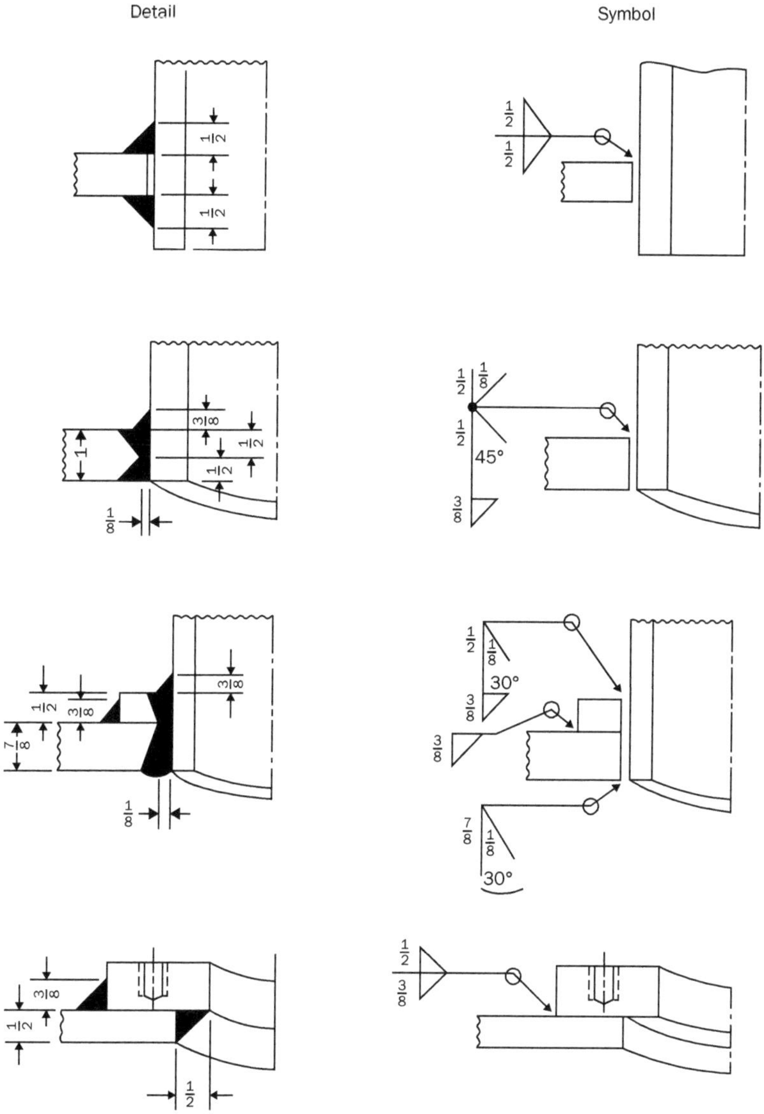

FIGURE 6.11 Welding details and symbols for nozzles and connections.

overemphasized. A good weld also cannot be obtained if the plate adjacent to the welding groove is covered with paint, rust, scale, grease, or oil. The Code states that if flame cutting is used to cut plates, the edges must be uniformly smooth and free of all loose scale and slag accumulations.

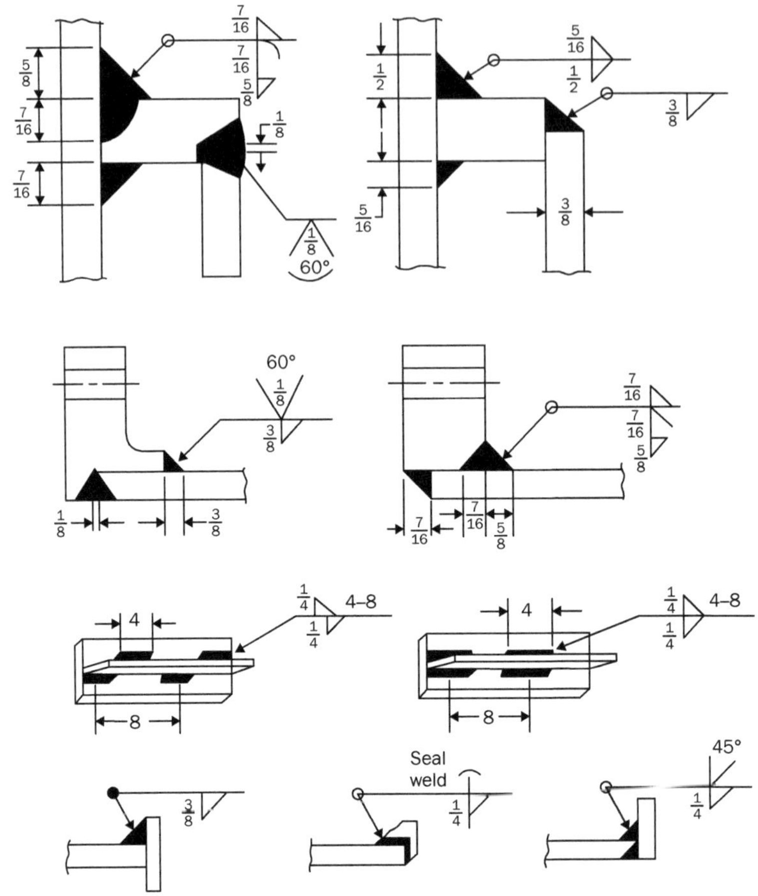

FIGURE 6.12 Miscellaneous welding details and symbols.

Before welding is undertaken, all joint surfaces for a distance of at least ½ in. from the edge of the welding groove on ferrous material and at least 2 in. on nonferrous material should be free of all paint, scale, rust, grease, and oil. The Code permits a maximum offset, after welding, of one-quarter the plate thickness at the joint, with a maximum allowable offset of the lesser of 1⁄16 the plate thickness or ⅜ in. in the longitudinal joints. For circumferential seams, the Code permits a maximum offset, after welding, of one-quarter the plate thickness at the joint, with a maximum allowable offset of the lesser of one-eighth the plate thickness or ¾ in. (Section VIII, Division 1, Table UW-33).

In pressure vessels of larger diameter in which a weld may be chipped to clean metal from the opposite side and welded the full thickness of the plate, the above permitted offsets are satisfactory.

If single-welded butt joints are used, however, special care must be taken in aligning and separating the edges to allow sufficient clearance for complete penetration and fusion at the bottom of the joint.

On smaller vessels and pipe that cannot be welded from the inside, special care must be taken when fitting up the abutting sections. This is especially true in a pipe of uneven thickness, in which the proper fit-up of inside diameters is almost impossible to obtain. However, if there is sufficient thickness in the pipe wall, machining can provide the proper fit-up. If the inside diameters are not equal, it is difficult to achieve full penetration at the root of a weld. As such penetration is needed to prevent root pass defects and cracks, alignment of these diameters is very important. Good joint preparation is especially essential in vessels that are intended for high-pressure and critical service.

The probes of welds that appear sound often reveal cracks that failed to show up in the radiographs made when the joint was first welded. As time went on, the small, unseen crack opened up and became larger—the probe showing the crack to have started at the edge of the plate at the backing ring because of a poor fit or a tack weld that was not properly removed. This crack formation has happened so often that large companies regularly examine pipe that is in service at critical pressures and temperatures. They have also set up more rigid standards of inspection for new construction, requiring a fit-up tolerance of 0.005 in. and the use of heliarc welding on the first pass to eliminate any incipient cracking that might develop into a major defect. However, not all companies have inspection departments to make such field inspections, and often the unit cannot be shut down long enough to permit such examinations.

When the investigation reveals poor fit-up or a cracked tack weld before welding as the cause of an accident, the manufacturer of the vessel or pipe is often unaware that such practices are permitted in the shop. Although care may be taken to prevent similar accidents in the future, preventive measures should always be in practice beforehand.

Effects of Welding Heat

Welding metallurgy is the study of the properties of welded metals and the thermal and mechanical effects occurring during the welding process, including the movements of the atoms in the liquid state and the formation of the grain structure of the metal in changing from the liquid to the solid state. Some understanding of the behavior of metal during welding is useful and necessary to the designer and fabricator of pressure vessels in order to help prevent dangerous residual stresses or defects in the welded joints.

Many metallurgists think of metals in terms of a "life cycle," that is, a metal is "born" in the melting operation, "dies" when it fails in service or is no longer used, and is even subject in the interim to "diseases."

It is known that heat treatment can change grain structure, which in turn may affect the physical properties. A metal can be hardened or softened, toughened or made more ductile. Welding is an operation in which separate pieces of metal are fused together to produce an integral section. Some metals are more difficult to weld than others. However simple the welding operation may appear, it is noteworthy that certain phenomena that occur in the making of steel also take place in the electric arc welding of steel products (plate, pipe, and the like), but more rapidly and on a smaller scale. Substantially, the same processes and metallurgical changes are involved. The heat of the arc melts a quantity of plate metal and the electrode. Chemical ingredients may be added by the coating on the electrode or by fluxes. The heat of the arc also melts or vaporizes the flux or coating, causing a gaseous shield to surround the arc stream and the pool of molten metal and to perform the important job of keeping out all the elements of the air (including nitrogen, hydrogen, and oxygen).

During the cooling and solidification process, the behavior of the molten metal in the weld joint is similar to that of molten metal poured into a mold to form a steel casting. The mold in the welded joint is the solid parent metal, and the melt consists of the molten metal of the electrode plus the liquified part of the parent metal—the two combining to form a casting within the solid parent metal. When the metal is in the liquid state, its atoms can move about at random. When it cools to the temperature at which it starts to solidify, the atoms are arranged into a regular pattern (for iron and steel, the pattern is a cube). This atomic arrangement and the corresponding effect on grain size are influenced by the rate of cooling from the liquid to the solid state and also by the composition.

Mr. H. S. Blumberg, a New York consulting metallurgist, had prepared two charts on the changes that take place at different temperatures in ferritic (carbon and low-alloy steels) and austenitic (chromium–nickel stainless steels that are nonmagnetic) materials (see Figs. 6.13 and 6.14). These charts show some of the simple fundamentals of atom arrangement and the grain size effects that take place when the metals change from the liquid to the solid state. They also show the important grain size changes in the "critical range" (the temperature range, with transition points, in which the structural changes take place) of ferritic carbon and low-alloy steels, including the temperatures used to normalize, anneal, draw, and stress-relieve these steels.

Grain size is influenced by the maximum temperature reached by the metal, the length of time the metal is held at a specific temperature, and the rate of cooling. For example, annealing, which involves heating to temperatures above the critical range followed by slow cooling, is used to refine the grain size to provide ductility and softness. Stresses can be relieved by heating beneath the critical range, but

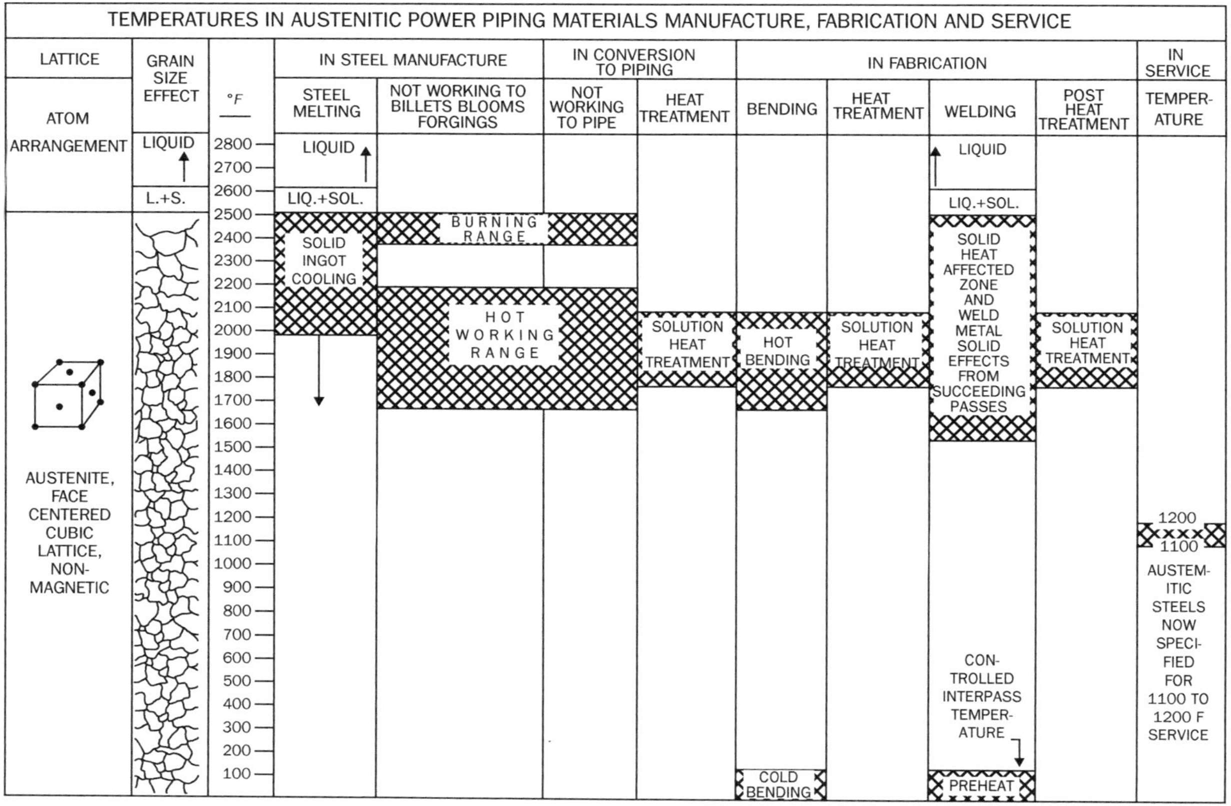

FIGURE 6.13 Metallurgical changes in austenitic materials at different fabrication and service temperatures.

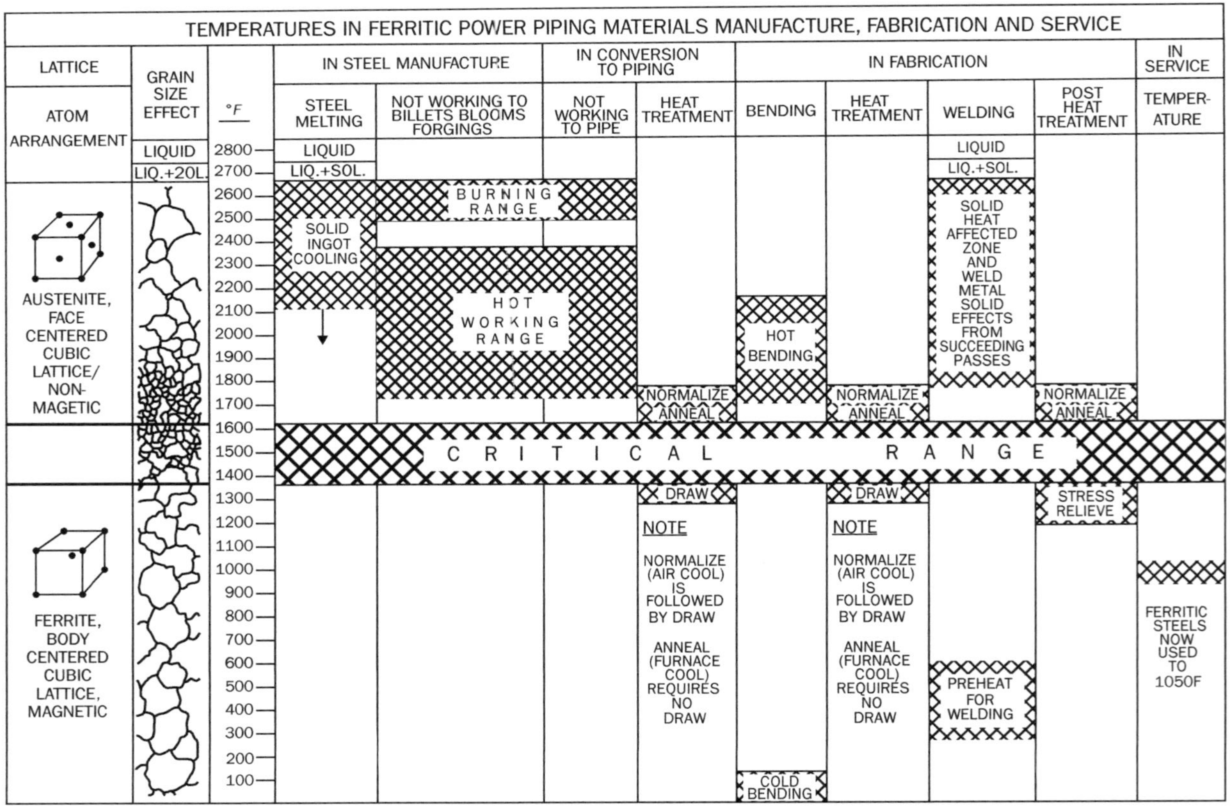

FIGURE 6.14 Metallurgical changes in ferritic materials at different fabrication and service temperatures.

unless temperatures exceed that range, full grain refinement does not take place. Normalizing is the process by which the steel is heated to approximately 100°F above the critical range (as in annealing) and then cooled in still air. This heat treatment also refines the grain size and improves certain mechanical properties. The refinement of ferritic steel can be easily achieved by heat treatment because of the movement of the atoms in the solid metal. Note that the austenitic steels do not have a critical range and that consequently their grain size cannot be changed by heat treatment alone.

The production of a fusion weld naturally results in the heating of some of the solid metal adjacent to the weld. The temperature reached by the metal in this zone varies from just below the melting point at the line of fusion to just above room temperature at a distance of a few inches from the weld. The effect of the welding heat on the parent metal depends on the temperature reached, the time this temperature is held, and the rate of cooling after welding. It is therefore necessary to understand the changes that take place in the zone where the metal is heated to very high temperatures without ever reaching the molten condition.

Usually, the weld metal zone is stronger than the parent metal but has less ductility. The heat-affected zone is the region where most cracking defects are likely to occur because the grain structure becomes coarse just beyond the fusion line and the ductility is lowest at this point. This reduction of ductility must be taken into consideration and measures taken to minimize the effects of total welding heat. For example:

1. The correct choice of parent material and welding electrode
2. Preheating before welding
3. Use of the proper welding procedure
4. Postweld heat treatment after welding

Preheating

Preheating prior to welding will eliminate the danger of formation of cracks, reduce hardness, reduce distortion, and reduce or prevent shrinkage stresses. Basically, preheating is a means of raising the temperature of a metal to a desired level above the surrounding ambient temperature before welding. It is usually applied between 125°F and 500°F but may involve temperatures up to 1200°F depending upon the material to be welded.

Preheating reduces the thermal strains set up during welding that may crack a weld. Preheating is especially necessary in thicker carbon steel plates in order to guard against the high thermal conductivity of steel when the heat of the welding arc is absorbed rapidly by the metal adjacent to the weld. The rate of cooling is slowed by preheating and thus the hardness is decreased. This slowing of the cooling rate allows more time for the metal to transform to a more favorable structure, minimizing hard zones and cracks; thus preheating retards the formation of undesirable metallurgical structures in metals having high hardness and susceptibility to cracking. Preheating dries up moisture, reducing the incidence of hydrogen and oxygen accumulation in a weld zone; consequently, it minimizes porosity, embrittlement, and cracking in the weld and heat-affected zone. Even though most medium- and low-carbon steels are welded without preheat, the metal adjacent to the start of a weld should be preheated thoroughly.

Preheating is an essential requirement when welding low-alloy steels in the P-No. 3, P-No. 4, and P-No. 5 material groups. Most alloy steels permit higher stress values at high temperatures, have greater strength at low temperatures, and allow higher notch toughness. In these steels, the required preheat temperature is stated and may be mandatory in order to comply with the Code requirements. The higher yield strength of these metals allows the use of higher stress values. These groups require careful planning of the welding procedure specifications in order to prevent cracking in the heat-affected zone—where the microstructure of the steel has been modified by the heat of welding. This heat-affected zone is comparatively hard and is formed during the welding operation; in some steels it may be necessary to hold the preheat temperature until the weld can be postweld heat-treated to avoid cracking.

Preheating the metal before welding has proved a dependable method of securing a sound weld. Preheating helps equalize and distribute the stresses evenly by distributing the expansion and contraction forces over an enlarged area to which the stress differential is applied due to the heat of welding (see Fig. 6.15).

Appendix R, Preheating, in Section VIII, Division 1 refers the reader to other Code paragraphs and to Section IX of the Code, P Numbers and Groups, where preheating is required.

Postweld Heat Treatment

The process of postweld heat treatment consists of the uniform heating of a vessel or part of a vessel to a suitable temperature for the material below the critical range of the base metal, followed by uniform cooling. This process is used to release the locked-up stresses in a structure or weld in order to "stress-relieve" it. Because of the great ductility of steel at high temperatures, usually above 1100°F, heating the material to such

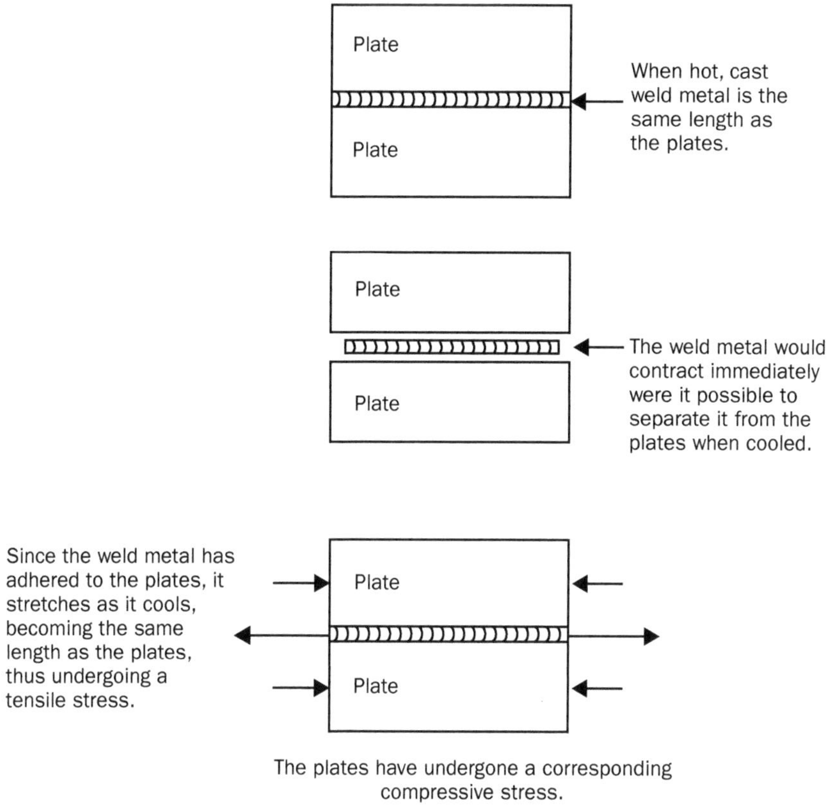

FIGURE 6.15 Good design provides for expansion and contraction.

a temperature permits the stresses caused by the deformation or straining of the metal to be released. Thus, postweld heat treatment provides more ductility in the weld metal and a lowering of hardness in the *heat-affected zone* (HAZ). It also improves the resistance to corrosion and caustic embrittlement. The proper heat treatment before and after fabrication is one that renews the material as near as possible to its original state.

Heat treatment procedures require careful planning and will depend on a number of factors—temperatures required and time control for material, the thickness of material and weldment sizes, and contour (or shape), and heating facilities. To be effective, the heat treatment must be carefully watched and recorded. Code Par. UW-49 requires that the Inspectors themselves be assured that postweld heat treatment is properly performed.

Furnace charts should be reviewed to determine if the rate of heating, holding time, and cooling time are correct. If a vessel is to be fully postweld heat-treated, the furnace should be checked to make sure that the heat will be applied uniformly and without flame impingement on the vessel.

The vessels with thin walls or large diameters should be protected against deformation by internal bracing. The vessel should be evenly blocked in the furnace to prevent deformation at the blocks and sagging during the postweld heat treatment operation.

During this operation, both increase and decrease in temperature must be gradual in order to allow uniform temperatures throughout. Where the difference in thickness of materials is greater, the rate of temperature change should be slower. The applicable requirements of Code Pars. UW-10, UW-40, UW-49, UCS-56, UCS-85, Table UCS-56, Par. UHT-56 or Table UHT-56, Pars. UHT-79, UW-80, Table UHA-32, Pars. UHA-32, UHA-105, UCL-34, UNF-56, ULW-26, ULT-56, UF-31, UF-32, and UF-52 must be followed in the postweld heat treatment of a Code vessel.

Conditions requiring postweld heat treatment

The Code requires postweld heat treatment of pressure vessels for certain design and service conditions (see Code Pars. UW-40 and UCS-56 for prescribed procedures). Postweld heat treatment requirements (according to materials, thicknesses, service, etc.) include the following:

1. All carbon-steel weld joints must be postweld heat-treated if their thickness exceeds 1¼ in. or 1½ in. if the material has been preheated to a minimum temperature of 200°F during welding [Code Par. UCS-56 and Table UCS-56-1 General Note (b)(2)
2. The vessels containing lethal substances must be postweld heat-treated (Code Par. UW-2).
3. Carbon and low-alloy steel vessels must be postweld heat-treated when the minimum design metal temperatures are below –55°F, unless exempted from the impact tests (Code Par. UCS-68).
4. Ferritic steel flanges must be given a normalizing or fully annealing heat treatment if the thickness of the flange section exceeds 3 in. (Code Par. 2-2(b)).
5. Unfired steam boilers with a design pressure exceeding 50 psi and fabricated with carbon or low-alloy steels must be postweld heat-treated (Code Par. UW-2(c)(3)).
6. Castings repaired by welding to obtain a 90 or 100 percent casting factor must be postweld heat-treated (Code Par. UG-24).
7. Carbon and low-alloy steel plates formed by blows at a forging temperature must be postweld heat-treated (Code Par UCS-79).
8. Some vessels of integrally clad or applied corrosion-resistance lining material must be postweld heat-treated when the base plate is required to be postweld heat-treated (Code Par. UCL-34).

9. Some high-alloy chromium steel vessels must be postweld heat-treated (Code Par. UHA-32 and Table UHA-32).
10. Some welded repairs on forgings require postweld heat treatment (Code Par. UF-37). (See Table 6.2.)

Remarks	Code references
All carbon and low-alloy steel seams must be postweld heat-treated if nominal thickness exceeds 1¼ in. or 1½ in. if preheated to 200°F before welding	Par. UCS-56
Some materials must be postweld heat-treated at lower thickness	Table UCS-56
Vessels containing lethal substances must be postweld heat-treated	Par. UW-2
Unfired steam boilers	Par. UW-2
Carbon steel vessels for service at lowered temperature must be postweld heat-treated unless exempted from impact test	Pars. UCS-66, UCS-67, UCS-68
Welded vessels	Pars. UW-10, UW-40
Carbon- and low-alloy steel vessels	Pars. UCS-56, UCS-66, UCS-67, UCS-79, UCS-85
High-alloy steel vessels	Par. UHA-32
Cast-iron vessels	Par. UCI-3
Clad-plate vessels	Par. UCL-34
Bolted flange connections	Appendix 2
Castings	Par. UG-24
Forgings	Par. UF-31
When only part of the vessel is postweld heat-treated, the extent of postweld heat treatment must be stated on Manufacturer's Data Report Form U-1	Par. UG-116
The letters HT must be stamped under the Code symbol when the entire vessel has been postweld heat-treated	Par. UG-116
The letters PHT must be stamped under the Code symbol when only part of the vessel has been postweld heat-treated	Par. UG-116
Low-temperature vessels	Par. ULT-56
Ferritic steel vessels	Pars. UHT-80, UHT-81
Nonferrous vessels	Par. UNF-56
Layered vessels	Par. ULW-26
Vessels or parts subject to direct firing, when the thickness exceeds ⅝ in.	Par. UW-2

Table 6.2 Postweld Heat Treatment

11. When the thickness of welded joints for vessel parts or pressure vessels constructed with carbon (P-No. 1) steels exceeds ⅝ in. as well as for all thicknesses of low-alloy steels (other than P-No. 1 steels) which are subject to direct firing, postweld heat treatment must be performed (Code Par. UW-2).

CHAPTER 7

Nondestructive Examination

Nondestructive examinations allow trained personnel to evaluate the integrity of pressure vessel parts without damaging them. Various sections of the *ASME Boiler and Pressure Vessel Code* now refer to the ASME Code Section V, Nondestructive Examinations, for nondestructive examination methods to detect internal and surface discontinuities in welds, materials, components, and fabricated parts.

Subsection A of Section V describes the methods of nondestructive examination to be used if referenced by other Code sections. They include the following:

- General requirements
- Radiographic examination (RT)
- Ultrasonic examination (UT)
- Liquid penetrant examination (PT)
- Magnetic particle examination (MT)
- Eddy current examination (ET)
- Visual examination (VT)
- Leak testing (LT)
- Acoustic emission examination (AE)

Subsection B lists the American Society of Testing Material standards covering nondestructive examination methods which have been accepted as Code standards. These are included for use or for reference, or as sources of technique details which may be used in the preparation of manufacturer's procedures.

The acceptance standards for these methods and procedures are stated in the referencing Code sections. Subsection B includes radiographic, ultrasonic, liquid penetrant, magnetic particle, eddy current, leak testing, acoustic emission, and visual examination standards.

Subsections A and B of Section V, Nondestructive Examination, provide rules and instruction in the selection of nondestructive test methods, development of inspection procedures, and evaluation of test indications. With the growing emphasis upon quality, nondestructive evaluation is becoming increasingly important in the construction of pressure vessels.

These examinations should be made according to a procedure that satisfies the requirements of the referencing Code and should be acceptable to the Authorized Inspector.

For the pressure vessel fabricator, nondestructive examinations provide a means of spotting faulty welds, plates, castings, and forgings before, during, and after fabrication. For the user, it offers a maintenance technique that more or less assures the safety of a vessel until the next scheduled inspection period. A serious flaw in a welded seam or a dangerously diminished plate thickness can cause a costly shutdown or, worse, a serious accident. Such defects must be located as soon as possible. Visual inspection is not enough. A weld may look perfect on the outside and yet reveal cracks, lack of penetration, heavy porosity, or undercutting at the backing strip when the weld is given a nondestructive examination by radiographic or ultrasonic tests. The internal condition of a material or weld can be accurately appraised only by the use of proper testing techniques.

Accident prevention requires many skills. Nondestructive testing is one that should be used more often. Since no weld or material is perfect, nondestructive testing is an excellent way to examine the internal conditions of a material. Nondestructive testing is often used on materials supplied to the fabricator of pressure vessels. Additional tests are made during fabrication and are also used during a vessel's operating life to assure safe operation and to gage plate thicknesses, thus enabling an accurate prediction of the life of the vessel. Flaws are not uncommon in plate material and welds during any stage of fabricating a vessel. The point is to find them as soon as possible in order to prevent difficulties later. For example, a serious flaw in material or a welded seam may cause an accident. Many such accidents can be prevented with nondestructive tests.

In this chapter, some nondestructive test procedures are explained with regard to Code pressure vessel fabrication. The understanding of their use in addition to visual inspection is essential to safe pressure vessel fabrication and operation.*

*For more complete instructions on the methods of nondestructive examinations, refer to Section V of the *ASME Boiler and Pressure Vessel Code.*

Table 7.1 and Figs. 7.1 to 7.5 analyze and illustrate some nondestructive testing procedures commonly used in Code pressure vessel fabrication. An authorized Code vessel manufacturer's quality control manual must describe the system that is employed for assuring that all nondestructive examination personnel are qualified in accordance with the applicable Code.

There should be written procedures to assure that nondestructive examination (NDE) requirements are maintained in accordance with the Code. If the nondestructive examination is subcontracted, the name of the person responsible for checking the subcontractor's site, equipment, and qualification of procedures to assure their conformance with the applicable Code should be given. In a nondestructive examination certification program, the manufacturer has the responsibility for the qualification and certification of NDE personnel. To obtain these goals, the manufacturer may train and certify personnel or may hire an independent laboratory that has qualified NDE procedures and personnel in accordance with the reference Code section and Section V to perform these duties. If others do the work, it must be stated in the quality control manual which department and person (by name and title) are responsible for ensuring that the necessary quality control requirements are met, because the manufacturer's responsibility is the same as if the parent company had done the work.

When required by the referencing section of the pressure vessel Code, personnel performing nondestructive examinations must be qualified in accordance with publication SNT-TC-1A of the American Society for Nondestructive Testing. SNT-TC-1A, *Recommended Practice for Nondestructive Examining Personnel Qualification and Certification,* is published by the American Society for Nondestructive Testing, Inc., 1711 Arlingate Lane, P.O. Box 28518, Columbus, OH 43228.

It is the responsibility of the manufacturer to conduct training and qualification tests of nondestructive examination personnel in accordance with the requirements of SNT-TC-1A, to arrange for eye examinations for the personnel interpreting nondestructive examinations, and to have audits made of the system in order to assure that qualified procedures and personnel are being used. The manufacturer must also prepare and maintain test examinations as well as test procedures which comply with Section V of the Code and SNT-TC-1A. Only personnel qualified to perform examinations should be assigned this task.

Adequate nondestructive examining procedures are an essential part of any well-managed operation. Procedure format and content must be appropriate to the specific requirements of the shop. Without a written procedure, there is no guarantee as to the effectiveness or repeatability of the test. Nondestructive testing should be performed in accordance with a written procedure. Each manufacturer should

Applications	Advantages	Remarks
	X-ray radiographic inspection*	
For examination of internal soundness of welds, castings, forgings, and plate material (see Fig. 7.1)	1. Gives sharper definition and greater contrast for thicknesses up to 3 in. 2. Gives a graphic and permanent record indicating size and nature of defects 3. Offers established standards of interpretation for guidance 4. Radiation source can be turned off when not in use	1. Protective precautions necessary to protect personnel in surrounding area 2. Trained technicians required to take and interpret film
	Gamma-ray radiographic inspection†	
For examination of internal soundness of welds, castings, forgings, and plate material (see Fig. 7.2)	1. More suitable for heavy thicknesses 2. Portable equipment 3. Source holder permits use through small openings 4. Lower initial cost 5. Requires no cooling mechanism 6. Gives a graphic and permanent record 7. No tube replacement	1. Government license is required for possession and use of isotopes 2. Protective precautions necessary 3. Sensitivity is not as high as x-ray testing, and persons not properly trained in interpreting film may underestimate the seriousness of a defect 4. Energy cannot be adjusted, so isotope must be chosen to meet sensitivity requirements and the thickness of material

*See Code Pars. UW-51 and UW-52.
†See Code Section V, Article 2.

Table 7.1 Analysis of Nondestructive Testing Techniques

follow the procedure in accordance with the requirements of the referencing section of the Code and ASME Code Section V, Nondestructive Examinations.

The manufacturer is responsible for seeing that all personnel performing nondestructive examinations for Code vessels are competent and knowledgeable in applicable nondestructive examining requirements. They must be qualified in accordance with the requirements of

Applications	Advantages	Remarks
Magnetic particle inspection*		
For locating surface and subsurface defects that are not too deep (see Figs. 7.3 and 7.4) Technique employs either electric coils wound around the part or prods to create a magnetic field. A magnetic powder applied to the surface will show defects as local magnetic fields. The nature of the defects will be revealed by the way the powder is attracted	1. Useful for the inspection of nozzle and manhole welds for which radiography would be difficult at best and most often impossible 2. Reveals small surface defects that cannot be detected radiographically 3. Useful in weld repair to assure the removal of defect before rewelding 4. Can be used to detect laminations at plate edges	1. Useful only for magnetic material 2. Not suitable for defects parallel to the magnetic field 3. Training needed to evaluate visual indications of defects 4. Prod methods of testing may mar the surface of a machined or smooth part 5. Magnetic particle examination methods (see Code Appendix 6)
Liquid penetrant inspection†		
For locating surface defects A liquid dye penetrant is applied to a dry, clean surface and allowed to soak long enough to penetrate any surface defects. After a time interval up to an hour, the excess penetrant is wiped off and a thin coating of developer applied. The penetrant entrapped in a defect will be drawn to the surface by the developer, and the defect will be indicated by the contrast between the color of the penetrant and that of the developer	1. Especially applicable for nonmagnetic material 2. Useful for the inspection of nozzle and manhole welds when radiography is impossible 3. Easy to apply, accurate, fast, and low cost	1. Only detects the defects that are open to the surface 2. Not practical on rough surface 3. Liquid penetrant examination methods (see Code Appendix 8) 4. Take care in applying dye penetrant to ensure that excessive amounts of vapors are not inhaled

*See Code Appendix 6.
†See Code Appendix 8.

Table 7.1 *(Contd)*

Applications	Advantages	Remarks
Fluorescent penetrant inspection*		
For locating defects that run through to surface Penetrant is applied by spraying, dipping, or brushing. Excess penetrant is washed from the surface with a water spray and the surface allowed to dry. Dry powder or water-suspension developer is applied to part to draw penetrant to surface. Penetrant glows under black (ultraviolet) light For leak test, penetrant is applied to one side and the other side is examined under black light for the indications of glow	1. Easy to perform 2. Defects show clearly under black light 3. Especially applicable for nonmagnetic material 4. Can be used on rough surfaces 5. Detects porosity and defects that would otherwise be hard to find	1. Only detects the defects that are open to the surface 2. Black light requires a source of electricity 3. Not very effective as a leak test for plates more than ¼ in. thick

*See Code Appendix 8.

Table 7.1 *(Contd)*

SNT-TC-1A. The intent of SNT-TC-1A is to serve as a guide for the qualification and certification of nondestructive test personnel. This document establishes the general framework for a qualification and certification program. It explains the various levels of qualification. It gives the requirements and shows the type of examination required. Certification may take any one of several forms.

The manufacturer may employ previously trained persons and certify these people. If this is done, the responsibility would be the same as if the manufacturer had trained and qualified these persons. The manufacturer may conduct a training program to meet the requirements of SNT-TC-1A. This responsibility would be carried out through a designated employee who meets Level III requirements. This employee will train the persons, examine them, and certify them as to their ability to do nondestructive examinations. Through the examinations, this employee will be able to judge the qualifications of these persons for Level I and/or Level II certification.

The *American Society of Nondestructive Testing* (ASNT) has training course programs to help a manufacturer write a training program to conduct the courses, which are usually given by a qualified Level III individual who uses questions derived from the ASNT course.

In addition to the required general examination, a specific written examination is necessary. The specific examination should reflect the procedures and equipment used in the plant.

Applications	Advantages	Remarks
	Ultrasonic inspection*	
For detecting defects in welds and plate material and for gaging thickness of plate (see Fig. 7.5) High-frequency sound impulses are transmitted through a search unit, usually a quartz crystal. This search unit is held in intimate contact with the part being tested, using an intermediary such as oil or glycerine to exclude air. The sound waves pass through the part being tested and are reflected from the opposite side or from a defect. The time of travel shows the defect on a cathode-ray tube or scope. This scope can also measure the depth of crack or defect Although the impulse instrument used to detect defects will also give thickness measurements, the resonance instrument is actually an electronic micrometer designed to measure thickness from one side of the material within the tolerances of ±0.002 in.	1. Portable equipment 2. Access from one side of the part being tested is possible 3. Reveals small root cracks and defects not indicated by radiographic film, especially in thick-walled vessels 4. Thickness measurements are rapidly made 5. Good for the detection of laminated plates 6. Can be used in nuclear vessels, where induced radiation eliminates the use of radiography 7. Remote inspection can be made in hostile environment 8. Can be used to determine whether use or corrosion has affected the thickness of vessel walls and piping	1. Training required for the interpretation of visual indications of defects 2. Rough surfaces must be made smooth if crystal is to make contact with part 3. Photographs must be taken to provide permanent records 4. Ultrasonic test methods when required (see Code Appendix 12) 5. Not very effective on welds with backing rings

*See Code Appendix 12.

Table 7.1 (*Contd*)

In the practical examination, the test objects should be similar to those fabricated by the manufacturer, applying the written procedures and the same equipment used in production. The examinee must be able to use the written procedures and equipment.

Applications	Advantages	Remarks
	Eddy current testing	
Eddy current testing is based upon the principles involving circulating currents into an electrically conductive article and observing the interaction between the article and the currents	1. Detects small discontinuities 2. High-speed testing 3. Accurate measurement of conductivity 4. Checks variation in wall thickness 5. Noncontacting	1. Limited to use with conductive materials or conductive-base materials 2. Depth of penetration restricts testing to depths of less than ¼ in. in most cases 3. The presence of strong magnetic fields will cause erroneous readings 4. Testing of ferromagnetic metals is sometimes difficult 5. Training required for the interpretations of visual defects and indications

Table 7.1 *(Contd)*

In summarizing, a manufacturer must have qualified personnel. In addition, the manufacturer should have written procedures if the shop does nondestructive examinations. The Code requires that nondestructive examining personnel be trained and qualified, and also that their qualifications be documented as required in the referencing Code and SNT-TC-1A with its supplements and appendixes. It cannot be emphasized too strongly that the nondestructive examination of the ASME Code pressure vessels must be performed by qualified personnel.

Leak Testing

Several leak-detecting methods are used in testing pressure vessels. One is the *halogen diode detector testing method*, which is generally performed by pressurizing the vessel being tested with an inert gas, usually Freon or a trace of Freon, and examining the areas to be tested with a device that is sensitive to Freon. The presence of a leak will give an audible alarm, will be indicated on a meter, or may light up an indicator light.

Another method consists of pressurizing the vessel to be tested with air containing Freon gas. A halide torch, whose flame changes color when a leak appears, is employed to search for leaks.

Often a helium mass spectrometer leak detector is used. It measures a tracer medium, usually helium gas. If a leak is present, the leak detector detects the helium gas and gives an electrical signal. In the helium mass spectrometer test, the tracer medium helium gas passes through a

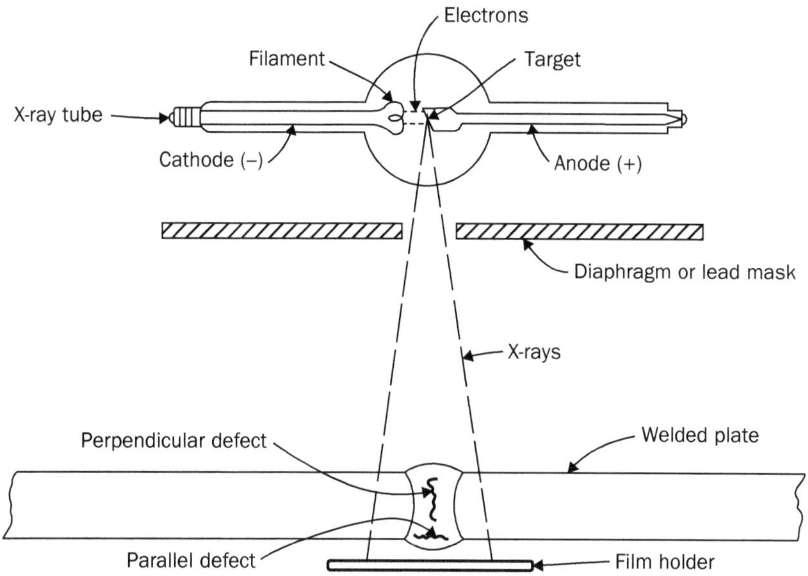

FIGURE 7.1 X-ray inspection of welds.

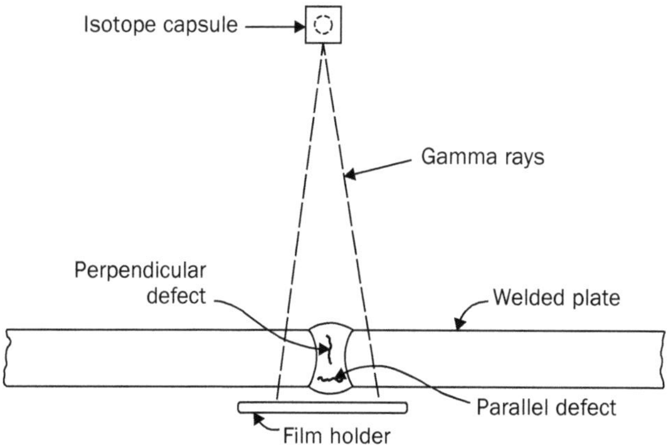

FIGURE 7.2 Gamma-ray inspection.

leak from the pressurized side to the nonpressurized side. The test can be made under a vacuum, or pressure test, or a combination of both.

A common method in pressure vessel testing is the *gas and bubble formation test*. Its purpose is to detect leaks in a vessel by applying a solution which will form bubbles at the leak area when gas passes through it.

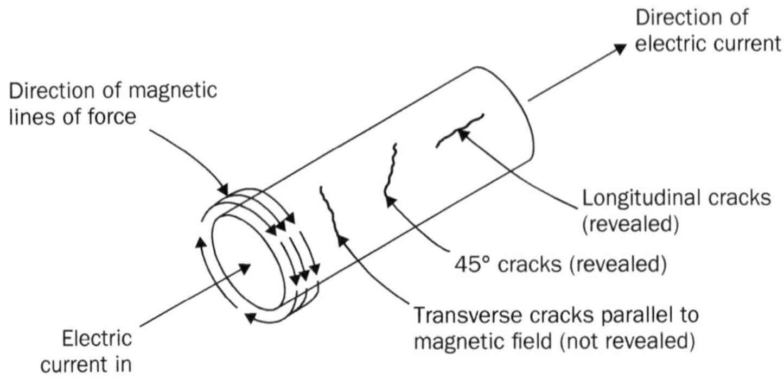

FIGURE 7.3 Magnetic particle inspection (circumferential magnetic field).

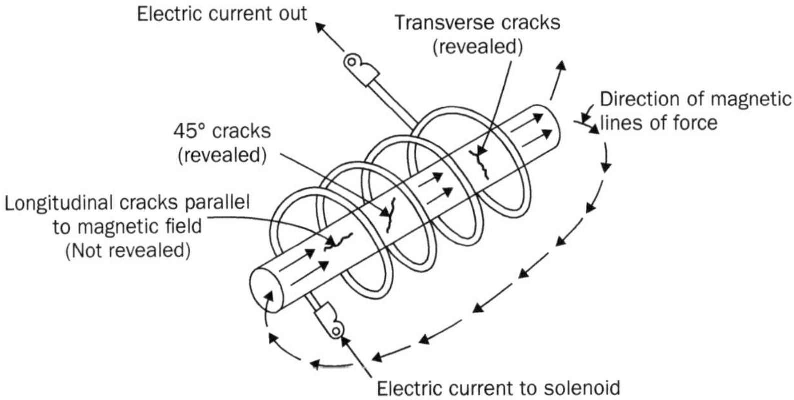

FIGURE 7.4 Magnetic particle inspection (longitudinal magnetic field).

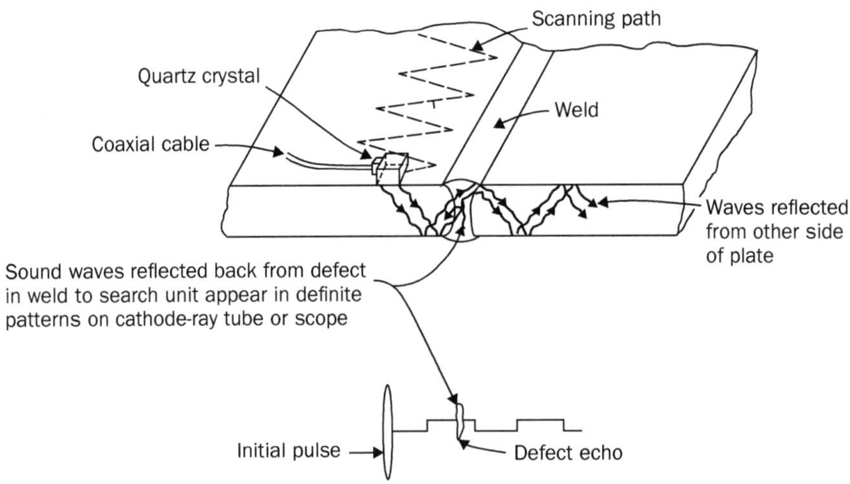

FIGURE 7.5 Ultrasonic weld inspection.

Acoustic Emission Testing

Acoustic emission testing is a recent NDE method that has been successfully used in the industry with hydrostatic tests and in-service examination of pressure vessels and piping. Acoustic emission systems use piezoelectric transducers attached directly to the structure being evaluated to detect the stress waves generated by discontinuities in the materials under stress. The acoustic emission device measures the emission of these minute pulses of elastic energy; when properly detected and analyzed, the test provides information concerning the discontinuities contained in the structure under stress. Once detected, a follow-up analysis of these signals verifies the location and meaning of a particular discontinuity.

The number of transducers needed to test a structure is determined by its complexity and not its size. Twelve to fifteen transducers may be required to test a storage vessel. Sixty to ninety may be required to test the pressure containment system of a nuclear plant, which includes the reactor vessel, pumps, steam generator, pressurizer, main piping, and valves. Pipeline testing needs at least 1 transducer every 1000 ft.

Acoustic emission technology is most beneficial where the accessibility to a pressure containment system is limited, because of its ability to conduct a complete structural integrity analysis with a minimum of shutdown time by sensing the presence of potential problems.

CHAPTER 8

ASME Code Section VIII, Division 2, Alternative Rules, and Division 3, Alternative Rules for High Pressure Vessels

The *ASME Boiler and Pressure Vessel Code*, Section VIII, Pressure Vessels, is published in three parts—Division 1; Division 2, Alternative Rules; and Division 3, Alternative Rules for Construction of High Pressure Vessels. In comparison to Division 1, the vessels in Division 2 require more complex design procedures and have more restrictive requirements on design, materials, and nondestructive examinations; however, higher design stress values are permitted in Division 2. It should be noted that in this chapter, all the Code paragraph references are to the specific parts of Division 2. The alternative rules may be applied to fixed vessels, to vessels used in transport (on ships, barges, rail, and roadways), to vessels used in direct firing from the combustion of fuel (and not within the scope of Sections I, III, or IV), and to some vessels in which steam is generated. Unfired steam boilers are not included in Division 2. See Code Par. 1.2.1.2 for details.

The requirements of Section VIII are more stringent in Division 2 vessels as compared with Division 1. There are many restrictions in the choice of materials, more exacting designs, and stress analyses. A closer inspection of the required fabrication details, material inspection, welding procedures, and welding as well as more nondestructive examinations will be necessary since the design stresses are larger than those permitted in Division 1. With larger allowable stresses, a vessel can be built for the same pressure and diameter using thinner shells and heads, resulting in less weld metal and lighter vessels. However, the cost of engineering and quality control will be greater. The owner must weigh the savings in materials and fabrication labor versus the extra costs of engineering and quality control to decide up the choice between Divisions 1 and 2.

In Division 2, full radiography is required. The spot radiography or no radiography options of Division 1 are not allowed.

The Division 2 Code has been divided into 9 parts which represent very logical sequences in the design and fabrication of a pressure vessel. They are as follows:

Part 1: General requirements. It includes items such as Scope, Referenced Standards, and Definitions.

Part 2: Responsibilities and duties. The User, Manufacturer, and Inspector involved in the production and certification of vessels in accordance with Division 2 have definite responsibilities or duties in meeting the requirements of this Division.

Division 2 requires that the owner furnish a User's Design Specification to the vessel fabricator. The User's Design Specification must include the following:

- *Installation site*—This includes the vessel intended location, jurisdictional authority, and environmental conditions.
- *Vessel identification*—Equipment name and item number.
- *Vessel configuration and dimensions*—This includes the drawings, orientation, openings, and support methods, as well as any included components.
- *Design conditions*—Design pressure and temperature, MDMT, and design loads on the vessels.
- *Operating conditions*—Operating pressure(s) and temperature(s), fluid transients, and flow.
- *Design fatigue life*—Sufficient operating information must be provided such that a determination may be made as to the necessity for a cyclic service analysis.
- *Materials of construction*—The user must either identify the specific materials for the construction of the vessel or provide sufficient information such that the fabricator can propose a specific material. The corrosion and/or erosion allowance must be specified.
- *Loads and load cases*—The user must identify the loads and load case combinations.

- *Overpressure protection*—It is the user's responsibility to provide overpressure protection for the vessel. The type of overpressure protection shall be specified in the User's Design Specification.

The User's Design Specification must be signed by a Registered Professional Engineer. The minimum requirements for the User's Design Specification are listed in Code Par. 2.2.2. The Code offers a guide for certifying the User's Design Specification in Code Annex 2-A.

Division 2 requires that the vessel fabricator supply a Manufacturer's Design Report. The Manufacturer's Design Report must include the following:

- Final as-built drawings of the vessel.
- The complete listing of material specifications for each component of the vessel.
- Documentation for all Part 4 Design-by-Rule calculations.
    - Include information about the name and version of any software used to perform the analyses.
    - Loading and boundary conditions used to address the load cases specified in the User's Design Specification.
    - References for all material physical properties used in the analyses.
    - Details showing the results of calculations and their comparisons to Code acceptance criteria.
    - A summary of all results.
- Documentation for any Design-by-Analysis calculations per Code Part 5.
    - A detailed description of the numerical method used—include the name and version of the software used.
    - A description of model geometry.
    - The loading and boundary conditions used to address the load cases defined in the User's Design Specification.
    - A summary of analysis results showing a comparison to Code acceptance criteria.
    - Graphical representation of results.
- The results of cyclic service analysis, if applicable.

The Manufacturer's Design Report must be signed by a Registered Professional Engineer. Note that the User's Design Specification and the Manufacturer's Design Report cannot both be signed by the same engineer. The minimum requirements for the Manufacturer's Design

Report are listed in Code Par. 2.3.3. The Code offers a guide for certifying the Manufacturer's Design Report in Code Annex 2-B.

Part 3: Materials requirements. The requirements for materials used in the construction of pressure vessel parts according to the rules of Division 2 are defined in Part 3. General rules and supplemental requirements are defined for different material types and product forms. Supplemental requirements for the several classes of materials are included. Material testing requirements and toughness requirements are defined. Allowable stresses for the various materials are addressed, with ultimate reference to the tables of Section II, Part D. Also included are strength parameters, physical properties, and design fatigue curves.

Part 4: Design-by-rule requirements. Part 4 includes the rules, criteria, and equations for the commonly used shapes and configurations of pressure vessels. A great many of the vessels purchased as Division 2 manufacture are engineered using only Part 4 (Design by Rule). The calculation equations and methods in Part 4 are very similar to, and in some cases identical to, their counterparts in Division 1. Further, most of the commercially available pressure vessel design software include the option to prepare calculations as per the Division 2 criteria and equations. In such cases, the only additional engineering expense is the engagement of a Registered Professional Engineer to certify the final Manufacturer's Design Report package.

Part 5: Design-by-analysis requirements. The design requirements for the application of the design-by-analysis methodology of Division 2 are described in Part 5. Detailed design procedures utilizing the results from a stress analysis are provided to evaluate the components for plastic collapse, local failure, buckling, and cyclic loading. Supplemental requirements are provided for the analysis of bolts, perforated plates, and layered vessels. Procedures are also provided for design, using the results from experimental stress analysis, and for fracture mechanics evaluations.

Part 6: Fabrication requirements. This part addresses the general fabrication requirements, beginning with materials, material identification, repair of defects, and forming. Included are specific requirements for welding, including special requirements for tube-to-tubesheet welds, for preheating, and for the heat treatment of weldments, along with special requirements for clad or lined vessels.

Part 7: Inspection and examination requirements. The requirements for examination including *nondestructive examination* (NDE) during the fabrication of pressure vessels of welded construction to be marked with Certification Mark are provided in Part 7. The requirements for the examination of materials are provided in Part 3. Within this part are included the responsibilities, duties, and qualification of NDE Personnel. Instructions are given for the examination of welded joints as well as for the examination methods and acceptance criteria. Also included are sections on leak testing and acoustic emission testing.

Part 8: Pressure testing requirements. The primary subjects addressed in this part are the hydrostatic and pneumatic testing of completed pressure vessels. Also included is the subject of alternative testing wherein there is an air pocket over a partially liquid-filled vessel.

Part 9: Pressure vessel overpressure protection. Part 9 provides the requirements for pressure relief devices used to protect against overpressure in pressure vessels constructed to the requirements of Division 2. It establishes the type, quantity, and settings of acceptable devices along with the relieving capacity requirements. Also provided are the requirements for capacity certification testing, as well as for obtaining and using the Certification Mark for pressure relief devices. In addition, this part provides the requirements for the installation of the pressure relief devices.

Division 1 specifies that the purchaser shall address the issue of cyclic service, but does not provide a specific set of criteria by which to define the need for a fatigue analysis. However, Division 2 provides a very detailed set of criteria which must be addressed for every vessel. In most instances, the owners of Division 1 vessels requiring fatigue analysis will utilize the criteria, methodology, and acceptance criteria presented in Section 5.5 of Division 2.

Division 2 of the Code does not hold the Authorized Inspector responsible for approving design calculations. The Inspector verifies the existence of the User's Design Specifications and the Manufacturer's Design Reports and that these reports have each been certified by Registered Professional Engineers.

It is the Authorized Inspector's duty to review the design, with respect to the materials used, to see that the manufacturer has qualified welding procedures and welders to weld the material and to review the vessel geometry, weld details, nozzle attachment details, and nondestructive testing requirements. The manufacturer should obtain the Inspector's review of these details before ordering materials and starting fabrication.

With the growing use of increased pressures in the operation of vessels it became apparent that a third division limiting itself to higher-pressure vessels, generally above 10,000 psi, would be helpful. Section VIII does not establish minimum pressure limits for Division 3 vessels or maximum pressure limits for Division 1 or 2 vessels.

Division 3 is limited to fixed vessels (Code Par. KG-104). Vessels fabricated to Division 3 may be transported from work site to work site between pressurizations. Division 3 explicitly excludes cargo tanks mounted on transport vehicles (see Chapter 11). If a direct-fired vessel is not within the scope of Section I, it may be constructed to the rules of Division 3.

As is required for Division 2, a Professional Engineer registered in the United States or Canada must certify the design, as well as the final design report, of a Division 3 vessel. In general, Division 3 provides very sophisticated methods of analysis, manufacture, and control, and therefore, any more than a cursory discussion falls outside the boundaries of this book's simplified approach to the Code. If the reader is contemplating constructing or designing a Division 3 vessel, he or she should use the Code itself as a source.

CHAPTER 9

Power Boilers: Guide to the ASME Code Section I, Power Boilers

The *ASME Boiler and Pressure Vessel Code* Section I, Power Boilers, includes the rules and general requirements for all the methods of construction of power, electric, and miniature boilers and high-temperature water boilers used in stationary service. It also includes the power boilers used in locomotive, portable, and traction service.

The rules of Section I are applicable to the boilers in which steam or other vapor is generated at a pressure more than 15 psig and high-temperature water boilers intended for operation at pressures exceeding 160 psig and/or temperatures exceeding 250°F. Superheaters, economizers, and other pressure parts connected directly to the boiler without intervening valves are considered as part of the scope of Section I (see Code Section I, Power Boilers Preamble).

References in this chapter are for Code Section I, Power Boilers.

The 1914 ASME Boiler Code was the original pressure vessel code, which now includes 11 sections. Section VIII, Division I, Pressure Vessel Code, is used to build more vessels than all the others combined. This book was written mainly for the users of Section VIII, Division I. However, many engineers, fabricators, and inspectors must work with all the sections of the Code.

Those who use the ASME Code, Section I, Power Boilers, regularly in construction and inspection may not have difficulty in finding the information needed. It is the engineer, fabricator, or Inspector—who consults it infrequently—who may have difficulty.

A Quick Reference Guide (Fig. 9.1) illustrates some of the types of boiler construction which are provided under Section I of the ASME

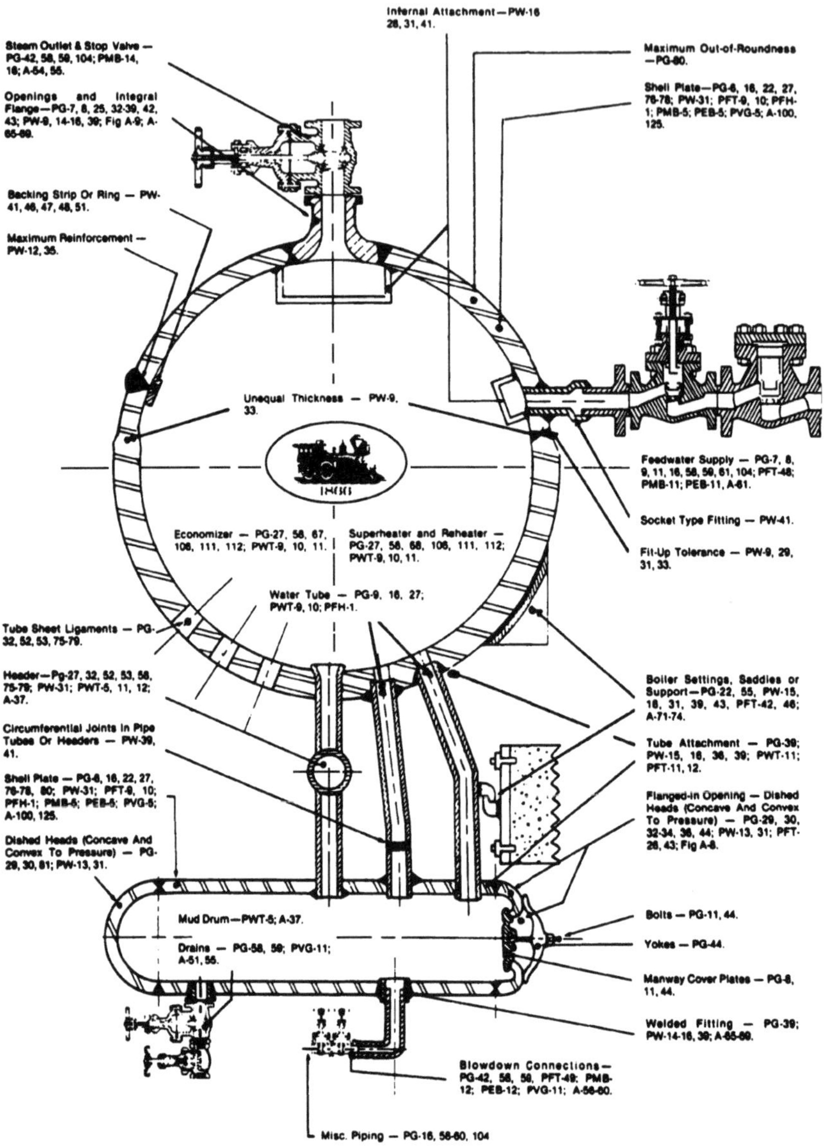

FIGURE 9.1 Quick Reference Guide: *ASME Boiler and Pressure Vessel Code*, Section I, Power Boilers. (*Courtesy of The Hartford Steam Boiler Inspection and Insurance Company.*)

Code and furnishes direct reference to the applicable rule in the Code. In the event of a discrepancy, the rules in the current edition of the Code shall govern. This chart should be used only as a quick reference. The current edition of the Code should always be referenced.

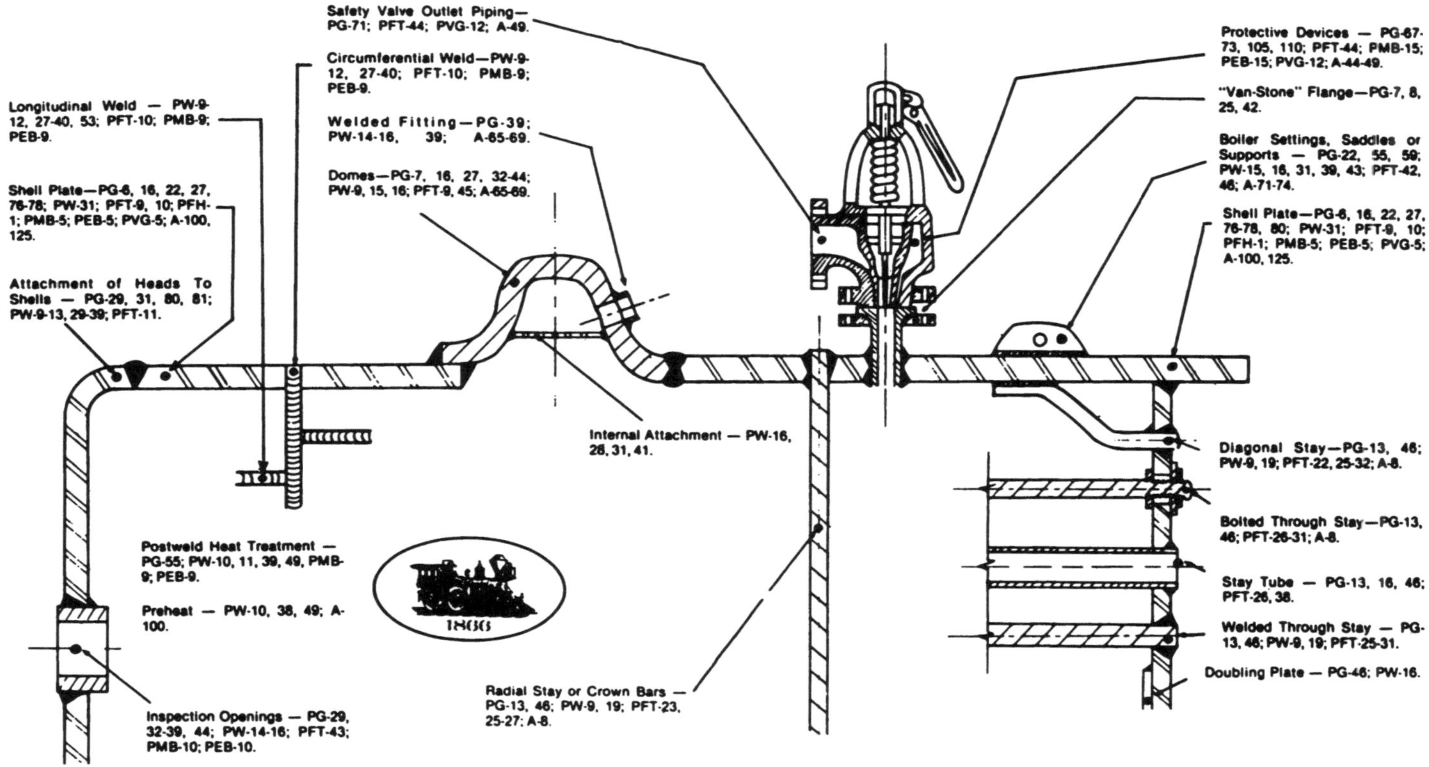

FIGURE 9.1 *(Contd)*

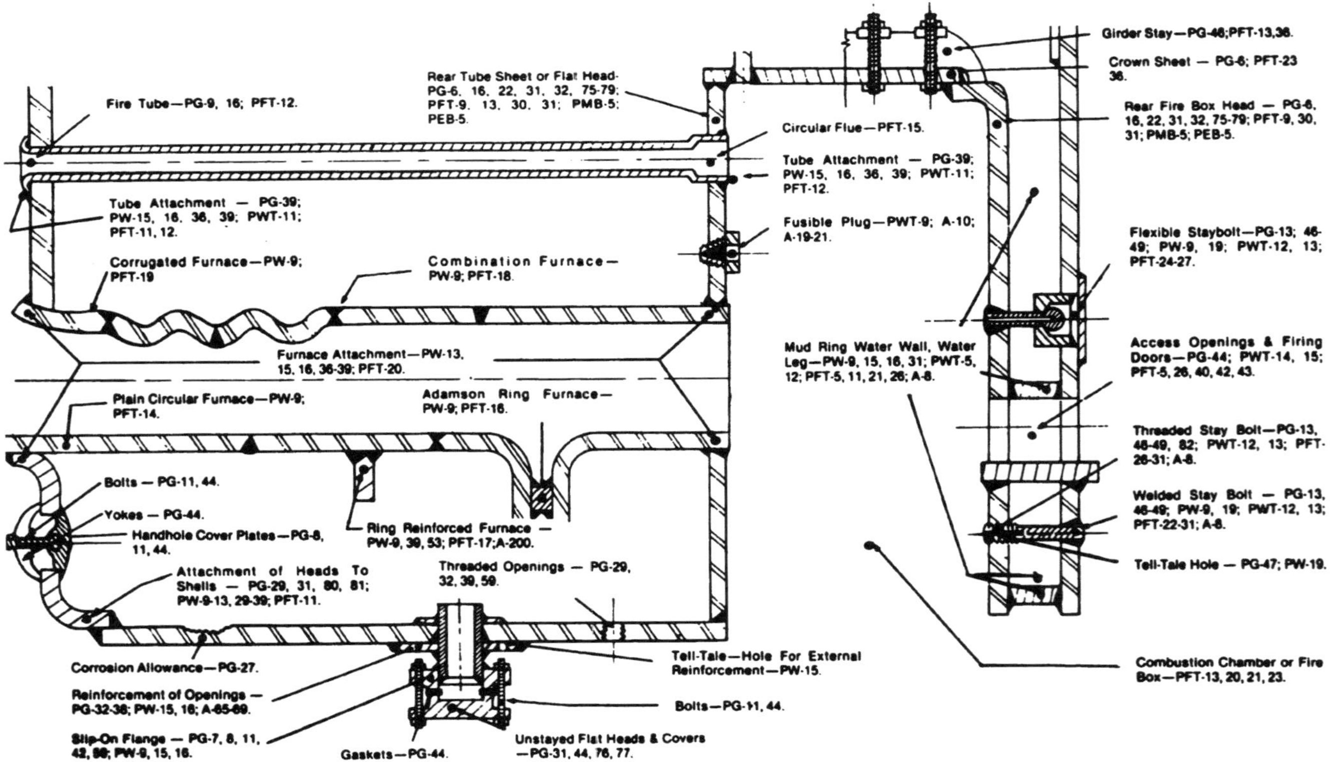

Figure 9.1 *(Contd)*

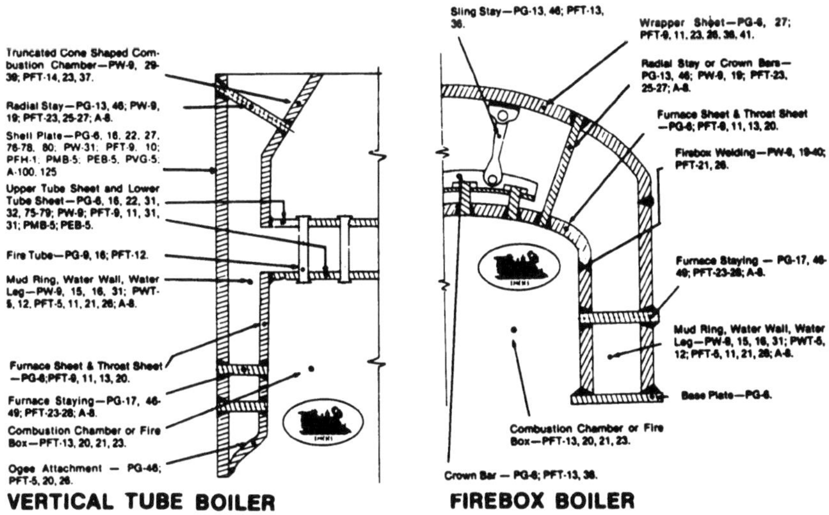

Figure 9.1 *(Contd)*

There are also many conditions in the boiler Code for which all boiler manufacturers, assemblers, and Authorized Inspectors have to take special precautions, especially when large boilers are fabricated partly in the shop and partly in the field by one manufacturer and the piping and components are manufactured by others. This is particularly important if all the parts are being assembled in the field by the boiler manufacturer or an assembler who is certified by the ASME to assemble boilers. When the construction of a boiler is to be completed in the field, either the manufacturer of the boiler or the assembler has the responsibility of designing the boiler unit, fabricating or obtaining all the parts that make up the entire boiler unit including all piping and piping components within the scope of the boiler code, and including these components in the master data report to be completed for the boiler unit (see Code Par. PG-107).

Possible arrangements include the following:

1. The boiler manufacturer assumes full responsibility by supplying the personnel to complete the job.
2. The boiler manufacturer assumes full responsibility by completing the job with personally supervised local labor.
3. The boiler manufacturer assumes full responsibility but engages an outside erection company to assemble the boiler to Code procedures specified by the manufacturer. With this arrangement, the manufacturer must make inspections to be sure that these procedures are being followed.

General Notes

Quality Control System	—A-300
Material General:	—PG-5
a. Plate	—PG-6, 77
b. Forgings	—PG-7
c. Castings	—PG-8, 25
d. Pipe, Tubes & Pressure Containing Parts	—PG-9
e. Misc. Pressure Parts	—PG-11
f. Stays	—PG-13
Loadings	—PG-22; PW-43; A-71 - 74
Stress—Max. Allowable	—Tables PG-23.1, 23.2 and 23.3; A-150
Manufacturer's Responsibility	—PG-90. 104; PW-1; PEB-18
Inspector's Responsibility	—PG-90, 91
Pressure Tests	—PG-99, 100; PW-54; PMB-21; PEB-17; A-22, 63
Safety Devices	—PG-67 - 73; PMB-15; PEB-15; PVG-12, A-44-49
Trim:	
a. Gage Glasses	—PG-60; PFT-47; PMB-13; PEB-13; A-50 - 52
b. Pressure Gages	—PG-60; PEB-14; A-53
c. Gage Cocks	—PG-60; A-51
d. Water Columns	—PG-8, 60; PW-42; A-52
Service Restrictions	—PG-2; PMB-2; PEB-2
Welded Construction	—Part PW
Nameplates, Stamping, & Reports	—PG-104 - 113; PFH-1; PEB-19 ; A-350
NDE Requirements:	
a. MT & PT	—PG-25
b. RT	—PG-25; PW-51
c. UT	—PW-52
Porosity Charts	—A-250
Code Jurisdiction	—PG-58
Minimum Thickness	—PG-16;PFT-9
Feedwater Heater (Optional Requirements)	—PFH
Miniature Boilers	—PMB
Electric Boilers	—PEB
Organic Fluid Vaporizer Generators	—PVG

Special Reference to Particular Boiler Types

Traction, Stationary, or Portable Return Tube Firetube (Typically the Locomotive Type)	PG-60.1.5, 111.3; PFT-5.2. 23.2. 24.4, 29, 41, 42, 43.2, 43.3, 44, 45, 45.3. 45.7, 46.6
Horizontal Tube & Horizontal Return Tube	PG-59.3.7, 60.1.4. 111.1, 111.2; PFT-11.4.7, 42, 43.1, 43.2. 43.3, 44, 45.3, 46.3, 46.4, 46.5, 46.6, 47, 48.1
Vertical Tube	PG-111.4 PFT-5.2, 11.4.5, 20.5, 21.2, 21.3, 23.3, 23.4, 23.5, 42, 43.1, 43.5, 48.2
Fire Box	PFT-21, 42, 43.1
Scotch Marine	PG-111.6 PFT-20.2, 43.1, 43.4
Waste Heat	PG-106.4.1 PFT-43.1, 44
Forced Flow Steam Generator	PG-16.2, 21.1, 21.2, 58.3.3, 58.3.5, 58.3.6, 59.3.3, 60.1.2, 60.6.2, 61.5, 67.4, 106.3, 111.5.2, 112.1

Riveted Boilers
ASME Section I, Part PR, 1971 Edition

FIGURE 9.1 (*Contd*)

4. The boiler may be assembled by an assembler with the required ASME certification, Code symbol stamp, and quality control system. This indicates that the organization is familiar

with the boiler Code and that the welding procedures and operators are Code qualified. Such an assembler, if maintaining an approved quality control system, may assume full responsibility for the completed boiler assembly.

Other arrangements are also possible. If more than one organization is involved, the Authorized Inspector should find out on the first visit which one will be responsible for the completed boiler. If a company other than the shop fabricator or certified assembler is called to the job, the extent of association must be determined. This is especially important with a power boiler when an outside contractor is hired to install the external piping. This piping, often entirely overlooked, lies within the jurisdiction of the Code and must be fabricated in accordance with the applicable rules of the ASME B31.1 Code for pressure piping by a certified manufacturer or contractor. It must be inspected by the Authorized Inspector, who conducts the final tests before the master data report is signed.

Once the responsible company has been identified, the Inspector must directly contact the person responsible for the erection to agree upon a schedule that will not interfere with the erection but will still permit a proper examination of the important fit-ups, chip-outs, and the like.

The first inspection will include a review of the company's quality control manual, welding procedures, and test results to make sure they cover all the positions and materials used on the job. The performance qualification records of all the welders working on the parts of the boiler and piping under the jurisdiction of the Code must also be examined.

It is the manufacturer's, contractor's, or assembler's responsibility to submit the drawings, calculations, and all pertinent design and test data so that the Inspector may make comments on the design or inspection procedures before the field work is started and thereby expedite the inspection schedule.

New material received in the field for field fabrication must be acceptable Code material and must carry proper identification markings for checking against the original manufacturer's certified mill test reports. At this time, the material may be examined for thicknesses, tolerances, and possible defects.

The manufacturer who completes the boiler has the responsibility for the structural integrity of the whole or the part fabricated. Included in the responsibilities of the manufacturer are the documentation of the quality control program to the extent specified in the Code.

CHAPTER 10

Nuclear Vessels and Required Quality Assurance Systems

What is the difference between a fossil-fuel power plant and a nuclear power plant? In both types of plants, heat is used to produce steam in order to drive a turbine that turns a generator that, in turn, creates electricity. The basic difference between fossil-fuel power plants and nuclear power plants is the source of heat.

In a fossil-fuel power plant, coal, oil, or gas is burned in a furnace to heat the water in a boiler in order to create steam. In nuclear plants, no burning or combustion takes place. Nuclear fission is used in these plants; the fission reaction generates heat and this heat is transferred, sometimes indirectly, to the water that provides the steam.

In a nuclear energy plant, the furnace is replaced by a reactor which contains a core of nuclear fuel consisting primarily of uranium. Energy is produced by fission in the reactor. A tiny particle called a neutron strikes the center or nucleus of a uranium atom splitting it into fragments releasing more neutrons. These fragments fly apart at great speed, generating heat as they collide with other atoms. The free neutrons collide with other uranium atoms causing fission; the collision creates more neutrons that continue to split more atoms in a controlled chain reaction.

The fission chain reaction heats the water that moderates the reaction in the reactor creating steam, the force of which turns the turbinegenerator. From this point on, nuclear energy plants operate essentially the same as fossil-fuel steam electric plants.

Power Plant Cycles

The purpose of this chapter is not to discuss nuclear technology but to let the reader know more about the heat cycles of an atomic power plant and have a better understanding of the classifications and rules for the construction and inspection of nuclear plant components. Code Classes 1, 2, 3, and MC allow a choice of rules that provide the different levels of structural integrity and quality and explain why these components are subjected to a rigorous quality assurance inspection, bearing upon the relative importance assigned to the individual components of the nuclear power system to ensure their lasting performance.

Several types of reactor coolant systems are used. The two plant cycles most often found in the utility service at present are the pressurized water reactor and the boiling water reactor.

Pressurized Water Reactor Plants

The *pressurized water reactor* (PWR), which is inside the containment vessel, is subjected to a high coolant water pressure. Here the pressurized water is heated. A pump circulates the heated water through a heat exchanger where the steam for the turbine is generated (see Fig. 10.1).

The part of the pressurized water nuclear power plant that contains the reactor coolant is called the primary circuit. Included in the primary circuit is an important vessel called the pressurizer. The coolant volume varies when the load changes require coolant temperature

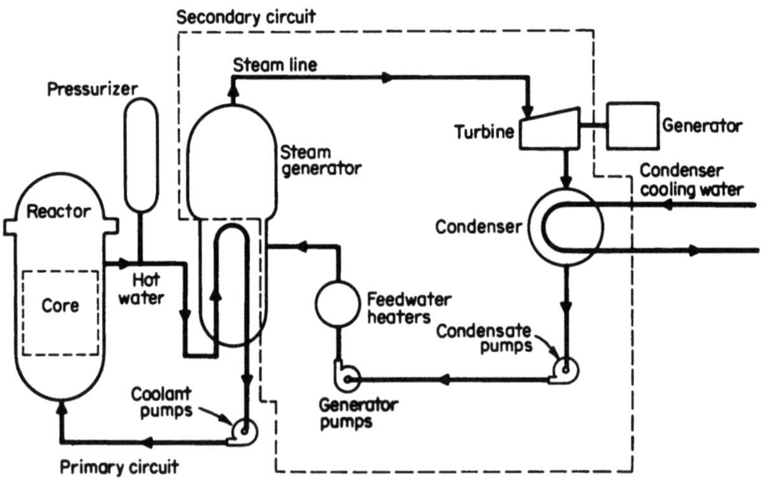

Figure 10.1 PWR

changes. When this occurs, the pressurizer serves as the expansion tank in the primary system, allowing the water to undergo thermal expansion and contraction and holding the primary circuit pressure nearly constant. If pressures are allowed to fluctuate too far, steam bubbles may form at the reactor heating surfaces; these bubbles, or voids, if formed inside the reactor core, greatly alter the reactor power output. The pressurizer has electric heating elements located low inside to provide the vapor needed to cushion the flowing liquid coolant. All of the above-mentioned items are located in the primary circuit.

The rest of the plant, called the secondary circuit, is essentially the same as a fossil-fuel steam electric plant. The heat exchangers make steam that passes through the turbine, condenser, condensate pump, feed pump, feedwater heater, and back to the heat exchanger.

Boiling Water Reactor Plants

In the *boiling water reactor* (BWR), the cooling water is inside the reactor core. The reactor makes steam that passes through the turbine, condenser, condensate pump, feedwater heater, and back to the reactor vessel.

In a BWR plant, the steam generators and the pressurizer are not needed, as the plant has no secondary circuit, only a primary circuit. No primary-to-secondary heat exchange loss takes place (see Fig. 10.2).

The above explanation of the composition and operation of the BWR and the PWR is meant to give the reader a basic understanding of the nuclear power plant systems.

In addition to the two described above, there are other types of plants, such as gas-cooled and liquid-cooled reactors, the number of which in operation is small.

Nuclear energy will only be viable if the plants that make more nuclear fuel than they consume are built. Such plants are called *breeder reactors*. If the breeder reactor is soon perfected and put in use,

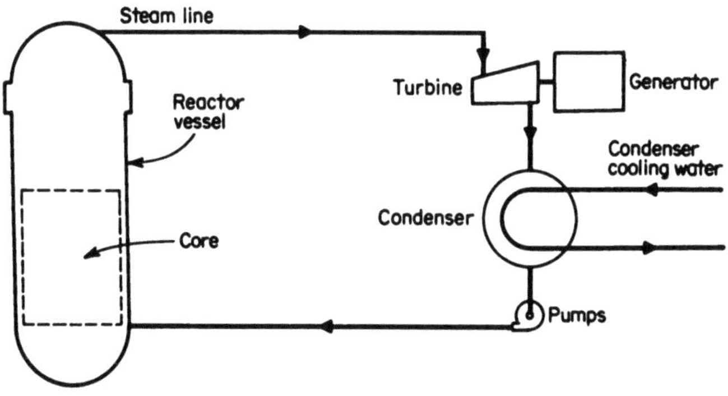

FIGURE 10.2 BWR

the energy bank of the United States uranium can be extended until a system is developed whereby energy is produced by fusion similar to the heat production method of the sun and other stars. If not, our uranium resources may be used up within a century.

Nuclear fusion is the joining together or fusion of atoms to form an atomic nucleus of higher atomic number and weight. *Nuclear fission* is the splitting apart of atoms. Both release enormous amounts of energy. All nuclear plants today operate by nuclear fission. Extensive research is being done to make electricity by nuclear fusion. It represents our best alternative for an abundant, accessible, and long-term energy source. When we know the secret of fusion energy, we will be able to duplicate the heat production methods of the sun.

All nuclear vessels must be manufactured under the appropriate part of the ASME Code Section III, Nuclear Vessels, Division 1, Nuclear Power Plant Components, or Division 2, Concrete Reactor Vessels and Containments. Both divisions are broken down into subsections which are designated by capital letters preceded by the letter N for Division 1 and by the letter C for Division 2.

The ASME Code Section III, Division 1, Nuclear Power Plant Components, is divided into seven subsections covering all the aspects of safety-related design and construction:

Subsection NCA: General Requirements for Division 1 and Division 2

Subsection NB: Class 1 Components

Subsection NC: Class 2 Components

Subsection ND: Class 3 Components

Subsection NE: Class MC (Metal Containment) Components

Subsection NF: Component Supports

Subsection NG: Core Support Structures

Section III: Appendixes The appendices contain both mandatory and non-mandatory appendices referenced by all the Divisions of Section III. The mandatory appendices contain the requirements for construction while the non-mandatory appendices provide additional information or guidance for the use of Section III.

Section III, Division 2: Addresses concrete containment structures, pre-stressed or reinforced. These are applicable only to those components that are designed to provide a pressure retaining or containing barrier.

Section III, Division 3: Addresses the design and construction of the containment systems, including internal support structures used for the transportation and/or storage of spent nuclear fuel and high-level radioactive material.

Section III, Division 5: Provides construction rules for high-temperature reactors, including both high-temperature gas-cooled reactors and liquid-metal reactors.

These Code classes are intended to be applied to the classification of items of a nuclear power system and containment system. Within these systems the Code recognizes the different levels of importance associated with the functions of each item as related to the safe operation of the nuclear power plant.

The Code classes allow a choice of rules that provide an assurance of structural integrity and quality commensurate with the relative importance assigned to the individual items of the nuclear power plant. Each subsection is intended to be an independent document where no crossreferencing of subsections is necessary, with the exception of Subsection NCA, General Requirements, which is a mandatory requirement for those who intend to use one or more of the other subsections of Section III, Division 1 and Division 2. It includes Subsections NB through NG. Subsection NCA comprises all the general requirements for the manufacturers, installers, material manufacturers, suppliers, and owners of nuclear power plant components including the requirements for quality assurance, Inspectors' duties, and stamping.

Section III, Division 2, Concrete Reactor Vessels and Containments, defines the requirements for the design and construction of concrete reactor vessels and concrete containments and Division 2 Appendices. It covers the requirements for materials, parts, and appurtenances of components that are designed to provide a pressure-retaining or -containing barrier.

The following explanation covers only Section III, Division 1, and does not include Division 2. To avoid confusion it should be noted that all Code references in this chapter, unlike those in the rest of this book, refer to the ASME Code Section III, Division 1, Nuclear Vessels, rather than Section VIII, Division 1, Pressure Vessels.

These seven subsections are intended to be applied to the classification of pressure-retaining or -containing components of a nuclear power system and containment system, component supports, and core support structures.

The *ASME Boiler and Pressure Vessel Code* Section III, Division 1, Nuclear Power Plant Components, differs from the other sections of the Code in that it requires a very detailed stress analysis of vessels classified as Class 1, Class 2, Class 3, and Class MC.

Class 1 includes reactor and primary circuit vessels, containing coolant for the core that cannot be separated from the core and its radioactivity.

Class 2 components are those which are not a part of the reactor–coolant boundary but are used to remove the heat from the reactor–coolant pressure boundary.

Class 3 components support Class 2 without being part of them. Class MC includes containment vessels.

All nuclear vessels and components require constant vigilance during fabrication and operation. The manufacturer is required to prepare, or have prepared, design specifications, drawings, and a complete stress analysis of pressure parts. The design report that will include all the materials has to be certified by a registered professional engineer. The copies of the certified stress report are required to be made available to the Authorized Nuclear Inspector at the manufacturing site as well as to the enforcement authority at the point of installation.

It is the duty of the Authorized Nuclear Inspector to verify that the equipment is fabricated according to the design specifications and drawings as submitted. Another duty is to make all inspections as specified by the Code rules.

Quality controls that were adequate in the past are unacceptable today. Pressure vessel fabricators view this period as the age of quality assurance. Today, the industry faces heavy demand for test reports and other supporting data, especially for the construction of nuclear vessels and parts. Fundamentally, there should be no difference in the level of quality between a high-pressure, high-temperature fossil-fuel boiler and a nuclear vessel. The reality is that nuclear fabrication requires significantly more testing and documentation than power boiler fabrication.

The nuclear code requires that the manufacturers of nuclear equipment have a more rigorous quality assurance program. Excellence of each item produced must be the aim of each manufacturer.

The responsibility for the program lies with the manufacturer. A manufacturer who wants to fabricate nuclear vessels or parts covered by Section III obligates the company with respect to quality assurance and documentation. The manufacturer must have an understanding of the many Code details, a written quality assurance manual, and written implementing procedures describing the system of design, material, procurement, fabrication, testing, and documentation. The manufacturer must also have a clear understanding of the nuclear quality assurance standards being used in every aspect of the nuclear power plant construction—from the design and construction phases to full operation.

The nuclear industry's good safety record is due to the rigorous standards of the quality assurance programs applied to all the phases of information, creation, planning, and fabrication as well as to the design, construction, testing, inspection, and corrective action taken when nonconformities are discovered during construction and later when implementing the quality assurance programs in the operation and maintenance of a nuclear power plant.

Quality Assurance

The quality assurance control program must be evaluated and approved for compliance by the ASME before issuance and upon each renewal of the Certificate of Authorization. This is done at the manufacturer's request and expense. The evaluation of the manufacturer's quality assurance capability will be made by a survey team designated by the ASME. The team will be made up of one or more consultants to the ASME as designated by the secretary of the Boiler and Pressure Vessel Committee, which includes the ASME team leader, a team member, a representative of the National Board of Boiler and Pressure Vessel Inspectors, a representative of the jurisdictional authority in which the manufacturing facility is located, the Authorized Nuclear Inspector Supervisor from the manufacturer's Authorized Inspection Agency, the Authorized Nuclear Inspector performing the Code inspection at the plant being surveyed, and a representative of a utility using nuclear energy.

The survey team's purpose will be to compile a report of the manufacturer's quality assurance capabilities at the plant or site. This report will be submitted to the ASME Subcommittee on Nuclear Accreditation and forwarded to the secretary of the Boiler and Pressure Vessel Committee, who will carry out the necessary procedures leading to the issuance of a Certificate of Authorization to the manufacturer for the use of an N Code symbol for the types of fabrication to be manufactured.

The intent of the ASME Code survey team will be to verify that the manufacturer has an acceptable quality assurance program describing the controls and manner in which the nuclear items will be produced. In addition, the team will verify that an inspection contract or agreement with an Authorized Inspection Agency is in effect to review the manufacturer's quality assurance program and its compliance with the intent of Section II, Section III, NQA-1, Section V, and Section IX. As with Section VIII, Division 1, Section III also references Section V for the qualifications of nondestructive examination procedures and Section IX for the welding qualification procedures. The team will examine the means and methods for the documentation of the manufacturer's quality assurance program, review the personnel involved in the manufacturer's quality assurance activities, and examine their status with respect to the production and management functions. The team will also examine the manufacturer's relationship with the Authorized Nuclear Inspector with respect to the organization being audited and the means of keeping the Inspector informed and up-to-date on the quality control matters, review all the procedures as required by the Code with respect to the qualifications of processes and individuals, and examine the equipment and facilities employed for testing and carrying out the quality assurance work.

It is the manufacturer's responsibility to aid the ASME survey team so that it can do its work. It would be advisable for the manufacturer to provide qualified personnel who are able to reply to the necessary inquiries generally made by the individuals of the survey team. The manufacturer should also provide a suitable workroom for the survey team. Facilities must be made accessible to the survey team so that an understanding of the manufacturer's operational capabilities is possible.

A management organization chart as well as the organization chart of the quality assurance department should be available. Also available to the survey team should be the manufacturer's qualification records of welding procedures and welders, nondestructive examination procedures and testing personnel, traveler or process sheet check-off lists, and other exhibits showing the methods used with respect to the quality control of production. All nondestructive examining equipment and measuring and testing facilities must be available for examination. A written procedure of the training and indoctrination program for the personnel performing work that affects quality control with the names of persons involved and the documentation records of examinations should be included.

Most importantly, the quality assurance manual, in which the company explains how its personnel control fabrication to meet the Code requirements, will form the basis for understanding the manufacturer's system. It will be reviewed by the survey team. The manual will serve as the basis for continuous monitoring of the system by the Authorized Nuclear Inspector. In addition, the manual will serve as the document used by the Authorized Nuclear Inspector Supervisors during semiannual ANSI 626.0 audits and later survey reviews when the manufacturer's Certificate of Authorization is to be renewed. All manuals should be classified as either controlled or noncontrolled, as follows:

1. A controlled manual is one that is given a number, assigned to one individual, and kept up-to-date in accordance with the approved system described in the manual by the party responsible for manuals in the plant.
2. A noncontrolled manual is a copy of the manual distributed to the customers and other interested parties, which will only be considered up-to-date at the time of distribution and should be marked as such by rubber stamping on the first page as well as several other pages throughout the book, indicating that it will not be kept up-to-date. All the pages of the manual should have the name of the manufacturer and a revision block including a space for revision levels and the dates of these revisions.

An outline of the procedures to be used by the manufacturer in describing a quality assurance system should include the following items:

1. **Organization**
   a. A statement of authority should be specified in this section. This statement should identify management's participation in the program—that the quality assurance manager shall have the authority and responsibility to establish and maintain a quality assurance program and organizational freedom to identify the quality problems in order to provide solutions to these problems. This statement of authority should be signed by the president of the company or someone else in top management.
   b. The manufacturer's organization chart must be included, giving titles and showing the relationship between management, quality assurance, purchasing, engineering, fabrication, testing, and inspection.
   c. Assurance that quality is achieved and maintained by those assigned the responsibility for completing the task should be given.
   d. A description of how quality achievement is verified should be listed.

2. **Quality assurance program**
   a. The manual will document a quality assurance program which includes the procedures that will assure and provide that the plans and activities affecting the quality are carried out.
   b. The quality assurance manual should give a program for indoctrination and training, which includes the program descriptions, listing of personnel involved, and records of their performance. All personnel involved with quality control must be given training with the excellence of performance in mind.

3. **Design control**
   a. The design shall be controlled, defined, and verified. Verification is required by the manufacturer that the design specifications certified by a registered professional engineer are the owner's or the agent's certified design specifications, a copy of which is to be made available to the Authorized Nuclear Inspector before the construction begins and to the enforcement authority at the location of installation before the components are placed in service.
   b. Responsibility for structural integrity and the Design Reports for Class 1 components and component supports, Class 2

vessels designed under NC-3200, Class MC vessels and component supports, certain Class 2 and 3 components, and Class CS should be certified by the registered professional engineer who makes the stress report and the owner or agent.

c. The load capacity data sheets for Class 1 and Class MC component supports designed under NC-3200 must be certified by a registered professional engineer stating that the load capacity of the component support is rated in accordance with Code Subsection NF, with the data sheet specifying the organization responsible for retaining the data that substantiates stated load capacity.
d. When the class of a vessel or part does not require a design report, the design calculations must be made available to the Authorized Nuclear Inspector.
e. The complete certified stress report by the manufacturer and owner shall be made available to the Authorized Nuclear Inspector.
f. Design reviews and checking shall be made by individuals other than those who initiated the original design.

4. **Procurement document control**

a. The manufacturer should have an order entry system which provides for competent review and remarks by applicable persons to assure the correct translation of design specification requirements into specifications, drawings, procedures, and instructions.
b. The document control of revisions is needed to assure that the documents and changes are reviewed for adequacy, approved, documented, and released by the authorized personnel and distributed and used wherever required.

5. **Instructions, procedures, and drawings**

a. The activities affecting quality shall be described in accordance with the written instructions, procedures, and drawings.
b. The inclusion, generation, revision, and issuance of these documents within the quality assurance system should be explained.

6. **Document control**

a. Documents must be controlled so that only the correct documents are issued. This system should include how the documents are identified, controlled, prepared, reviewed, approved, issued, and corrected prior to issuance.
b. The procedures identifying responsible personnel to assure that the required documents are reviewed and approved before the release should be established.

c. The revision system must provide for engineering review by qualified individuals, including the registered professional engineer responsible for stress report or design calculations.
d. All the revisions and replacements must be controlled and documented.
e. Documents must be made available to the Authorized Nuclear Inspector.

7. **Control of purchased items and services**
    a. The manufacturer is responsible for surveying and qualifying the vendor quality system programs, or the material manufacturer must hold a quality system certificate issued by the ASME.
    b. The system of material control used in purchasing to assure the identification of all material used and to check compliance with specifications should be explained.
    c. The details of the vendor's evaluation should include a sample of the survey questions asked by the manufacturer, the vendor's responses to these questions, and how often the vendor is to be audited.
    d. The qualification and listing of a vendor should be described, as should the revisions to and control of the accepted vendor list. The procedure of removing a vendor from the accepted vendor's list shall be explained.
    e. The control of material at the source—who issues and approves the purchase documents—must be described.
    f. The way in which the material manufacturer's and subcontractor's corrective action is verified and controlled should be given.

8. **Identification and control of items**
    a. The procedure of receiving inspection control—identification of material with receiving checklist—must be explained.
    b. The control of nonconformity of purchased items should be discussed and the corrective action system should be stated.
    c. The manufacturer's certified materials test report, including welding and brazing materials, must be given. How the test report is examined and signed off to assure a correlation with material used for traceability must be stated.

9. **Control of processes**
    a. A description of the system for assuring that all welding procedures and welders or welding operators are properly qualified (Code Par. NCA-5250) must be stated.

b. The welding qualification records and identifying stamps, including the method of recording the continuance of valid performance qualification by production welding, should be presented.
c. The method of welding electrode control, that is, electrode ordering, inspection when received, storage, how electrodes are issued, and the in-process protection of the electrode, should be described.
d. The method used on drawings to tell the type and size of weld and electrode to use, that is, whether the drawings show weld detail in full cross section or use welding symbols, should be given.
e. The fit-up and weld inspection must be explained.
f. The system for assuring that all the nondestructive examination personnel are qualified should be described.
g. Written procedures to assure that the nondestructive examination requirements are maintained should be listed.
h. The required written training program should be explained.
i. The examination methods in the following required areas should be detailed:
   (1) general
   (2) specific
   (3) practical
   (4) physical
j. The personnel qualification record and Level III support documents for examinations, experience, education, and physical examinations should be presented.
k. The equipment qualification and calibration should be described.
l. For subcontracted NDE operations, similar controls as above and the contract terms with the NDE organization should be given.
m. The procedures for preheating, postweld heat treatment, hot forming, etc. should be listed.
n. How the furnace is loaded, the thermocouple placement, how and where thermocouples are placed, the recording system for time and how and where it is placed, the recording system for time and temperature, and how time and temperature are recorded for each thermocouple should be described.
o. The methods of heating and cooling, including temperature control, should be given.
p. The time and temperature recordings should be described, including their review by the Authorized Nuclear Inspector.

10. **Inspection**
    a. A statement which states that the personnel performing functions affecting the quality of the work shall not report directly to the persons responsible for the performance of that work shall be included.
    b. That each person who performs the work shall be properly qualified, indoctrinated, and trained to perform the work should be stated.
    c. This chapter could be included with the chapter titled "Inspections, Tests, and Operating Status."

11. **Test control**
    a. The tests that will require the verification of conformance of an item must be planned and executed and must be defined in the quality assurance program.
    b. The results must be documented as outlined in the manual.

12. **Control of measurement and test equipment**
    a. The established and documented procedures used for assuring that gages and measuring and testing devices are calibrated and controlled should be described.
    b. Calibration should be to national standards where such standards exist; other standards, when used, should be designated.
    c. The procedures to assure that measuring and test equipment are calibrated and adjusted at specified periods must be established.
    d. The control of nonconforming equipment should be discussed. Corrective action taken when discrepancies in the examination or testing equipment are found should be stated.

13. **Handling, storage, and shipping**
    a. The procedures used to control the methods of handling, storage, cleaning, and preservation of materials, packaging, and shipping in order to prevent deterioration, damage, and loss should be described.
    b. The procedures for examination and inspection should be included on the traveler check-off list.

14. **Inspection, test, and operating status**
    a. The identification and control of the material heat number transfer should be described.
    b. The use of a "traveler" or process control checklist indicating the fabricating processes is needed.

c. The preparation and use of a traveler or process control checklist with sign-off spaces for use by the manufacturer should be described.
d. Mandatory hold points (Code Par. NCA-5241) must be listed.
e. The control of nonconformities, material, parts, and components should be explained.
f. The methods used for corrective action and nonconformities should be presented.
g. The manufacturer should be carefully acquainted with the provisions in the Code that establish the duties of the Authorized Nuclear Inspector. The third party in the manufacturer's plant, by virtue of authorization by the state jurisdiction to do Code inspections, is the legal representative. This party's presence permits the manufacturer to fabricate under state laws. Under the ASME Code rules, the Authorized Nuclear Inspector has certain duties as are specified in the Code. It is the obligation of the manufacturer to see that the Inspector is given the opportunity to perform as required by law. The details of the Authorized Nuclear Inspector visits at the plant or site are to be worked out between the Inspector and the manufacturer. The manufacturer's quality assurance manual should state the provisions for assisting the Authorized Nuclear Inspectors performing of their duties.
h. The manual should state that the Authorized Nuclear Inspector, Authorized Nuclear Inspector Supervisor, and the Authorized Inspection Agency have free access to the areas and documents as necessary to carry out the duties specified in the Code, such as monitoring the manufacturer's quality assurance system, reviewing qualification records, verifying materials, inspecting and verifying in-process fabrication, witnessing final tests, and certifying data reports.
i. The Inspector should be given the opportunity to indicate "hold points" and to discuss these inspection points with the quality assurance manager as well as make arrangements for their desired inspections.
j. The Authorized Nuclear Inspector must be provided with the current quality assurance manual with the latest revisions which have been accepted by the Authorized Nuclear Inspector Supervisor. The final hydrostatic or pneumatic tests required shall be witnessed by the Authorized Nuclear Inspector, and any examinations that are necessary before and after these tests shall be made by the Inspector.
k. The manufacturer's data reports shall be carefully reviewed by a responsible representative of the manufacturer

or assembler. If the reports are properly completed, the representative shall sign them for the company. The Authorized Nuclear Inspector, after examining all the documents and feeling assured that the requirements of the Code have been met and the manufacturer's data sheets properly completed, may sign the data sheets with his or her name and the National Board commission number.

15. **Control of nonconforming items**
    a. The definition of a nonconformity should be given.
    b. Nonconforming materials, components, and parts that do not meet the required standards should be corrected, or they may not be used.
    c. The identification and inventory of parts should be explained.
    d. The methods of corrective action, return of material for use, and record keeping of traceability as well as material segregation to prevent inadvertent use should be included.
    e. The names of the appropriate levels of management involved with responsibilities and authority for the control of nonconformities should be given.
    f. When the disposition of nonconformity involves design, design report, or dimensional changes, review should be provided by a responsible engineering person or group. (See Correction of Nonconformities, Chap. 4, item 6.)

16. **Corrective action.** The way in which the nonconforming items will not continue to be a problem should be explained. An identification, cause, and corrective action plan shall be documented so that the conditions adverse to quality do not reoccur.

17. **Quality assurance records**
    a. A written procedure should be established for the handling of the records—permanent and nonpermanent—for the time specified in the Code, showing how the records will be kept. The procedure should be determined early on and a list developed to serve as index to the document file.
    b. Permanent records: The owner has the responsibility for the continued maintenance of all the records and the traceability of the records for the life of the plant. These records shall be stored at the power plant site, the manufacturer's plant, or other locations determined by mutual agreement.
    c. Nonpermanent records: State the responsibility for who will maintain, store and distribute; where they will be kept; and how long they will be kept.

d. The quality assurance system must verify the records of materials, manufacturing, examination, and tests before and during construction. The responsibility for and the location of all the records should be stated.
e. The storage and protection of records from damage should be explained.

18. **Audits**
    a. The system of planned, periodic audits to determine the effectiveness of the quality assurance program should be explained.
    b. Audits should be performed in accordance with a written procedure using a checklist developed for each audit in order to assure that a consistent approach is used by the audit personnel.
    c. A report of audit results should be given to the top management, including the corrective action to be taken if necessary.
    d. The corrective action taken by the quality assurance manager to resolve the deficiencies, if any, revealed by the audits should be monitored.
    e. The results of the audit must be made available to the Authorized Nuclear Inspector.
    f. The name and title of person responsible for the audit system should be listed.

When constructing nuclear pressure vessels according to the ASME Code, complete quality assurance is a must. In this chapter, an attempt has been made to acquaint the reader with the basic elements of a total quality assurance manual and system—from design specifications and control to auditing of the system, which includes document control; purchase control; vendor surveys and control; receiving inspection; identification and control of materials; control of manufacturing and/or installations; control of examination and tests; control of measuring and test equipment; handling, storage, shipping, and preservation; examination of process status; and nonconformity control of materials and items and corrective action to take, as well as quality assurance record retention and authorized inspection.

If a manufacturer uses this chapter as a guide, it will help in preparing for an ASME nuclear survey, or if one is contemplated, it will help in understanding the rigorous standards an organization and manufacturing system must meet to enable the nuclear power plant components to be fabricated. The pursuit of excellence in the manufacture of nuclear vessel components must be the aim of each manufacturer seeking ASME certification to fabricate nuclear vessel components.

CHAPTER 11

Section XII Rules for Construction and Continued Service of Transport Tanks

Overview of Section XII

The rules of Section XII constitute the requirements for the construction and continued service of pressure vessels for the transportation of dangerous goods via highway, rail, air, or water.

By way of definition:

- Construction comprises materials, design, fabrication, examination, inspection, testing, certification, and overpressure protection.
- Continued service refers to the inspection, testing, repair, alteration, and recertification of a transport tank that has been in service.
- Pressure vessel refers to the pressure boundary defined by the geometric scope of Section XII and includes, but is not limited to, the shell, heads, and openings.
- Tank refers to the pressure vessel, appurtenances, and additional components that are covered by the modal appendices.

The first edition of Section XII was published in 2004. Prior to that, time tank construction was primarily under DOT regulation, with a significant influence from Section VIII, Division 1. Section XII is an amalgam of the DOT regulations and the Section VIII, Division 1 rules. Section XII contains ten Parts applicable to all tanks, four Modal Appendices which address tanks in various distinct services, and Mandatory and Non-Mandatory Appendices. The following is a brief review of the Parts and Appendices. In those areas wherein the subject matter is similar to Section VIII, Division 1, the discussion will primarily cover the Section XII rules that are not a part of Section VIII, Division 1.

The ten Parts that comprise the main body of the Code are arranged in a very logical sequence and are generally simple to navigate, particularly for someone who has a familiarity with Section VIII, Division 1 pressure vessel code. Therefore, the following very brief introduction to these ten Parts serves only to assist the designer in locating the Code information, as necessary.

Part TG: General Requirements

Article TG-1 introduces the scope and jurisdiction of Section XII. This includes the definition of the set of pressure vessels to which this section is applicable, the geometric and physical scope of vessels, and the documents and product standards referenced within Section XII.

Article TG-3 defines the duties and responsibilities of the various entities involved in the manufacture and continued service of transport vessels. The article briefly addresses the owner and user responsibilities. It also provides more detailed information on the manufacturer's responsibilities and the Inspector's duties.

Article TG-4 discusses the general rules for inspection, including the definition of the entity responsible for inspection duties.

Part TM: Material Requirements

Article TM-1 presents information on the materials which may be used in the construction of transport vessels. The article includes the tables of approved ASME materials, along with special requirements. Article TM-2 provides a detailed set of rules for the notch toughness requirements for ferrous materials.

Part TD: Design Requirements

Article TD-2 defines the loadings that are to be considered in the design of transport tanks. Also included in this article are the allowables for the categories and combination of stresses encountered in the design and analysis.

Articles TD-3 and TD-4 present the rules for minimum thickness of heads and shells under internal and external pressure, respectively.

Openings and their reinforcement are covered in Article TD-6. This article addresses the size and shape of openings; the strength and design of openings under internal and external pressure; and the size, shape, and strength of reinforcement.

Part TW: Requirements for Tanks Fabricated by Welding

The first part of Article TW-1 includes service restrictions applied to transport tanks containing fluids with United Nations Hazard Classifications. The article further goes on to discuss the design of welded joints, welded joint classifications, joint efficiencies at welded joints, and welded joint attachment details.

Part TF: Fabrication Requirements

Article TF-1 addresses subjects related to general fabrication procedures such as material identification, repair of defects, forming and tolerances, and inspection during fabrication.

Article TF-2 covers the requirements for welding fabrication. Included here are topics such as welding qualifications, records, and stamps; qualification of welding procedures; tests of welders and welding operators; and requirements for production welding.

Articles TF-3 thru TF-6 are the specialized requirements pertaining to particular classes of materials; namely, carbon and low-alloy steels, high-alloy steel, nonferrous materials, and ferritic steels that are heat treated to enhance tensile properties.

The procedures and requirements for postweld heat treatment are contained in Article TF-7.

And finally, Article TF-8 pertains to vessels that have been internally lined for erosion or corrosion control.

Part TE: Examination Requirements

The subject of Article TE-1 is requirements for examination procedures and personnel qualification. Attention is given to examination by radiography, by ultrasonics, and by magnetic particle and liquid penetrant crack detection.

Article TE-2 goes into detail regarding the examination of welds and their acceptance criteria.

Part TT: Testing Requirements

The requirements for pressure testing are in Article TT-2. This article addresses both hydrostatic and pneumatic testing. For those cases wherein accurate pressures cannot be computed for testing purposes, Article TT-3 covers proof testing requirements.

And Article TT-4 discusses the requirements for elastomeric lining testing.

Part TR: Pressure Relief Devices

All transport tanks covered by Section XII must be equipped with protection against overpressure. Part TR provides the requirements and rules for overpressure protection devices. Article TR-2 outlines the requirements for pressure relief valves. Included are the design, mechanical and material requirements along with general requirements for manufacturers, and production testing by manufacturers.

Article TR-3 presents the requirements for nonreclosing pressure relief devices. The article addresses both rupture disk devices and pin devices. As above, included are the design, mechanical and material requirements along with general requirements for manufacturers, and production testing by manufacturers.

The requirements for the certification of the capacity of pressure relief devices is covered in Article TR-4. And the marking and certification of pressure relief devices is in Article TR-5.

Part TS: Stamping, Marking, Certification, Reports, and Records

Article TS-1 contains the information related to nameplates and nameplate content. Transport tanks built according to Section XII rules are labeled with the ASME Certification Mark and the T Designator. A sample nameplate is given along with the various additional information required to be applied to the nameplate.

Article TS-2 provides the information needed for a manufacturer to apply for the ASME Certificate of Authorization and the ongoing requirements to use the Certification Mark. The discussion includes the Inspection Agreement along with subsequent renewals of the Certificate of Authorization.

Article TS-3 addresses report forms and record maintenance. Particularly addressed are the Manufacturer's Data Reports and Partial Data Reports.

Part TP: Requirements for Repair, Alteration, Testing, and Inspection for Continued Service

Part TP is issued to cover the requirements for the continued use of transport tanks once they are put into service. Articles TP-2 and TP-3 are for general requirements for repairs and alterations and for tests and inspections, respectively.

Article TP-4 discusses the specific tests, including visual, pressure, and leak tests. Criteria for tests and inspections are given in Article TP-5.

Modal Appendices

The above ten Parts contain Code rules and regulations applicable to all transport tanks. The four Modal Appendices provide the rules and regulations for specific types of tanks.

Modal Appendix 1: Cargo Tanks

Modal Appendix 1 applies to cargo tanks carrying hazardous materials with MAWPs not exceeding 34.5 bar (500 psi).

A cargo tank is defined in Article 1 as "bulk packaging, namely a tank intended primarily for the carriage of liquids or gases. It includes appurtenances, reinforcements, fittings, and closures to a tank permanently attached to or forming a part of a motor vehicle. Because of its size, construction, or attachment to a motor vehicle, it is loaded or unloaded without being removed from the motor vehicle and is not fabricated under another specification. A cargo tank is a highway transport tank."

There are five categories of tanks under this classification, which are circular in cross section except as noted:

Category 406 tanks

- circular or noncircular cross section
- MAWP ≤ 0.276 bar (4 psi)
- minimum design pressure = 0.183 bar (2.65 psi)
- no external pressure loading
- Category 406 tanks are Class 3 tanks
- Article 2 of Modal Appendix 1 covers the specific requirements for Category 406 tanks

Category 407 tanks

- MAWP ≤ 6.90 bar (100 psi)
- minimum MAWP = 1.72 bar (25 psi)
- if subject to vacuum loading, shall be designed for at least 1.03 bar (15 psi) external pressure
- tanks with MAWP > 2.41 bar (35 psi) are Class 1 tanks
- tanks with MAWP ≤ 2.41 bar (35 psi) are Class 3 tanks
- Article 2 of Modal Appendix 1 covers the specific requirements for Category 407 tanks

Category 412 tanks

- MAWPs ≤ 6.90 bar (100 psi)
- minimum MAWP = 0.345 bar (5 psi)
- circular cross section except that for MAWP ≤ 51.03 bar (15 psi), they may be of noncircular cross section
- tanks with MAWP >1.03 bar (15 psi) are Class 1 tanks
- tanks with MAWP ≤ 51.03 bar (15 psi) are Class 3 tanks
- Article 2 of Modal Appendix 1 covers the specific requirements for Category 412 tanks

Category 331 tanks

- Category 331 is for tanks primarily for the transportation of compressed gases
- MAWP ≤ 34.5 bar (500 psi)
- minimum MAWP = 6.90 bar (100 psi)
- All Category 331 tanks are Class 1 tanks
- Article 3 of Modal Appendix 1 covers the specific requirements for Category 331 tanks

Category 338 tanks

- Category 338 is for vacuum insulated tanks
- MAWP ≤ 34.5 bar (500 psi)
- minimum design MAWP = 1.75 bar (25.3 psi)
- Category 338 tanks are enclosed in an outer structural vacuum jacket
- All Category 338 inner tanks are Class 1 tanks
- Article 4 of Modal Appendix 1 covers the specific requirements for Category 338 tanks
- Sections 1-1.3 through 1-1.11 of Article 1 shall not be used for Category 338 tanks

Section 1-1.4, Design Requirements Common to More Than One Category includes several options for the allowable stress levels to be used in the design of the DOT 400 tanks. Those options include: the allowable stresses from Section II, Part D for materials listed in Part TM; derivations of allowable stress using the specified yield stress and tensile stress from Section II, Part D, Subpart 1, Table 1A; and derivations of allowable stress using actual material properties as established by the *Material Test Report* (MTR). The section also includes examples to aid in the calculation of allowable stress levels.

Figure 1-1.4 shows a number of static and dynamic loading conditions that must be considered in the tank design. These include a static loading case, four dynamic loading cases, and one extreme load condition. They are described in Pars. 1-1.4(c)(1) thru (6).

A few additional requirements specific to Category 406, 407, and 412 cargo tanks are provided in Article 2. Article 3 has an extensive listing of additional requirements related to Category 331 cargo tanks. A partial list of these is as follows:

- The tank must have a reflective outer surface. It can be clad with a bright nontarnishing material or painted white, aluminum, or similar reflecting color on the upper two-thirds of the vessel.

- Impact tests in accordance with Article TM-2 are required.
- Maximum allowable stress values are further restricted as defined in Par. 1-3.5.1(a)(1).
- Shell design stresses shall be determined for the normal operating and extreme dynamic loading conditions [see Par. 1-3.5.1(c)].
- Fatigue analysis—The design of tank vessels constructed by the rules of Article 3 shall take into account the effect of cyclic loads [see Par. 1-3.5.1(i)].
- Loads transferred into the tank vessel by damage protection devices and their resulting stresses shall be considered according to the maximum allowable stress design criteria [see Par. 1-3.5.5(a)].
- After postweld heat treatment and hydrostatic tests, each tank vessel shall be subjected to a wet fluorescent magnetic particle inspection of all welds in or on the tank shell and heads, inside and out [see Par. 1-3.6.2(b)].
- Paragraph 1-3.11.1 is a presentation of the construction requirements for cargo tank vessels used to transport specific hazardous materials.

Article 4 contains specific requirements for cargo tanks in Category 338. These are vacuum insulated cargo tanks for transporting refrigerated fluids. A partial list of specific requirements for these tanks is as follows:

- Table 1-4.1 is a listing of refrigerated nontoxic gases with respect to Category 338 tanks.
- A Category 338 cargo tank consists of one or more suitably supported welded pressure vessels enclosed within an outer vacuum jacket with insulation between the inner vessel and the outer shell or jacket.
- The jacket covering the insulation shall be of steel if the refrigerated fluid is to be transported by water or if the fluid is oxygen or a flammable.
- The pressure vessel, its attachments, and supports shall be designed for the static forces obtained by multiplying the load factors applicable for the transportation modes given in Table 1-4.4-1 with the maximum pressure vessel weight plus the design pressure [see Par. 1-4.4(c)(1)].
- The outer jacket, its attachments, and supports shall be designed for the static forces obtained by multiplying the load factors applicable for the transportation modes given in Table 1-4.4-1 with the maximum weight imposed on the outer

jacket in addition to 2 bar (29 psi) external pressure [see Par. 1-4.4(c)(2)].

- The design shall include a fatigue analysis to account for cyclic loads on the inner vessel and its attachments during normal operation. The mechanical loads shall include the load factors specified in Table 1-4.4-2 as well as the design pressure. The total life shall consider 109 cycles for the loads in Table 1-4.4-2 and 3 × 105 pressure cycles. Fatigue life analysis shall be in accordance with ASME Code Section VIII, Division 2, Appendix 5 [see Par. 1-4.4(d)].
- Maximum allowable stresses are as defined in Pars. 1-4.4(e) (1) thru (3).
- For shells with vacuum insulation, the test pressure shall not be less than 1.3 times the sum of the MAWP and 1 bar (14.5 psi) [see Par. 1-4.4(f)(2)].
- Every Category 338 tank shall be provided with two independent pressure relief systems that are not connected in series, a primary system of one or more spring-loaded pressure relief valves, and a secondary system of one or more frangible disks or pressure relief valves [see Par. 1-4.6(a)].

Modal Appendix 2: Rail Tank Cars

(As of the 2017 edition of Section XII, this appendix is in the course of preparation.)

Modal Appendix 3: Portable Tanks

Article 1 provides the minimum requirements for the design, construction, and continued service for portable tanks transporting cryogenic fluids. For the purpose of this modal appendix, cryogenic fluids are listed in Table 3-1.1. It may be noted that much of the presentation in this article and in Article 4 of Modal Appendix 1 (Category 338 tanks) is very similar in that they both address tanks for the transport of cryogenic fluids. In this article, the tanks are constructed as separate, stand-alone structures, and not as a permanent attachment to a truck or trailer chassis. Further, the tanks under consideration in this article may be designed for shipment by road, by water, or by rail.

A partial list of specific requirements for Article 1 tanks is as follows:

- A cryogenic portable tank is a multimodal tank having a capacity of more than 450 L (120 gal) used for the transport of the cryogenic fluids listed in Table 3-1.1. Each portable tank shall consist of one or more suitably supported welded pressure vessels enclosed within an outer jacket with insulation between the inner vessel and the outer shell or jacket.

- The jacket covering the insulation is mandated to be constructed of steel.
- The pressure vessel, its attachments, and supports shall be designed for the static forces obtained by multiplying the load factors applicable for the transportation modes given in Table 3-1.5-1 with the maximum pressure vessel weight plus the design pressure. Note that the table lists three specified transportation modes—road and water, rail without cushioning devices, and rail with cushioning devices [see Par. 3-1.5(a)(1)].
- The design shall take into account the effect of pressure and thermal and mechanical cyclic loads on the inner vessel and its attachments. The mechanical loads shall include the load factors specified in Table 3-1.5-2 as well as the design pressure. Fatigue life analysis shall be based on 109 cycles, in accordance with ASME Code Section VIII, Division 2, Appendix 5 [see Par. 3-1.5(b)].
- Maximum allowable stresses are as defined in Par. 3-1.5(c).
- Every tank shall be provided with not less than two independent spring-loaded reclosing pressure relief devices (see Par. 3-1.7).

Modal Appendix 4: Category 106A500-X, 106A800-X, 110A500-W, 110A600-W, 110A800-W, 110A1000-W, and 110A2000-W Ton Containers

As per Par. 4-2, a ton container is a cylindrical steel tank having a water capacity of more than 680 L (181 gal) but not more than 1180 L (313 gal) and intended to transport compressed gases or liquids. It shall not possess stabilizing members external to the shell and shall be capable of being rolled about its longitudinal axis with all openings protected within the heads. It shall be capable of being lifted filled with product. It shall be designed to be loaded onto a transport vehicle or container and so secured that it can be shipped by road, rail, or sea.

The internal pressure of Modal Appendix 4 tanks is in the range from full vacuum to 138 bar (2000 psi). Table 4-6.1-1 lists the minimum thicknesses, the test pressures, the start-to-discharge or burst pressures, and the minimum vapor-tightness pressures of relief devices as related to the various categories.

APPENDIX A

Cylinder Volume Tables and Diagrams

The most economical method of constructing cylindrical pressure vessels is to keep the length-to-diameter ratio between 2 and 5 while using a diameter that will provide the lowest number of courses. Occasionally a customer will order a vessel of a given capacity but allow the fabricator some leeway in its diameter and length. The fabricator may then select the size that will be easiest and least expensive to construct. If a 1000-gal tank has been ordered, for example, a course 8 ft long with two hemispherical heads 48 in. in diameter might be used. The cylinder would then have a capacity of 752 gal and the two heads a capacity of 125 gal each, for a total of 1002 gal. By fabricating the vessel in one course rather than two, moreover, the manufacturer will save time in laying out the shell and in handling as well as the cost of joint preparation and welding on the girth seam. It is important to choose a length and diameter that will give the required capacity with as few seams as possible. The volume tables for cylinders and heads in this appendix, Tables A.1 and A.2, are based on the diameters and lengths most used in the pressure vessel industry. To use the cylinder tables, locate the inside diameter of the vessel in the first column on the left-hand side and move across horizontally to the column headed by the length of the cylinder. Read the capacity in gallons at this point. If the vessel has dished heads, the length of the cylinder is measured between the tangent lines of each head. The accompanying volume diagram allows for larger diameters and longer lengths and also for rapid estimation of tank lengths and diameters for a known capacity. To use the diagram, simply rotate a straight edge around the required capacity in the middle column and read the desired length in the right-hand column and the corresponding diameter for this length and capacity in the left-hand column. To illustrate: A tank is wanted that will hold 8000 gal. If the shell is made 20 ft long from tangent line to tangent line of two torispherical heads and 96 in. in inside diameter, the capacity will be 8016 gal (7520 gal for the cylinder and 298 gal for each head). The volume diagram, Fig. A.1, can also be used to determine the length of a tank of which the capacity and the diameter are both fixed. In fact, as long as any two factors are known, the third can easily be found. For this reason, many prefer to use the diagram rather than the tables.

Inside diameter, in.	Cubic feet per foot of cylinder	Gallons per inch of cylinder	Gallons per foot of cylinder	Length of shell, ft										
				3	4	5	6	7	8	9	10	11	12	13
				Volume, gal										
12	0.79	0.49	5.88	17.64	23.52	29.40	35.28	41.16	47.04	52.92	58.80	64.68	70.56	76.44
18	1.77	1.10	13.22	39.66	52.88	66.10	79.32	92.54	105.7	118.9	132.2	145.4	158.6	171.8
24	3.14	1.96	23.58	70.74	94.32	117.9	141.4	165.0	188.6	212.2	235.8	259.3	282.9	306.5
30	4.9	3.06	38.72	116.1	154.8	193.6	232.3	271.0	309.7	348.4	387.2	425.9	464.6	503.3
36	7.1	4.41	52.88	158.6	211.5	264.4	317.2	370.1	423.0	475.9	528.8	581.6	634.5	687.4
42	9.6	6.00	71.97	215.9	287.8	359.8	431.8	503.7	575.7	647.3	719.7	791.6	863.6	935.6
48	12.6	7.83	94.00	282.0	376.0	470.0	564.0	658.0	752.0	846.0	940.0	1,034	1,128	1,222
54	15.9	9.91	118.9	356.9	475.8	594.8	713.8	832.7	951.7	1,070	1,189	1,308	1,427	1,546
60	19.6	12.24	146.8	440.6	587.5	734.4	881.2	1,028	1,175	1,321	1,468	1,615	1,762	1,909
66	23.8	14.81	177.7	533.1	710.8	888.6	1,066	1,244	1,421	1,599	1,777	1,954	2,132	2,310
72	28.3	17.63	211.5	634.5	846.0	1,057	1,269	1,480	1,692	1,903	2,115	2,326	2,538	2,749
78	33.2	20.69	248.2	744.6	992.8	1,241	1,489	1,737	1,985	2,233	2,482	2,730	2,978	3,226
84	38.5	23.99	287.8	863.6	1,151	1,439	1,727	2,015	2,303	2,591	2,878	3,166	3,454	3,742
90	44.2	27.54	330.4	991.4	1,321	1,652	1,982	2,313	2,643	2,974	3,304	3,635	3,965	4,296
96	50.3	31.33	376.0	1,128	1,504	1,880	2,256	2,632	3,008	3,384	3,760	4,136	4,512	4,888
102	56.8	35.37	424.4	1,273	1,697	2,122	2,546	2,971	3,395	3,820	4,244	4,669	5,093	5,518
108	63.6	39.66	475.8	1,427	1,903	2,379	2,855	3,331	3,807	4,283	4,758	5,234	5,710	6,186
114	70.9	44.19	530.2	1,590	2,120	2,651	3,181	3,711	4,241	4,772	5,302	5,832	6,362	6,893
120	78.5	48.96	587.5	1,762	2,350	2,937	3,525	4,112	4,700	5,287	5,875	6,462	7,050	7,637
126	86.3	53.98	647.7	1,943	2,590	3,238	3,886	4,534	5,181	5,829	6,477	7,125	7,772	8,420
132	95.	59.24	710.9	2,132	2,843	3,554	4,265	4,976	5,687	6,398	7,109	7,819	8,530	9,241
138	103.6	64.75	777.0	2,331	3,108	3,885	4,662	5,439	6,216	6,993	7,770	8,547	9,324	10,101
144	113.1	70.50	846.0	2,538	3,384	4,230	5,076	5,922	6,768	7,614	8,460	9,306	10,152	10,998

TABLE A.1 Cylinder Volume

Volume = $D^2 \times L \times 0.0034$, in which D =internal diameter, in inches, and L = length, in inches

Inside diameter, in.	Cubic feet per foot of cylinder	Gallons per inch of cylinder	Gallons per foot of cylinder	Length of shell, ft										
				14	15	16	17	18	19	20	25	30	35	40
				Volume, gal										
12	0.79	0.49	5.88	82.32	88.20	94.08	99.96	105.8	111.7	117.6	147.0	176.4	205.8	235.2
18	1.77	1.10	13.22	185.0	198.3	211.5	224.7	237.9	251.1	264.4	330.5	396.6	462.7	528.8
24	3.14	1.96	23.58	330.1	353.7	377.2	400.8	424.4	448.0	471.6	589.5	707.4	825.3	943.2
30	4.9	3.06	38.72	542.0	580.8	619.5	658.2	696.9	735.6	774.4	968.0	1,161	1,355	1,548
36	7.1	4.41	52.88	740.3	793.2	846.0	898.9	951.8	1,004	1,057	1,322	1,586	1,850	2,115
42	9.6	6.00	71.97	1,007	1,079	1,151	1,223	1,295	1,367	1,439	1,799	2,159	2,518	2,878
48	12.6	7.83	94.00	1,316	1,410	1,504	1,598	1,692	1,786	1,880	2,350	2,820	3,290	3,760
54	15.9	9.91	118.9	1,665	1,784	1,903	2,022	2,141	2,260	2,379	2,974	3,569	4,163	4,758
60	19.6	12.24	146.8	2,056	2,203	2,350	2,496	2,643	2,790	2,937	3,672	4,406	5,140	5,875
66	23.8	14.81	177.7	2,488	2,665	2,843	3,021	3,198	3,376	3,554	4,443	5,331	6,220	7,108
72	28.3	17.63	211.5	2,961	3,172	3,384	3,595	3,807	4,018	4,230	5,287	6,345	7,402	8,460
78	33.2	20.69	248.2	3,475	3,723	3,971	4,219	4,467	4,716	4,964	6,205	7,446	8,687	9,928
84	38.5	23.99	287.8	4,030	4,318	4,606	4,894	5,182	5,469	5,757	7,197	8,636	10,076	11,515
90	44.2	27.54	330.4	4,626	4,957	5,287	5,618	5,948	6,279	6,609	8,262	9,914	11,566	13,219
96	50.3	31.33	376.0	5,264	5,640	6,016	6,392	6,768	7,144	7,520	9,400	11,280	13,160	15,040
102	56.8	35.37	424.4	5,942	6,367	6,791	7,216	7,640	8,489	8,489	10,612	12,734	14,856	16,979
108	63.6	39.66	475.8	6,662	7,138	7,614	8,090	8,566	9,041	9,517	11,897	14,276	16,656	19,035
114	70.9	44.19	530.2	7,432	7,953	8,483	9,014	9,544	10,074	10,604	13,256	15,907	18,558	21,209
120	78.5	48.96	587.5	8,225	8,812	9,400	9,987	10,575	11,162	11,750	14,688	17,625	20,563	23,500
126	86.3	53.98	647.7	9,068	9,716	10,363	11,011	11,659	12,307	12,954	16,193	19,432	22,670	25,909
132	95.	59.24	710.9	9,952	10,663	11,374	12,085	12,796	13,507	14,218	17,772	21,327	24,881	28,436
138	103.6	64.75	777.0	10,878	11,655	12,432	13,209	13,986	14,763	15,540	19,425	23,310	27,195	31,080
144	113.1	70.50	846.0	11,844	12,690	13,536	14,382	15,228	16,074	16,920	21,150	25,380	29,611	33,841

TABLE A.1 Cylinder Volume *(Continued)*

Volume = $D^2 \times L \times 0.0034$, in which D =internal diameter, in inches, and L = length, in inches

Torispherical heads (flanged and dished)

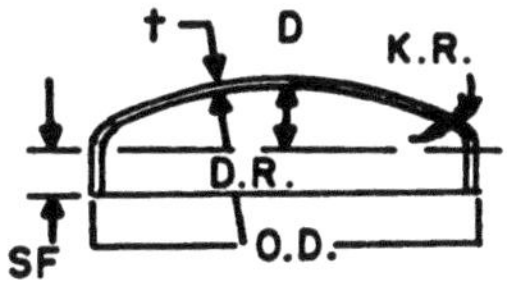

$$\text{Volume} = \left(\frac{\text{I.D.}}{12}\right)^3 \times 0.582$$

Inside Diameter, in	*Radius of dish, in*	*Volume, gal*
12	12	0.58
18	18	1.96
24	24	4.65
30	30	9.2
36	36	15.7
42	42	25
48	48	37
54	54	53
60	60	73
66	66	97
72	72	126
78	78	160
84	84	200
90	84	246
96	90	298
102	96	357
108	102	424
114	108	499
120	114	582
126	120	674
132	130	775
138	132	885
144	132	1,006

Elliptical heads

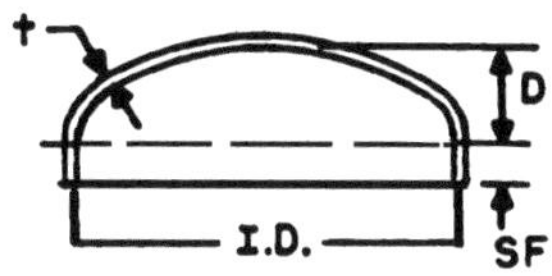

$$\text{Volume} = \left(\frac{\text{I.D.}}{12}\right)^3 \times 0.954$$

Inside Diameter, in	*Volume, gal*
12	0.95
18	3.2
24	7.6
30	15
36	26
42	41
48	61
54	87
60	119
66	159
72	206
78	262
84	327
90	403
96	489
102	586
108	696
114	818
120	954
126	1,110
132	1,270
138	1,452
144	1,649

Hemispherical heads

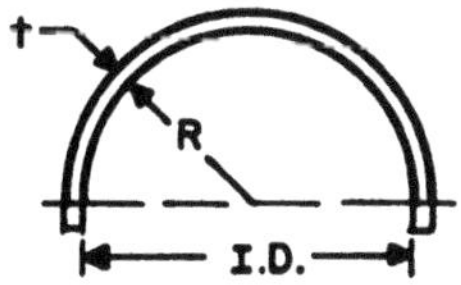

$$\text{Volume} = \left(\frac{\text{I.D.}}{12}\right)^3 \times 1.958$$

Inside Diameter, in	*Volume, gal*
12	1.95
18	6.6
24	15.6
30	30.5
36	53
42	84
48	125
54	178
60	245
66	326
72	423
78	538
84	672
90	826
96	1,003
102	1,203
108	1,428
114	1,679
120	1,958
126	2,267
132	2,606
138	2,978
144	3,384

TABLE A.2 Head Volume

Volumes listed are approximate volumes of dished portion of head; straight flanges are not included

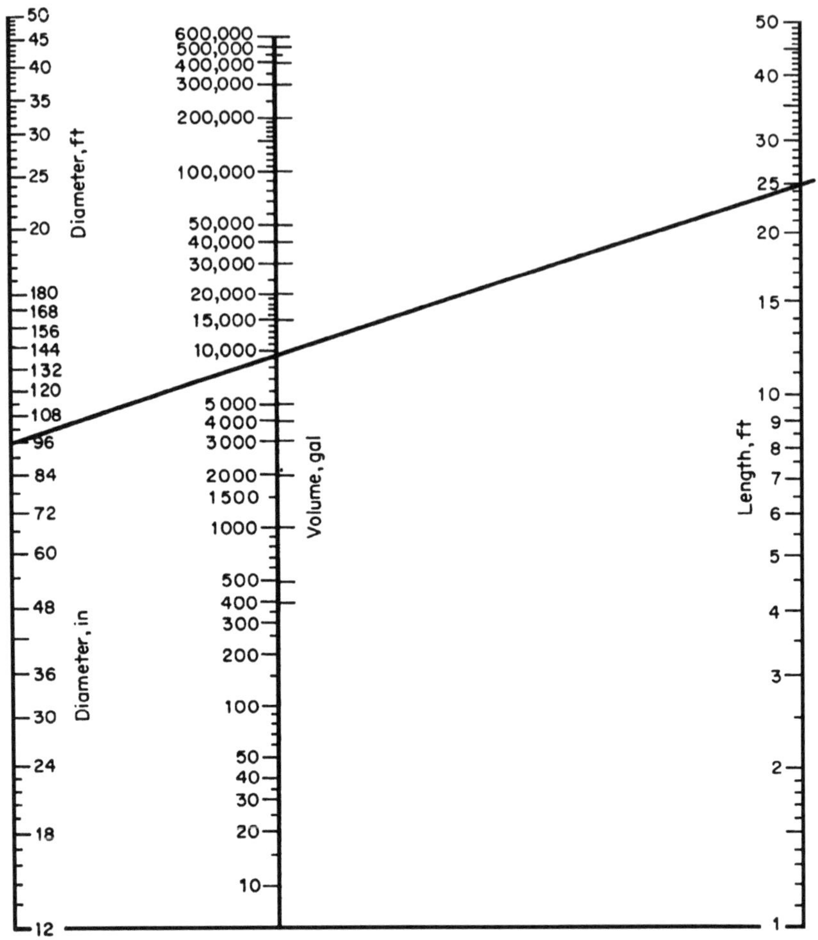

Figure A.1 Cylinder volume diagram. A cylinder 25 ft long and 96 in. in diameter will hold 9400 gal.

APPENDIX B

Circumferences of Cylinders

The length of a shell plate to be cut for a cylindrical vessel is measured by the mean or bend-line circumference of the shell so that the plate will equal a required inside diameter after being rolled. The circumferences given in Table B.1 are for the inside diameter.

When plates are to be cut to provide a given inside diameter, an amount equal to the plate thickness multiplied by 3.1416 must be added to the circumference shown in the table for that diameter. For example, suppose a vessel is to be 60 in. in inside diameter and is to have a shell thickness of 1 in. The 60-in. inside diameter generates a circumference of 188.496 in. To this figure must be added the plate thickness multiplied by 3.1416, or 3.1416 in. The length of plate to be cut to make a cylinder 60 in. in inside diameter is therefore 188.496 plus 3.1416, or 191.6376 in.

A simpler method is to add the thickness of plate to the inside diameter and obtain the circumference directly from the table. For example, for a vessel requiring a 60-in. inside diameter and a 1-in. shell plate, add 1 in. to 60 in. for a total of 61 in. The circumference given in the table for a 61-in. diameter equals 191.637 in.

The plate for cylinders with dished heads should never be cut until the heads have been taped if a good fit is to be secured at the head joint. To illustrate: If the inside circumference of a head has been taped at 188.5 in. and the thickness of the shell and head is 1 in., an amount equal to the plate thickness multiplied by 3.1416 will have to be added to the taped measurement for the shell to meet the inside diameter of the head when the plate is cut. Therefore, the plate will have to be cut to 191.64 in.

(*Text continues on page 262.*)

Dia.	Circum.	Area	Dia.	Circum.	Area	Dia.	Circum.	Area	Dia.	Circum.	Area
3.	9.4248	7.0686	11.	34.558	95.033	19.	59.690	283.53	27.	84.823	572.56
1/8	9.8175	7.6699	1/8	34.950	97.205	1/8	60.083	287.27	1/8	85.216	577.87
1/4	10.210	8.2958	1/4	35.343	99.402	1/4	60.476	291.04	1/4	85.608	583.21
3/8	10.603	8.9462	3/8	35.736	101.62	3/8	60.868	294.83	3/8	86.001	588.57
1/2	10.996	9.6211	1/2	36.128	103.87	1/2	61.261	298.65	1/2	86.394	593.96
5/8	11.388	10.321	5/8	36.521	106.14	5/8	61.654	302.49	5/8	86.786	599.37
3/4	11.781	11.045	3/4	36.914	108.43	3/4	62.046	306.35	3/4	87.179	604.81
7/8	12.174	11.793	7/8	37.306	110.75	7/8	62.439	310.24	7/8	87.572	610.27
4.	12.566	12.566	12.	37.699	113.10	20.	62.832	314.16	28.	87.965	615.75
1/8	12.959	13.364	1/8	38.092	115.47	1/8	63.225	318.10	1/8	88.357	621.26
1/4	13.352	14.186	1/4	38.485	117.86	1/4	63.617	322.06	1/4	88.750	626.80
3/8	13.744	15.033	3/8	38.877	120.28	3/8	64.010	326.05	3/8	89.143	632.36
1/2	14.137	15.904	1/2	39.270	122.72	1/2	64.403	330.06	1/2	89.535	637.94
5/8	14.530	16.800	5/8	39.663	125.19	5/8	64.795	334.10	5/8	89.928	643.55
3/4	14.923	17.728	3/4	40.055	127.68	3/4	65.188	338.16	3/4	90.321	649.18
7/8	15.315	18.665	7/8	40.448	130.19	7/8	65.581	342.25	7/8	90.713	654.84
5.	15.708	19.635	13.	40.841	132.73	21.	65.973	346.36	29.	91.106	660.52
1/8	16.101	20.629	1/8	41.233	135.30	1/8	66.366	350.50	1/8	91.499	666.23
1/4	16.493	21.648	1/4	41.626	137.89	1/4	66.759	354.66	1/4	91.892	671.96
3/8	16.886	22.691	3/8	42.019	140.50	3/8	67.152	358.84	3/8	92.284	677.71
1/2	17.279	23.758	1/2	42.412	143.14	1/2	67.544	363.05	1/2	92.677	683.49
5/8	17.671	24.850	5/8	42.804	145.80	5/8	67.937	367.28	5/8	93.070	689.30
3/4	18.064	25.967	3/4	43.197	148.49	3/4	68.330	371.54	3/4	93.462	695.13
7/8	18.457	27.109	7/8	43.590	151.20	7/8	68.722	375.83	7/8	93.855	700.98
6.	18.850	28.274	14.	43.982	153.94	22.	69.115	380.13	30.	94.248	706.86
1/8	19.242	29.465	1/8	44.375	156.70	1/8	69.508	384.46	1/8	94.640	712.76
1/4	19.635	30.680	1/4	44.768	159.48	1/4	69.900	388.82	1/4	95.033	718.69
3/8	20.028	31.919	3/8	45.160	162.30	3/8	70.293	393.20	3/8	95.426	724.64
1/2	20.420	33.183	1/2	45.553	165.13	1/2	70.686	397.61	1/2	95.819	730.62
5/8	20.813	34.472	5/8	45.946	167.99	5/8	71.079	402.04	5/8	96.211	736.62
3/4	21.206	35.785	3/4	46.338	170.87	3/4	71.471	406.49	3/4	96.604	742.64
7/8	21.598	37.122	7/8	46.731	173.78	7/8	71.864	410.97	7/8	96.997	748.69

TABLE B.1 Circumferences and Areas of Circles
Areas are approximate but sufficiently accurate for estimating

7.	21.991	38.485	15.	47.124	176.71	23.	72.257	415.48	31.	97.389	754.77
1/8	22.384	39.871	1/8	47.517	179.67	1/8	72.649	420.00	1/8	97.782	760.87
1/4	22.776	41.282	1/4	47.909	182.65	1/4	73.042	424.56	1/4	98.175	766.99
3/8	23.169	42.718	3/8	48.302	185.66	3/8	73.435	429.13	3/8	98.567	773.14
1/2	23.562	44.179	1/2	48.695	188.69	1/2	73.827	433.74	1/2	98.960	779.31
5/8	23.955	45.664	5/8	49.087	191.75	5/8	74.220	438.36	5/8	99.353	785.51
3/4	24.347	47.173	3/4	49.480	194.83	3/4	74.613	443.01	3/4	99.746	791.73
7/8	24.740	48.707	7/8	49.873	197.93	7/8	75.006	447.69	7/8	100.138	797.98
8.	25.133	50.265	16.	50.265	201.06	24.	75.398	452.39	32.	100.531	804.25
1/8	25.525	51.849	1/8	50.658	204.22	1/8	75.791	457.11	1/8	100.924	810.54
1/4	25.918	53.456	1/4	51.051	207.39	1/4	76.184	461.86	1/4	101.316	816.86
3/8	26.311	55.088	3/8	51.444	210.60	3/8	76.576	466.64	3/8	101.709	823.21
1/2	26.704	56.745	1/2	51.836	213.82	1/2	76.969	471.44	1/2	102.102	829.58
5/8	27.096	58.426	5/8	52.229	217.08	5/8	77.362	476.26	5/8	102.494	835.97
3/4	27.489	60.132	3/4	52.622	220.35	3/4	77.754	481.11	3/4	102.887	842.39
7/8	27.882	61.862	7/8	53.014	223.65	7/8	78.147	485.98	7/8	103.280	848.83
9.	28.274	63.617	17.	53.407	226.98	25.	78.540	490.87	33.	103.673	855.30
1/8	28.667	65.397	1/8	53.800	230.33	1/8	78.933	495.79	1/8	104.065	861.79
1/4	29.060	67.201	1/4	54.192	233.71	1/4	79.325	500.74	1/4	104.458	868.31
3/8	29.452	69.029	3/8	54.585	237.10	3/8	79.718	505.71	3/8	104.851	874.85
1/2	29.845	70.882	1/2	54.978	240.53	1/2	80.111	510.71	1/2	105.243	881.41
5/8	30.238	72.760	5/8	55.371	243.98	5/8	80.503	515.72	5/8	105.636	888.00
3/4	30.631	74.662	3/4	55.763	247.45	3/4	80.896	520.77	3/4	106.029	894.62
7/8	31.023	76.589	7/8	56.156	250.95	7/8	81.289	525.84	7/8	106.421	901.26
10.	31.416	78.540	18.	56.549	254.47	26.	81.681	530.93	34.	106.814	907.92
1/8	31.809	80.516	1/8	56.941	258.02	1/8	82.074	536.05	1/8	107.207	914.61
1/4	32.201	82.516	1/4	57.334	261.59	1/4	82.467	541.19	1/4	107.600	921.32
3/8	32.594	84.541	3/8	57.727	265.18	3/8	82.860	546.35	3/8	107.992	928.06
1/2	32.987	86.590	1/2	58.119	268.80	1/2	83.252	551.55	1/2	108.385	934.82
5/8	33.379	88.664	5/8	58.512	272.45	5/8	83.645	556.76	5/8	108.778	941.61
3/4	33.772	90.763	3/4	58.905	276.12	3/4	84.038	562.00	3/4	109.170	948.42
7/8	34.165	92.886	7/8	59.298	279.81	7/8	84.430	567.27	7/8	109.563	955.25

Dia.	Circum.	Area	Dia.	Circum.	Area	Dia.	Circum.	Area	Dia.	Circum.	Area
35.	109.956	962.11	43.	135.088	1452.2	51.	160.221	2042.8	59.	185.354	2734.0
1/8	110.348	969.00	1/8	135.481	1460.7	1/8	160.614	2052.8	1/8	185.747	2745.6
1/4	110.741	975.91	1/4	135.874	1469.1	1/4	161.007	2062.9	1/4	186.139	2757.2
3/8	111.134	982.84	3/8	136.267	1477.6	3/8	161.399	2073.0	3/8	186.532	2768.8
1/2	111.527	989.80	1/2	136.659	1486.2	1/2	161.792	2083.1	1/2	186.925	2780.5
5/8	111.919	996.78	5/8	137.052	1494.7	5/8	162.185	2093.2	5/8	187.317	2792.2
3/4	112.312	1003.8	3/4	137.445	1503.3	3/4	162.577	2103.3	3/4	187.710	2803.9
7/8	112.705	1010.8	7/8	137.837	1511.9	7/8	162.970	2113.5	7/8	188.103	2815.7
36.	113.097	1017.9	44.	138.230	1520.5	52.	163.363	2123.7	60.	188.496	2827.4
1/8	113.490	1025.0	1/8	138.623	1529.2	1/8	163.756	2133.9	1/8	188.888	2839.2
1/4	113.883	1032.1	1/4	139.015	1537.9	1/4	164.148	2144.2	1/4	189.281	2851.0
3/8	114.275	1039.2	3/8	139.408	1546.6	3/8	164.541	2154.5	3/8	189.674	2862.9
1/2	114.668	1046.3	1/2	139.801	1555.3	1/2	164.934	2164.8	1/2	190.066	2874.8
5/8	115.061	1053.5	5/8	140.194	1564.0	5/8	165.326	2175.1	5/8	190.459	2886.6
3/4	115.454	1060.7	3/4	140.586	1572.8	3/4	165.719	2185.4	3/4	190.852	2898.6
7/8	115.846	1068.0	7/8	140.979	1581.6	7/8	166.112	2195.8	7/8	191.244	2910.5
37.	116.239	1075.2	45.	141.372	1590.4	53.	166.504	2206.2	61.	191.637	2922.5
1/8	116.632	1082.5	1/8	141.764	1599.3	1/8	166.897	2216.6	1/8	192.030	2934.5
1/4	117.024	1089.8	1/4	142.157	1608.2	1/4	167.290	2227.0	1/4	192.423	2946.5
3/8	117.417	1097.1	3/8	142.550	1617.0	3/8	167.683	2237.5	3/8	192.815	2958.5
1/2	117.810	1104.5	1/2	142.942	1626.0	1/2	168.075	2248.0	1/2	193.208	2970.6
5/8	118.202	1111.8	5/8	143.335	1634.9	5/8	168.468	2258.5	5/8	193.601	2982.7
3/4	118.596	1119.2	3/4	143.728	1643.9	3/4	168.861	2269.1	3/4	193.993	2994.8
7/8	118.988	1126.7	7/8	144.121	1652.9	7/8	169.253	2279.6	7/8	194.386	3006.9
38.	119.381	1134.1	46.	144.513	1661.9	54.	169.646	2290.2	62.	194.779	3019.1
1/8	119.773	1141.6	1/8	144.906	1670.9	1/8	170.039	2300.8	1/8	195.171	3031.3
1/4	120.166	1149.1	1/4	145.299	1680.0	1/4	170.431	2311.5	1/4	195.564	3043.5
3/8	120.559	1156.6	3/8	145.691	1689.1	3/8	170.824	2322.1	3/8	195.957	3055.7
1/2	120.951	1164.2	1/2	146.084	1698.2	1/2	171.217	2332.8	1/2	196.350	3068.0
5/8	121.344	1171.7	5/8	146.477	1707.4	5/8	171.609	2343.5	5/8	196.742	3080.3
3/4	121.737	1179.3	3/4	146.869	1716.5	3/4	172.002	2354.3	3/4	197.135	3092.6
7/8	122.129	1186.9	7/8	147.262	1725.7	7/8	172.395	2365.0	7/8	197.528	3104.9

TABLE B.1 Circumferences and Areas of Circles (*Continued*)

39.	122.522	1194.6	47.	147.655	1734.9	55.	172.788	2375.8	63.	197.920	3117.2
1/8	122.915	1202.3	1/8	148.048	1744.2	1/8	173.180	2386.6	1/8	198.313	3129.6
1/4	123.308	1210.6	1/4	148.440	1753.5	1/4	173.573	2397.5	1/4	198.706	3142.0
3/8	123.700	1217.7	3/8	148.833	1762.7	3/8	173.966	2408.3	3/8	199.098	3154.5
1/2	124.093	1225.4	1/2	149.226	1772.1	1/2	174.358	2419.2	1/2	199.491	3166.9
5/8	124.486	1233.2	5/8	149.618	1781.4	5/8	174.751	2430.1	5/8	199.884	3179.4
3/4	124.878	1241.0	3/4	150.011	1790.8	3/4	175.144	2441.1	3/4	200.277	3191.9
7/8	125.271	1248.8	7/8	150.404	1800.1	7/8	175.536	2452.0	7/8	200.669	3204.4
40.	125.664	1256.6	48.	150.796	1809.6	56.	175.929	2463.0	64.	201.062	3217.0
1/8	126.056	1264.5	1/8	151.189	1819.0	1/8	176.322	2474.0	1/8	201.455	3229.6
1/4	126.449	1272.4	1/4	151.582	1828.5	1/4	176.715	2485.0	1/4	201.847	3242.2
3/8	126.842	1280.3	3/8	151.975	1837.9	3/8	177.107	2496.1	3/8	202.240	3254.8
1/2	127.235	1288.2	1/2	152.367	1847.5	1/2	177.500	2507.2	1/2	202.633	3267.5
5/8	127.627	1296.2	5/8	152.760	1857.0	5/8	177.893	2518.3	5/8	203.025	3280.1
3/4	128.020	1304.2	3/4	153.153	1866.5	3/4	178.285	2529.4	3/4	203.418	3292.8
7/8	128.413	1312.2	7/8	153.545	1876.1	7/8	178.678	2540.6	7/8	203.811	3305.6
41.	128.805	1320.3	49.	153.938	1885.7	57.	179.071	2551.8	65.	204.204	3318.3
1/8	129.198	1328.3	1/8	154.331	1895.4	1/8	179.463	2563.0	1/8	204.596	3331.1
1/4	129.591	1336.4	1/4	154.723	1905.0	1/4	179.856	2574.2	1/4	204.989	3343.9
3/8	129.983	1344.5	3/8	155.116	1914.7	3/8	180.249	2585.4	3/8	205.382	3356.7
1/2	130.376	1352.7	1/2	155.509	1924.4	1/2	180.642	2596.7	1/2	205.774	3369.6
5/8	130.769	1360.8	5/8	155.902	1934.2	5/8	181.034	2608.0	5/8	206.167	3382.4
3/4	131.161	1369.0	3/4	156.294	1943.9	3/4	181.427	2619.4	3/4	206.560	3395.3
7/8	131.554	1377.2	7/8	156.687	1953.7	7/8	181.820	2630.7	7/8	206.952	3408.2
42.	131.947	1385.4	50.	157.080	1963.5	58.	182.212	2642.1	66.	207.345	3421.2
1/8	132.340	1393.7	1/8	157.472	1973.3	1/8	182.605	2653.5	1/8	207.738	3434.2
1/4	132.732	1402.0	1/4	157.865	1983.2	1/4	182.998	2664.9	1/4	208.131	3447.2
3/8	133.125	1410.3	3/8	158.258	1993.1	3/8	183.390	2676.4	3/8	208.523	3460.2
1/2	133.518	1418.6	1/2	158.650	2003.0	1/2	183.783	2687.8	1/2	208.916	3473.2
5/8	133.910	1427.0	5/8	159.043	2012.9	5/8	184.176	2699.3	5/8	209.309	3486.3
3/4	134.303	1435.4	3/4	159.436	2022.8	3/4	184.569	2710.9	3/4	209.701	3499.4
7/8	134.696	1443.8	7/8	159.829	2032.8	7/8	184.961	2722.4	7/8	210.094	3512.5

Dia.	Circum.	Area	Dia.	Circum.	Area	Dia.	Circum.	Area	Dia.	Circum.	Area
67.	210.487	3525.7	75.	235.619	4417.9	83.	260.752	5410.6	91.	285.885	6503.9
1/8	210.879	3538.8	1/8	236.012	4432.6	1/8	261.145	5426.9	1/8	286.278	6521.8
1/4	211.272	3552.0	1/4	236.405	4447.4	1/4	261.538	5443.3	1/4	286.670	6539.7
3/8	211.665	3565.2	3/8	236.798	4462.2	3/8	261.930	5459.6	3/8	287.063	6557.6
1/2	212.058	3578.5	1/2	237.190	4477.0	1/2	262.323	5476.0	1/2	287.456	6575.5
5/8	212.450	3591.7	5/8	237.583	4491.8	5/8	262.716	5492.4	5/8	287.848	6593.5
3/4	212.843	3605.0	3/4	237.976	4506.7	3/4	263.108	5508.8	3/4	288.241	6611.8
7/8	213.236	3618.3	7/8	238.368	4521.5	7/8	263.501	5525.3	7/8	288.634	6629.6
68.	213.628	3631.7	76.	238.761	4536.5	84.	263.894	5541.8	92.	289.027	6647.6
1/8	214.021	3645.0	1/8	239.154	4551.4	1/8	264.286	5558.3	1/8	289.419	6665.7
1/4	214.414	3658.4	1/4	239.546	4566.4	1/4	264.679	5574.8	1/4	289.812	6683.8
3/8	214.806	3671.8	3/8	239.939	4581.3	3/8	265.072	5591.4	3/8	290.205	6701.9
1/2	215.199	3685.3	1/2	240.332	4596.3	1/2	265.465	5607.9	1/2	290.597	6720.1
5/8	215.592	3698.7	5/8	240.725	4611.4	5/8	265.857	5624.5	5/8	290.990	6738.2
3/4	215.984	3712.2	3/4	241.117	4626.4	3/4	266.250	5641.2	3/4	291.383	6756.4
7/8	216.377	3725.7	7/8	241.510	4641.5	7/8	266.643	5657.8	7/8	291.775	6774.7
69.	216.770	3739.3	77.	241.903	4656.6	85.	267.035	5674.5	93.	292.168	6792.9
1/8	217.163	3752.8	1/8	242.295	4671.8	1/8	267.428	5691.2	1/8	292.561	6811.2
1/4	217.555	3766.4	1/4	242.688	4686.9	1/4	267.821	5707.9	1/4	292.954	6829.5
3/8	217.948	3780.0	3/8	243.081	4702.1	3/8	268.213	5724.7	3/8	293.346	6847.8
1/2	218.341	3793.7	1/2	243.473	4717.3	1/2	268.606	5741.5	1/2	293.739	6866.1
5/8	218.733	3807.3	5/8	243.866	4732.5	5/8	268.999	5758.3	5/8	294.132	6884.5
3/4	219.126	3821.0	3/4	244.259	4747.8	3/4	269.392	5775.1	3/4	294.524	6902.9
7/8	219.519	3834.7	7/8	244.652	4763.1	7/8	269.784	5791.9	7/8	294.917	6921.3
70.	219.911	3848.5	78.	245.044	4778.4	86.	270.177	5808.8	94.	295.310	6939.8
1/8	220.304	3862.2	1/8	245.437	4793.7	1/8	270.570	5825.7	1/8	295.702	6958.2
1/4	220.697	3876.0	1/4	245.830	4809.0	1/4	270.962	5842.6	1/4	296.095	6976.7
3/8	221.090	3889.8	3/8	246.222	4824.4	3/8	271.355	5859.6	3/8	296.488	6995.3
1/2	221.482	3903.6	1/2	246.615	4839.8	1/2	271.748	5876.5	1/2	296.881	7013.8
5/8	221.875	3917.5	5/8	247.008	4855.2	5/8	272.140	5893.5	5/8	297.273	7032.4
3/4	222.268	3931.4	3/4	247.400	4870.7	3/4	272.533	5910.6	3/4	297.666	7051.0
7/8	222.660	3945.3	7/8	247.793	4886.2	7/8	272.926	5927.6	7/8	298.059	7069.6

TABLE B.1 Circumferences and Areas of Circles (*Continued*)

71.	223.053	3959.2	79.	248.186	4901.7	87.	273.319	5944.7	95.	298.451	7088.2
1/8	223.446	3973.1	1/8	248.579	4917.2	1/8	273.711	5961.8	1/8	298.844	7106.9
1/4	223.838	3987.1	1/4	248.971	4932.7	1/4	274.104	5978.9	1/4	299.237	7125.6
3/8	224.231	4001.1	3/8	249.364	4948.3	3/8	274.497	5996.0	3/8	299.629	7144.3
1/2	224.624	4015.2	1/2	249.757	4963.9	1/2	274.889	6013.2	1/2	300.022	7163.0
5/8	225.017	4029.2	5/8	250.149	4979.5	5/8	275.282	6030.4	5/8	300.415	7181.8
3/4	225.409	4043.3	3/4	250.542	4995.2	3/4	275.675	6047.6	3/4	300.807	7200.6
7/8	225.802	4057.4	7/8	250.935	5010.9	7/8	276.067	6064.9	7/8	301.200	7219.4
72.	226.195	4071.5	80.	251.327	5026.5	88.	276.460	6082.1	96.	301.593	7238.2
1/8	226.587	4085.7	1/8	251.720	5042.3	1/8	276.853	6099.4	1/8	301.986	7257.1
1/4	226.980	4099.8	1/4	252.113	5058.0	1/4	277.246	6116.7	1/4	302.378	7276.0
3/8	227.373	4114.0	3/8	252.506	5073.8	3/8	277.638	6134.1	3/8	302.771	7294.9
1/2	227.765	4128.2	1/2	252.898	5089.6	1/2	278.031	6151.4	1/2	303.164	7313.8
5/8	228.158	4142.5	5/8	253.291	5105.4	5/8	278.424	6168.8	5/8	303.556	7332.8
3/4	228.551	4156.8	3/4	253.684	5121.2	3/4	278.816	6186.2	3/4	303.949	7351.8
7/8	228.944	4171.1	7/8	254.076	5137.1	7/8	279.209	6203.7	7/8	304.342	7370.8
73.	229.336	4185.4	81.	254.469	5153.0	89.	279.602	6221.1	97.	304.734	7389.8
1/8	229.729	4199.7	1/8	254.862	5168.9	1/8	279.994	6238.6	1/8	305.127	7408.9
1/4	230.122	4214.1	1/4	255.254	5184.9	1/4	280.387	6256.1	1/4	305.520	7428.0
3/8	230.514	4228.5	3/8	255.647	5200.8	3/8	280.780	6273.7	3/8	305.913	7447.1
1/2	230.907	4242.9	1/2	256.040	5216.8	1/2	281.173	6291.2	1/2	306.305	7466.2
5/8	231.300	4257.4	5/8	256.433	5232.8	5/8	281.565	6308.8	5/8	306.698	7485.3
3/4	231.692	4271.8	3/4	256.825	5248.9	3/4	281.958	6326.4	3/4	307.091	7504.5
7/8	232.085	4286.3	7/8	257.218	5264.9	7/8	282.351	6344.1	7/8	307.483	7523.7
74.	232.478	4300.8	82.	257.611	5281.0	90.	282.743	6361.7	98.	307.876	7543.0
1/8	232.871	4315.4	1/8	258.003	5297.1	1/8	283.136	6379.4	1/8	308.269	7562.2
1/4	233.263	4329.9	1/4	258.396	5313.3	1/4	283.529	6397.1	1/4	308.661	7581.5
3/8	233.656	4344.5	3/8	258.789	5329.4	3/8	283.921	6414.9	3/8	309.054	7600.8
1/2	234.049	4359.2	1/2	259.181	5345.6	1/2	284.314	6432.6	1/2	309.447	7620.1
5/8	234.441	4373.8	5/8	259.574	5361.8	5/8	284.707	6450.4	5/8	309.840	7639.5
3/4	234.834	4388.5	3/4	259.967	5378.1	3/4	285.100	6468.2	3/4	310.232	7658.9
7/8	235.227	4403.1	7/8	260.359	5394.3	7/8	285.492	6486.0	7/8	310.625	7678.3

Dia.	Circum.	Area	Dia.	Circum.	Area	Dia.	Circum.	Area	Dia.	Circum.	Area
99.	311.018	7697.7	107.	336.15	8992	115.	361.28	10387	123.	386.42	11882
1/8	311.410	7717.1	1/8	336.54	9014	1/8	361.68	10410	1/8	386.81	11907
1/4	311.803	7736.6	1/4	336.94	9035	1/4	362.07	10432	1/4	387.20	11931
3/8	312.196	7756.1	3/8	337.33	9056	3/8	362.46	10455	3/8	387.60	11956
1/2	312.588	7775.6	1/2	337.72	9077	1/2	362.86	10477	1/2	387.99	11980
5/8	312.981	7795.2	5/8	338.12	9098	5/8	363.25	10500	5/8	388.38	12004
3/4	313.374	7814.8	3/4	338.51	9119	3/4	363.64	10522	3/4	388.77	12028
7/8	313.767	7834.4	7/8	338.90	9140	7/8	364.03	10545	7/8	389.17	12052
100.	314.16	7854	108.	339.29	9161	116.	364.43	10568	124.	389.56	12076
1/8	314.55	7873	1/8	339.69	9183	1/8	364.82	10590	1/8	389.95	12101
1/4	314.95	7893	1/4	340.08	9204	1/4	365.21	10613	1/4	390.34	12125
3/8	315.34	7913	3/8	340.47	9225	3/8	365.60	10636	3/8	390.74	12150
1/2	315.73	7933	1/2	340.86	9246	1/2	366.00	10659	1/2	391.13	12174
5/8	316.12	7952	5/8	341.26	9268	5/8	366.39	10682	5/8	391.52	12199
3/4	316.52	7972	3/4	341.65	9289	3/4	366.78	10705	3/4	391.92	12223
7/8	316.91	7992	7/8	342.04	9310	7/8	367.18	10728	7/8	392.31	12248
101.	317.30	8012	109.	342.43	9331	117.	367.57	10751	125.	392.70	12272
1/8	317.69	8032	1/8	342.83	9353	1/8	367.96	10774	1/8	393.09	12297
1/4	318.09	8052	1/4	343.22	9374	1/4	368.35	10798	1/4	393.49	12321
3/8	318.48	8071	3/8	343.61	9396	3/8	368.75	10821	3/8	393.88	12346
1/2	318.87	8091	1/2	344.01	9417	1/2	369.14	10844	1/2	394.27	12370
5/8	319.27	8111	5/8	344.40	9439	5/8	369.53	10867	5/8	394.66	12395
3/4	319.66	8131	3/4	344.79	9460	3/4	369.92	10890	3/4	395.06	12419
7/8	320.05	8151	7/8	345.18	9481	7/8	370.32	10913	7/8	395.45	12444
102.	320.44	8171	110.	345.58	9503	118.	370.71	10936	126.	395.84	12469
1/8	320.84	8191	1/8	345.97	9525	1/8	371.11	10960	1/8	396.23	12494
1/4	321.23	8211	1/4	346.36	9546	1/4	371.49	10983	1/4	396.63	12518
3/8	321.62	8231	3/8	346.75	9568	3/8	371.89	11007	3/8	397.02	12543
1/2	322.01	8252	1/2	347.15	9589	1/2	372.28	11030	1/2	397.41	12568
5/8	322.41	8272	5/8	347.54	9611	5/8	372.67	11053	5/8	397.81	12593
3/4	322.80	8292	3/4	347.93	9633	3/4	373.07	11076	3/4	398.20	12618
7/8	323.19	8312	7/8	348.33	9655	7/8	373.46	11099	7/8	398.59	12643

TABLE B.1 Circumferences and Areas of Circles (*Continued*)

103.	323.59	8332	111.	348.72	9677	119.	373.85	11122	127.	398.98	12668
1/8	323.98	8352	1/8	349.11	9698	1/8	374.24	11146	1/8	399.38	12693
1/4	324.37	8372	1/4	349.50	9720	1/4	374.64	11169	1/4	399.77	12718
3/8	324.76	8393	3/8	349.90	9742	3/8	375.03	11193	3/8	400.16	12743
1/2	325.16	8413	1/2	350.29	9764	1/2	375.42	11216	1/2	400.55	12768
5/8	325.55	8434	5/8	350.68	9786	5/8	375.81	11240	5/8	400.95	12793
3/4	325.94	8454	3/4	351.07	9808	3/4	376.21	11263	3/4	401.34	12818
7/8	326.33	8474	7/8	351.47	9830	7/8	376.60	11287	7/8	401.73	12843
104.	326.73	8495	112.	351.86	9852	120.	376.99	11310	128.	402.13	12868
1/8	327.12	8515	1/8	352.25	9874	1/8	377.39	11334	1/8	402.52	12893
1/4	327.51	8536	1/4	352.65	9897	1/4	377.78	11357	1/4	402.91	12919
3/8	327.91	8556	3/8	353.04	9919	3/8	378.17	11381	3/8	403.30	12944
1/2	328.30	8577	1/2	353.43	9941	1/2	378.56	11404	1/2	403.70	12970
5/8	328.69	8597	5/8	353.82	9963	5/8	378.96	11428	5/8	404.09	12995
3/4	329.08	8618	3/4	354.22	9985	3/4	379.35	11451	3/4	404.48	13020
7/8	329.48	8638	7/8	354.61	10007	7/8	379.74	11475	7/8	404.87	13045
105.	329.87	8659	113.	355.00	10029	121.	380.13	11499	129.	405.27	13070
1/8	330.26	8679	1/8	355.39	10052	1/8	380.53	11522	1/8	405.66	13096
1/4	330.65	8700	1/4	355.79	10074	1/4	380.92	11546	1/4	406.05	13121
3/8	331.05	8721	3/8	356.18	10097	3/8	381.31	11570	3/8	406.44	13147
1/2	331.44	8741	1/2	356.57	10119	1/2	381.70	11594	1/2	406.84	13172
5/8	331.83	8762	5/8	356.96	10141	5/8	382.10	11618	5/8	407.23	13198
3/4	332.22	8783	3/4	357.36	10163	3/4	382.49	11642	3/4	407.62	13223
7/8	332.62	8804	7/8	357.75	10185	7/8	382.88	11666	7/8	408.02	13248
106.	333.01	8825	114.	358.14	10207	122.	383.28	11690	130.	408.41	13273
1/8	333.40	8845	1/8	358.54	10230	1/8	383.67	11714	1/8	408.80	13299
1/4	333.80	8866	1/4	358.93	10252	1/4	384.06	11738	1/4	409.19	13324
3/8	334.19	8887	3/8	359.32	10275	3/8	384.45	11762	3/8	409.59	13350
1/2	334.58	8908	1/2	359.71	10297	1/2	384.85	11786	1/2	409.98	13375
5/8	334.97	8929	5/8	360.11	10320	5/8	385.24	11810	5/8	410.37	13401
3/4	335.37	8950	3/4	360.50	10342	3/4	385.63	11834	3/4	410.76	13426
7/8	335.76	8971	7/8	360.89	10365	7/8	386.02	11858	7/8	411.16	13452

Dia.	Circum.	Area	Dia.	Circum.	Area	Dia.	Circum.	Area	Dia.	Circum.	Area
131.	411.55	13478	139.	436.68	15175	147.	461.82	16972	155.	486.95	18869
1/8	411.94	13504	1/8	437.08	15203	1/8	462.21	17000	1/8	487.34	18900
1/4	412.34	13529	1/4	437.47	15230	1/4	462.60	17029	1/4	487.73	18930
3/8	412.73	13555	3/8	437.86	15258	3/8	462.99	17058	3/8	488.13	18961
1/2	413.12	13581	1/2	438.25	15285	1/2	463.39	17087	1/2	488.52	18991
5/8	413.51	13607	5/8	438.65	15313	5/8	463.78	17116	5/8	488.91	19022
3/4	413.91	13633	3/4	439.04	15340	3/4	464.17	17145	3/4	489.30	19052
7/8	414.30	13659	7/8	439.43	15367	7/8	464.56	17174	7/8	489.70	19083
132.	414.69	13685	140.	439.82	15394	148.	464.96	17203	156.	490.09	19113
1/8	415.08	13711	1/8	440.22	15422	1/8	465.35	17232	1/8	490.48	19144
1/4	415.48	13737	1/4	440.61	15449	1/4	465.74	17262	1/4	490.88	19174
3/8	415.87	13763	3/8	441.00	15477	3/8	466.14	17291	3/8	491.27	19205
1/2	416.26	13789	1/2	441.40	15504	1/2	466.53	17321	1/2	491.66	19235
5/8	416.66	13815	5/8	441.79	15532	5/8	466.92	17350	5/8	492.05	19266
3/4	417.05	13841	3/4	442.18	15559	3/4	467.31	17379	3/4	492.45	19297
7/8	417.44	13867	7/8	442.57	15587	7/8	467.71	17408	7/8	492.84	19328
133.	417.83	13893	141.	442.97	15615	149.	468.10	17437	157.	493.23	19359
1/8	418.23	13919	1/8	443.36	15642	1/8	468.49	17466	1/8	493.62	19390
1/4	418.62	13946	1/4	443.75	15670	1/4	468.88	17496	1/4	494.02	19421
3/8	419.01	13972	3/8	444.14	15697	3/8	469.28	17525	3/8	494.41	19452
1/2	419.40	13999	1/2	444.54	15725	1/2	469.67	17555	1/2	494.80	19483
5/8	419.80	14025	5/8	444.93	15753	5/8	470.06	17584	5/8	495.20	19514
3/4	420.19	14051	3/4	445.32	15781	3/4	470.46	17614	3/4	495.59	19545
7/8	420.58	14077	7/8	445.72	15809	7/8	470.85	17643	7/8	495.98	19576
134.	420.97	14103	142.	446.11	15837	150.	471.24	17672	158.	496.37	19607
1/8	421.37	14130	1/8	446.50	15865	1/8	471.63	17702	1/8	496.77	19638
1/4	421.76	14156	1/4	446.89	15893	1/4	472.03	17731	1/4	497.16	19669
3/8	422.15	14183	3/8	447.29	15921	3/8	472.42	17761	3/8	497.55	19701
1/4	422.55	14209	1/2	447.68	15949	1/2	472.81	17790	1/2	497.94	19732
5/8	422.94	14236	5/8	448.07	15977	5/8	473.20	17820	5/8	498.34	19763
3/4	423.33	14262	3/4	448.46	16005	3/4	473.60	17849	3/4	498.73	19794
7/8	423.72	14288	7/8	448.86	16033	7/8	473.99	17879	7/8	499.12	19825

Table B.1 Circumferences and Areas of Circles (*Continued*)

135.	424.12	14314	143.	449.25	16061	151.	474.38	17908	159.	499.51	19856
1/8	424.51	14341	1/8	449.64	16089	1/8	474.77	17938	1/8	499.91	19887
1/4	424.90	14367	1/4	450.03	16117	1/4	475.17	17967	1/4	500.30	19919
3/8	425.29	14394	3/8	450.43	16145	3/8	475.56	17997	3/8	500.69	19950
1/2	425.69	14420	1/2	450.82	16173	1/2	475.95	18026	1/2	501.09	19982
5/8	426.08	14447	5/8	451.21	16201	5/8	476.35	18056	5/8	501.48	20013
3/4	426.47	14473	3/4	451.61	16229	3/4	476.74	18086	3/4	501.87	20044
7/8	426.87	14500	7/8	452.00	16258	7/8	477.13	18116	7/8	502.26	20075
136.	427.26	14527	144.	452.39	16286	152.	477.52	18146	160.	502.66	20106
1/8	427.65	14553	1/8	452.78	16314	1/8	477.92	18175	1/8	503.05	20138
1/4	428.04	14580	1/4	453.18	16342	1/4	478.31	18205	1/4	503.44	20169
3/8	428.44	14607	3/8	453.57	16371	3/8	478.70	18235	3/8	503.83	20201
1/2	428.83	14633	1/2	453.96	16399	1/2	479.09	18265	1/2	504.23	20232
5/8	429.22	14660	5/8	454.35	16428	5/8	479.49	18295	5/8	504.62	20264
3/4	429.61	14687	3/4	454.75	16456	3/4	479.88	18325	3/4	505.01	20295
7/8	430.01	14714	7/8	455.14	16485	7/8	480.27	18355	7/8	505.41	20327
137.	430.40	14741	145.	455.53	16513	153.	480.67	18385	161.	505.80	20358
1/8	430.79	14768	1/8	455.93	16542	1/8	481.06	18415	1/8	506.19	20390
1/4	431.19	14795	1/4	456.32	16570	1/4	481.45	18446	1/4	506.58	20421
3/8	431.58	14822	3/8	456.71	16599	3/8	481.84	18476	3/8	506.98	20453
1/2	431.97	14849	1/2	457.10	16627	1/2	482.24	18507	1/2	507.37	20484
5/8	432.36	14876	5/8	457.50	16656	5/8	482.63	18537	5/8	507.76	20516
3/4	432.76	14903	3/4	457.89	16684	3/4	483.02	18567	3/4	508.15	20548
7/8	433.15	14930	7/8	458.28	16713	7/8	483.41	18597	7/8	508.55	20580
138	433.54	14957	146.	458.67	16742	154.	483.81	18627	162.	508.94	20612
1/8	433.93	14984	1/8	459.07	16770	1/8	484.20	18658	1/8	509.33	20644
1/4	434.33	15012	1/4	459.46	16799	1/4	484.59	18688	1/4	509.73	20675
3/8	434.72	15039	3/8	459.85	16827	3/8	484.99	18719	3/8	510.12	20707
1/2	435.11	15067	1/2	460.24	16856	1/2	485.38	18749	1/2	510.51	20739
5/8	435.50	15094	5/8	460.64	16885	5/8	485.77	18779	5/8	510.90	20771
3/4	435.90	15121	3/4	461.03	16914	3/4	486.16	18809	3/4	511.30	20803
7/8	436.29	15148	7/8	461.42	16943	7/8	486.56	18839	7/8	511.69	20835

If the head has been taped on the outside circumference, subtract an amount equal to the plate thickness multiplied by 3.1416 from the total measurement before cutting the shell plate.

Some may find it easier to use Table B.1 to arrive at a measurement. For example, if the outside circumference of a head has been taped at 194.78 in., the diameter shown for this measurement is 62 in. If the shell plate is 1 in. thick, subtract 1 in. from 62 in., for a diameter of 61 in. The circumference given in the table for this diameter equals 191.637 in. This figure represents the mean or bend-line circumference and therefore the length of plate required to make a cylinder 60 in. in inside diameter.

APPENDIX C

Decimal Equivalents and Theoretical Weights of Steel Plates

Plate thickness		Weight, psf	Plate thickness		Weight, psf
Fractions of an inch	Decimal equivalent		Fractions of an inch	Decimal equivalent	
1/64	0.01562	0.637	33/64	0.51562	21.037
1/32	0.03125	1.274	17/32	0.53125	21.675
3/64	0.04687	1.911	35/64	0.54687	22.312
1/16	0.0625	2.55	9/16	0.5625	22.950
5/64	0.07812	3.1875	37/64	0.57812	23.587
3/32	0.09375	3.825	19/32	0.59375	24.225
7/64	0.10937	4.4625	39/64	0.60937	24.862
1/8	0.125	5.10	5/8	0.625	25.50
9/64	0.14062	5.7375	41/64	0.64062	26.137
5/32	0.15625	6.375	21/32	0.65625	26.775
11/64	0.17187	7.0125	43/64	0.67187	27.412
3/16	0.1875	7.65	11/16	0.6875	28.050
13/64	0.20312	8.2875	45/64	0.70312	28.687
7/32	0.21875	8.925	23/32	0.71875	29.325
15/64	0.23437	9.5625	47/64	0.73437	29.962
1/4	0.25	10.20	3/4	0.75	30.60
17/64	0.26562	10.837	49/64	0.76562	31.237
9/32	0.28125	11.475	25/32	0.78125	31.875
19/64	0.29687	12.112	51/64	0.79687	32.512
5/16	0.3125	12.75	13/16	0.8125	33.150
21/64	0.32812	13.387	53/64	0.82812	33.787
11/32	0.34375	14.025	27/32	0.84375	34.425
23/64	0.35937	14.662	55/64	0.85937	35.062
3/8	0.375	15.3	7/8	0.875	35.70
25/64	0.39062	15.937	57/64	0.89062	36.337
13/32	0.40625	16.575	29/32	0.90625	36.975
27/64	0.42187	17.212	59/64	0.92187	37.612
7/16	0.4375	17.85	15/16	0.9375	38.250
29/64	0.45312	18.487	61/64	0.95312	38.887
15/32	0.46875	19.125	31/32	0.96875	39.525
31/64	0.48437	19.762	63/64	0.98437	40.162
1/2	0.50	20.40	1	1.00	40.80

TABLE C.1

APPENDIX D

Pipe Dimensions and Weights

Properties of Commercial Steel Pipe

Nominal Pipe Size & Outside Diameter	Pipe Schedule			Nominal Wall Thickness (in.)	Inside Diameter (in.)	Inside Area (in.^{2})	Metal Area (in.^{2})	Weight (lb/ft)	Weight of Water (lb/ft)
1/8	5			0.035	0.335	0.0881	0.0407	0.1385	0.0382
0.405	10		10S	0.049	0.307	0.0740	0.0548	0.1865	0.0321
	40	Std	40S	0.068	0.269	0.0568	0.0720	0.2450	0.0246
	80	XS	80S	0.095	0.215	0.0363	0.0925	0.3149	0.0157
1/4	5			0.049	0.442	0.1534	0.0756	0.2572	0.0665
0.540	10		10S	0.065	0.410	0.1320	0.0970	0.3301	0.0572
	40	Std	40S	0.088	0.364	0.1041	0.1250	0.4253	0.0451
	80	XS	80S	0.119	0.302	0.0716	0.1574	0.5356	0.0310
3/8	5			0.049	0.577	0.2615	0.0964	0.3280	0.1133
0.675	10		10S	0.065	0.545	0.2333	0.1246	0.4239	0.1011
	40	Std	40S	0.091	0.493	0.1909	0.1670	0.5682	0.0827
	80	XS	80S	0.126	0.423	0.1405	0.2173	0.7396	0.0609
1/2	5		5S	0.065	0.710	0.3959	0.1583	0.5386	0.1715
0.840	10		10S	0.083	0.674	0.3568	0.1974	0.6718	0.1546
	40	Std	40S	0.109	0.622	0.3039	0.2503	0.8519	0.1316
	80	XS	80S	0.147	0.546	0.2341	0.3200	1.0892	0.1014
	160			0.187	0.466	0.1706	0.3836	1.3055	0.0739
		XXS		0.294	0.252	0.0499	0.5043	1.7162	0.0216
3/4	5		5S	0.065	0.920	0.6648	0.2011	0.6845	0.2880
1.050	10		10S	0.083	0.884	0.6138	0.2521	0.8581	0.2659
	40	Std	40S	0.113	0.824	0.5333	0.3326	1.1320	0.2310
	80	XS	80S	0.154	0.742	0.4324	0.4335	1.4753	0.1873
	160			0.218	0.614	0.2961	0.5698	1.9392	0.1283
		XXS		0.308	0.434	0.1479	0.7180	2.4434	0.0641
1	5		5S	0.065	1.185	1.1029	0.2553	0.8687	0.4778
1.315	10		10S	0.109	1.097	0.9452	0.4130	1.4054	0.4094
	40	Std	40S	0.133	1.049	0.8643	0.4939	1.6808	0.3744
	80	XS	80S	0.179	0.957	0.7193	0.6388	2.1740	0.3116
	160			0.250	0.815	0.5217	0.8364	2.8466	0.2260
		XXS		0.358	0.599	0.2818	1.0763	3.6630	0.1221
1 1/4	5		5S	0.065	1.530	1.8385	0.3257	1.1084	0.7965
1.660	10		10S	0.109	1.442	1.6331	0.5311	1.8075	0.7075
	40	Std	40S	0.140	1.380	1.4957	0.6685	2.2751	0.6479
	80	XS	80S	0.191	1.278	1.2828	0.8815	2.9998	0.5557
	160			0.250	1.160	1.0568	1.1074	3.7687	0.4578
		XXS		0.382	0.896	0.6305	1.5337	5.2195	0.2731
1 1/2	5		5S	0.065	1.770	2.4606	0.3747	1.2752	1.0659
1.900	10		10S	0.109	1.682	2.2220	0.6133	2.0872	0.9626
	40	Std	40S	0.145	1.610	2.0358	0.7995	2.7207	0.8819
	80	XS	80S	0.200	1.500	1.7671	1.0681	3.6351	0.7655
	160			0.281	1.338	1.4061	1.4292	4.8640	0.6091
		XXS		0.400	1.100	0.9503	1.8850	6.4149	0.4117
2	5		5S	0.065	2.245	3.9584	0.4717	1.6053	1.7148

TABLE D.1 Pipe Dimensions and Weights

Properties of Commercial Steel Pipe

Nominal Pipe Size & Outside Diameter	Pipe Schedule			Nominal Wall Thickness (in.)	Inside Diameter (in.)	Inside Area (in.^{2})	Metal Area (in.^{2})	Weight (lb/ft)	Weight of Water (lb/ft)
2.375	10		10S	0.109	2.157	3.6542	0.7760	2.6407	1.5830
	40	Std	40S	0.154	2.067	3.3556	1.0745	3.6568	1.4536
	80	XS	80S	0.218	1.939	2.9529	1.4773	5.0274	1.2792
	160			0.343	1.689	2.2405	2.1896	7.4517	0.9706
		XXS		0.436	1.503	1.7742	2.6559	9.0386	0.7686
2 1/2	5		5S	0.083	2.709	5.7638	0.7280	2.4776	2.4969
2.875	10		10S	0.120	2.635	5.4532	1.0386	3.5346	2.3623
	40	Std	40S	0.203	2.469	4.7878	1.7041	5.7992	2.0741
	80	XS	80S	0.276	2.323	4.2383	2.2535	7.6692	1.8360
	160			0.375	2.125	3.5466	2.9452	10.0233	1.5364
		XXS		0.552	1.771	2.4634	4.0285	13.7096	1.0671
3	5		5S	0.083	3.334	8.7301	0.8910	3.0322	3.7819
3.500	10		10S	0.120	3.260	8.3469	1.2742	4.3365	3.6159
	40	Std	40S	0.216	3.068	7.3927	2.2285	7.5839	3.2025
	80	XS	80S	0.300	2.900	6.6052	3.0159	10.2638	2.8614
	160			0.437	2.626	5.4160	4.2051	14.3109	2.3462
		XXS		0.600	2.300	4.1548	5.4664	18.6032	1.7998
3 1/2	5		5S	0.083	3.834	11.545	1.0214	3.4759	5.0013
4.000	10		10S	0.120	3.760	11.104	1.4627	4.9779	4.8101
	40	Std	40S	0.226	3.548	9.8868	2.6795	9.1190	4.2830
	80	XS	80S	0.318	3.364	8.8880	3.6784	12.5184	3.8503
		XXS		0.636	2.728	5.8449	6.7215	22.8744	2.5320
4	5		5S	0.083	4.334	14.753	1.1517	3.9196	6.3908
4.500	10		10S	0.120	4.260	14.253	1.6512	5.6194	6.1744
	40	Std	40S	0.237	4.026	12.730	3.1740	10.8019	5.5148
	60			0.281	3.938	12.180	3.7245	12.6752	5.2763
	80	XS	80S	0.337	3.826	11.497	4.4074	14.9994	4.9804
	120			0.437	3.626	10.326	5.5780	18.9830	4.4734
	160			0.531	3.438	9.2833	6.6210	22.5327	4.0215
		XXS		0.674	3.152	7.8030	8.1013	27.5703	3.3803
4 1/2		Std	40S	0.247	4.506	15.947	3.6882	12.5517	6.9081
5.000		XS	80S	0.355	4.290	14.455	5.1804	17.6300	6.2617
		XXS		0.710	3.580	10.066	9.5690	32.5651	4.3606
5	5		5S	0.109	5.345	22.438	1.8676	6.3559	9.7202
5.563	10		10S	0.134	5.295	22.020	2.2855	7.7779	9.5392
	40	Std	40S	0.258	5.047	20.006	4.2999	14.6333	8.6665
	80	XS	80S	0.375	4.813	18.194	6.1120	20.8003	7.8815
	120			0.500	4.563	16.353	7.9529	27.0655	7.0840
	160			0.625	4.313	14.610	9.6957	32.9965	6.3290
		XXS		0.750	4.063	12.965	11.340	38.5935	5.6166
6	5		5S	0.109	6.407	32.240	2.2313	7.5936	13.9665
6.625	10		10S	0.134	6.357	31.739	2.7325	9.2994	13.7494
	40	Std	40S	0.280	6.065	28.890	5.5814	18.9945	12.5153
	80	XS	80S	0.432	5.761	26.067	8.4049	28.6037	11.2921

Properties of Commercial Steel Pipe

Nominal Pipe Size & Outside Diameter	Pipe Schedule			Nominal Wall Thickness (in.)	Inside Diameter (in.)	Inside Area (in.^{2})	Metal Area (in.^{2})	Weight (lb/ft)	Weight of Water (lb/ft)
	120			0.562	5.501	23.767	10.705	36.4302	10.2958
	160			0.718	5.189	21.147	13.324	45.3449	9.1611
		XXS		0.864	4.897	18.834	15.637	53.2168	8.1590
7		Std	40S	0.301	7.023	38.738	6.9257	23.5696	16.7812
7.625		XS	80S	0.500	6.625	34.472	11.1919	38.0884	14.9331
		XXS		0.875	5.875	27.109	18.5550	63.1465	11.7434
8	5		5S	0.109	8.407	55.510	2.9162	9.9243	24.0470
8.625	10		10S	0.148	8.329	54.485	3.9414	13.4135	23.6028
	20			0.250	8.125	51.849	6.5777	22.3853	22.4608
	30			0.277	8.071	51.162	7.2646	24.7229	22.1632
	40	Std	40S	0.322	7.981	50.027	8.3993	28.5843	21.6717
	60			0.406	7.813	47.943	10.4832	35.6765	20.7689
	80	XS	80S	0.500	7.625	45.664	12.7627	43.4341	19.7814
	100			0.593	7.439	43.463	14.9633	50.9232	18.8281
	120			0.718	7.189	40.591	17.8355	60.6979	17.5839
	140			0.812	7.001	38.496	19.931	67.8283	16.6763
		XXS		0.875	6.875	37.122	21.3039	72.5015	16.0814
	160			0.906	6.813	36.456	21.9705	74.7699	15.7927
9		Std	40S	0.342	8.941	62.786	9.9739	33.9431	27.1988
9.625		XS	80S	0.500	8.625	58.426	14.3335	48.7798	25.3103
10	5		5S	0.134	10.482	86.294	4.4691	15.2091	37.3824
10.750	10		10S	0.165	10.420	85.276	5.4869	18.6729	36.9414
	20			0.250	10.250	82.516	8.2467	28.0651	35.7459
	30			0.307	10.136	80.691	10.0719	34.2769	34.9552
	40	Std	40S	0.365	10.020	78.854	11.9083	40.5263	34.1597
	60	XS	80S	0.500	9.750	74.662	16.1007	54.7938	32.3435
	80			0.593	9.564	71.840	18.9221	64.3958	31.1213
	100			0.718	9.314	68.134	22.6288	77.0104	29.5155
	120			0.843	9.064	64.525	26.2373	89.2909	27.9523
	140			1.000	8.750	60.132	30.6305	104.242	26.0492
	160			1.125	8.500	56.745	34.0176	115.769	24.5819
11		Std	40S	0.375	11.000	95.033	13.4009	45.6058	41.1684
11.750		XS	80S	0.500	10.750	90.763	17.6715	60.1395	39.3183
12			5S	0.156	12.438	121.50	6.1722	21.0051	52.6356
12.750	5			0.165	12.420	121.153	6.524	22.201	52.4833
	10		10S	0.180	12.390	120.57	7.1082	24.1905	52.2301
	20			0.250	12.250	117.86	9.8175	33.4108	51.0564
	30			0.330	12.090	114.80	12.8761	43.8201	49.7314
		Std	40S	0.375	12.000	113.10	14.5790	49.6151	48.9938
	40			0.406	11.938	111.93	15.7446	53.5820	48.4888
		XS	80S	0.500	11.750	108.43	19.2423	65.4852	46.9736
	60			0.562	11.626	106.16	21.5188	73.2329	45.9874
	80			0.687	11.376	101.64	26.0353	88.6032	44.0309
	100			0.843	11.064	96.142	31.534	107.317	41.6488
	120			1.000	10.750	90.763	36.9137	125.625	39.3183

TABLE D.1 Pipe Dimensions and Weights (Continued)

Properties of Commercial Steel Pipe

Nominal Pipe Size & Outside Diameter	Pipe Schedule			Nominal Wall Thickness (in.)	Inside Diameter (in.)	Inside Area (in.^{2})	Metal Area (in.^{2})	Weight (lb/ft)	Weight of Water (lb/ft)
	140			1.125	10.500	86.590	41.0861	139.824	37.5109
	160			1.312	10.126	80.531	47.145	160.443	34.8862
14			5S	0.156	13.688	147.15	6.7848	23.0900	63.7468
14.000			10S	0.188	13.624	145.78	8.1576	27.7621	63.1521
	10			0.250	13.500	143.14	10.7992	36.7519	62.0077
	20			0.312	13.376	140.52	13.4167	45.6596	60.8739
	30	Std	40S	0.375	13.250	137.89	16.0516	54.6267	59.7324
	40			0.437	13.126	135.32	18.6203	63.3687	58.6196
		XS	80S	0.500	13.000	132.73	21.206	72.1674	57.4996
	60			0.593	12.814	128.96	24.9768	85.0009	55.8660
	80			0.750	12.500	122.72	31.2196	106.246	53.1616
	100			0.937	12.126	115.48	38.453	130.864	50.0280
	120			1.093	11.814	109.62	44.320	150.828	47.4867
	140			1.250	11.500	103.87	50.069	170.395	44.9960
	160			1.406	11.188	98.309	55.629	189.316	42.5876
16			5S	0.165	15.670	192.85	8.2083	27.9344	83.5442
16.000			10S	0.188	15.624	191.72	9.3389	31.7821	83.0544
	10			0.250	15.500	188.69	12.3700	42.0977	81.7413
	20			0.312	15.376	185.68	15.3770	52.3311	80.4387
	30	Std	40S	0.375	15.250	182.65	18.4078	62.6453	79.1258
	40	XS	80S	0.500	15.000	176.71	24.3473	82.8589	76.5528
	60			0.656	14.688	169.44	31.6222	107.617	73.4013
	80			0.843	14.314	160.92	40.1412	136.609	69.7108
	100			1.031	13.938	152.58	48.4843	165.002	66.0966
	120			1.218	13.564	144.50	56.5627	192.494	62.5971
	140			1.437	13.126	135.32	65.744	223.741	58.6196
	160			1.593	12.814	128.96	72.101	245.373	55.8660
18			5S	0.165	17.670	245.22	9.2450	31.4626	106.231
18.000			10S	0.188	17.624	243.95	10.5201	35.8020	105.679
	10			0.250	17.500	240.53	13.9408	47.4434	104.197
	20			0.312	17.376	237.13	17.3374	59.0025	102.725
		Std	40S	0.375	17.250	233.71	20.7640	70.6639	101.241
	30			0.437	17.126	230.36	24.1118	82.0573	99.7907
		XS	80S	0.500	17.000	226.98	27.4889	93.5503	98.3278
	40			0.562	16.876	223.68	30.7881	104.778	96.8986
	60			0.750	16.500	213.82	40.6444	138.321	92.6288
	80			0.937	16.126	204.24	50.2279	170.936	88.4773
	100			1.156	15.688	193.30	61.172	208.181	83.7362
	120			1.375	15.250	182.65	71.815	244.400	79.1258
	140			1.562	14.876	173.80	80.6640	274.516	75.2923
	160			1.781	14.438	163.72	90.7482	308.834	70.9239
20			5S	0.188	19.624	302.46	11.7014	39.8220	131.025
20.000			10S	0.218	19.564	300.61	13.5480	46.1067	130.225
	10			0.250	19.500	298.65	15.5116	52.7891	129.374
	20	Std	40S	0.375	19.250	291.04	23.1202	78.6825	126.078
	30	XS	80S	0.500	19.000	283.53	30.6305	104.242	122.825
	40			0.593	18.814	278.00	36.1546	123.041	120.432

Properties of Commercial Steel Pipe

Nominal Pipe Size & Outside Diameter	Pipe Schedule			Nominal Wall Thickness (in.)	Inside Diameter (in.)	Inside Area (in.^{2})	Metal Area (in.^{2})	Weight (lb/ft)	Weight of Water (lb/ft)
	60			0.812	18.376	265.21	48.9481	166.580	114.889
	80			1.031	17.938	252.72	61.4403	209.093	109.478
	100			1.280	17.440	238.88	75.2776	256.185	103.484
	120			1.500	17.000	226.98	87.179	296.688	98.3278
	140			1.750	16.500	213.82	100.335	341.459	92.6288
	160			1.968	16.064	202.67	111.4856	379.408	87.7982
22			5S	0.188	21.624	367.25	12.8826	43.8420	159.093
22.000			10S	0.218	21.564	365.21	14.9178	50.7682	158.211
	10			0.250	21.500	363.05	17.0824	58.1349	157.273
	20	Std		0.375	21.250	354.66	25.4764	86.7011	153.637
	30	XS		0.500	21.000	346.36	33.7721	114.933	150.043
	60			0.875	20.250	322.06	58.0704	197.625	139.517
	80			1.125	19.750	306.35	73.7783	251.082	132.713
	100			1.375	19.250	291.04	89.0936	303.203	126.078
	120			1.625	18.750	276.12	104.016	353.988	119.614
	140			1.875	18.250	261.59	118.546	403.436	113.319
	160			2.125	17.750	247.45	132.6832	451.547	107.195
24			5S	0.218	23.564	436.10	16.2875	55.4297	188.919
24.000	10		10S	0.250	23.500	433.74	18.6532	63.4806	187.894
	20	Std	40S	0.375	23.250	424.56	27.8325	94.7197	183.918
		XS	80S	0.500	23.000	415.48	36.9137	125.625	179.984
	30			0.562	22.876	411.01	41.3815	140.830	178.049
	40			0.687	22.626	402.07	50.3158	171.235	174.178
	60			0.968	22.064	382.35	70.0417	238.366	165.633
	80			1.218	21.564	365.21	87.174	296.672	158.211
	100			1.531	20.938	344.32	108.071	367.787	149.159
	120			1.812	20.376	326.08	126.3067	429.847	141.259
	140			2.062	19.876	310.28	142.1136	483.641	134.411
	160			2.343	19.314	292.98	159.4118	542.510	126.918
26	10			0.312	25.376	505.75	25.1788	85.6884	219.091
26.000		Std	40S	0.375	25.250	500.74	30.1887	102.738	216.921
	20	XS	80S	0.500	25.000	490.87	40.0553	136.316	212.647
28	10			0.312	27.376	588.61	27.1391	92.3599	254.987
28.000		Std	40S	0.375	27.250	583.21	32.5449	110.757	252.645
	20	XS		0.500	27.000	572.56	43.1969	147.008	248.031
	30			0.625	26.750	562.00	53.7507	182.924	243.459
30			5S	0.250	29.500	683.49	23.3656	79.5178	296.089
30.000	10		10S	0.312	29.376	677.76	29.0995	99.0314	293.605
		Std	40S	0.375	29.250	671.96	34.9011	118.776	291.092
	20	XS	80S	0.500	29.000	660.52	46.3385	157.699	286.137
	30			0.625	28.750	649.18	57.6777	196.289	281.225
32	10			0.312	31.376	773.19	31.0598	105.703	334.945
32.000		Std	40S	0.375	31.250	766.99	37.2573	126.794	332.260
	20	XS	80S	0.500	31.000	754.77	49.4801	168.391	326.965
	30			0.625	30.750	742.64	61.6047	209.653	321.713

TABLE D.1 Pipe Dimensions and Weights (Continued)

Properties of Commercial Steel Pipe

Nominal Pipe Size & Outside Diameter	Pipe Schedule			Nominal Wall Thickness (in.)	Inside Diameter (in.)	Inside Area (in.^{2})	Metal Area (in.^{2})	Weight (lb/ft)	Weight of Water (lb/ft)
	40			0.688	30.624	736.57	67.6782	230.323	319.082
34	10			0.312	33.376	874.90	33.0202	112.374	379.007
34.000		Std	40S	0.375	33.250	868.31	39.6135	134.813	376.150
	20			0.500	33.000	855.30	52.6217	179.082	370.515
	30			0.625	32.750	842.39	65.5317	223.017	364.923
	40			0.688	32.624	835.92	72.0011	245.034	362.120
36	10			0.312	35.376	982.90	34.9806	119.046	425.790
36.000		Std	40S	0.375	35.250	975.91	41.9697	142.831	422.763
		XS	80S	0.500	35.000	962.11	55.7633	189.774	416.787
	30			0.625	34.750	948.42	69.4587	236.382	410.854
	40			0.750	34.500	934.82	83.0559	282.656	404.964
42		Std	40S	0.375	41.250	1336.4	49.0383	166.887	578.930
42.000		XS	80S	0.500	41.000	1320.3	65.1880	221.848	571.934
48		Std	40S	0.375	47.250	1753.5	56.1069	190.943	759.595
48.000		XS	80S	0.500	47.000	1734.9	74.6128	253.922	751.578

APPENDIX E
Dry Saturated Steam Temperatures

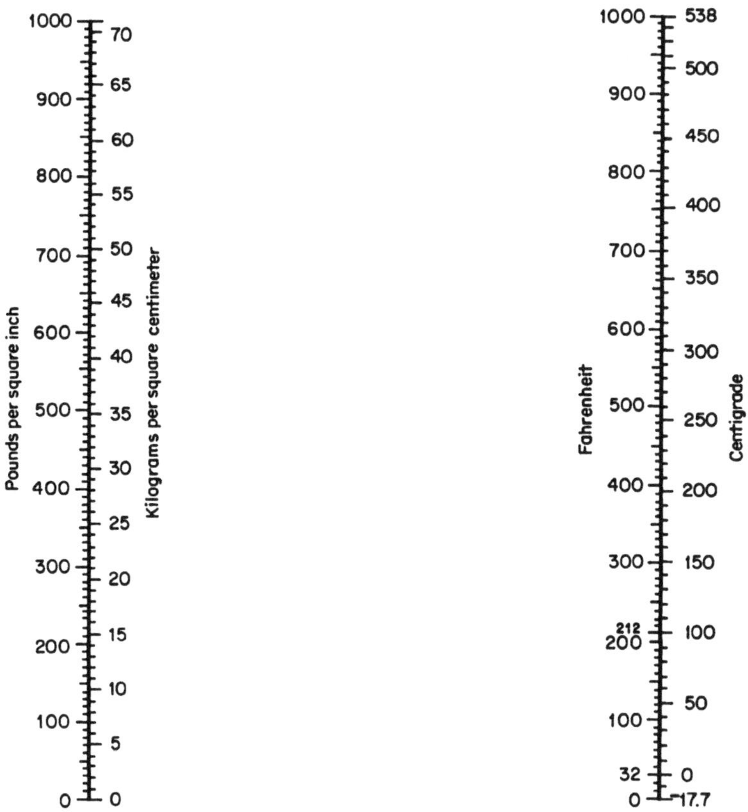

FIGURE E.1 U.S. customary versus metric conversions.

Temperature, deg F	Absolute pressure, psi	Specific volume: Saturated liquid	Specific volume: Evaporation	Specific volume: Saturated vapor	Enthalpy: Saturated liquid	Enthalpy: Evaporation	Enthalpy: Saturated vapor	Entropy: Saturated liquid	Entropy: Evaporation	Entropy: Saturated vapor
32	0.08854	0.01602	3306	3306	0.00	1075.8	1075.8	0.0000	2.1877	2.1877
35	0.09995	0.01602	2947	2947	3.02	1074.1	1077.1	0.0061	2.1709	2.1770
40	0.12170	0.01602	2444	2444	8.05	1071.3	1079.3	0.0162	2.1435	2.1597
45	0.14752	0.01602	2036.4	2036.4	13.06	1068.4	1081.5	0.0262	2.1167	2.1429
50	0.17811	0.01603	1703.2	1703.2	18.07	1065.6	1083.7	0.0361	2.0903	2.1264
60	0.2563	0.01604	1206.6	1206.7	28.06	1059.9	1088.0	0.0555	2.0393	2.0948
70	0.3631	0.01606	867.8	867.9	38.04	1054.3	1092.3	0.0745	1.9902	2.0647
80	0.5069	0.01608	633.1	633.1	48.02	1048.6	1096.6	0.0932	1.9428	2.0360
90	0.6982	0.01610	468.0	468.0	57.99	1042.9	1100.9	0.1115	1.8972	2.0087
100	0.9492	0.01613	350.3	350.4	67.97	1037.2	1105.2	0.1295	1.8531	1.9826
110	1.2748	0.01617	265.3	265.4	77.94	1031.6	1109.5	0.1471	1.8106	1.9577
120	1.6924	0.01620	203.25	203.27	87.92	1025.8	1113.7	0.1645	1.7694	1.9339
130	2.2225	0.01625	157.32	157.34	97.90	1020.0	1117.9	0.1816	1.7296	1.9112
140	2.8886	0.01629	122.99	123.01	107.89	1014.1	1122.0	0.1984	1.6910	1.8894
150	3.718	0.01634	97.06	97.07	117.89	1008.2	1126.1	0.2149	1.6537	1.8685
160	4.741	0.01639	77.27	77.29	127.89	1002.3	1130.2	0.2311	1.6174	1.8485
170	5.992	0.01645	62.04	62.06	137.90	996.3	1134.2	0.2472	1.5822	1.8293
180	7.510	0.01651	50.21	50.23	147.92	990.2	1138.1	0.2630	1.5480	1.8109
190	9.339	0.01657	40.94	40.96	157.95	984.1	1142.0	0.2785	1.5147	1.7932
200	11.526	0.01663	33.62	33.64	167.99	977.9	1145.9	0.2938	1.4824	1.7762
210	14.123	0.01670	27.80	27.82	178.05	971.6	1149.7	0.3090	1.4508	1.7598
212	14.696	0.01672	26.78	26.80	180.07	970.3	1150.4	0.3120	1.4446	1.7566
220	17.186	0.01677	23.13	23.15	188.13	965.2	1153.4	0.3239	1.4201	1.7440
230	20.780	0.01684	19.365	19.382	198.23	958.8	1157.0	0.3387	1.3901	1.7288
240	24.969	0.01692	16.306	16.323	208.34	952.2	1160.5	0.3531	1.3609	1.7140
250	29.825	0.01700	13.804	13.821	216.48	945.5	1164.0	0.3675	1.3323	1.6998
260	35.429	0.01709	11.746	11.763	228.64	938.7	1167.3	0.3817	1.3043	1.6860
270	41.858	0.01717	10.044	10.061	238.84	931.8	1170.6	0.3958	1.2769	1.6727
280	49.203	0.01726	8.628	8.645	249.06	924.7	1173.8	0.4096	1.2501	1.6597
290	57.556	0.01735	7.444	7.461	259.31	917.5	1176.8	0.4234	1.2238	1.6472

TABLE E.1 Dry Saturated Steam Temperatures

Temperature, deg F	Absolute pressure, psi	Specific volume			Enthalpy			Entropy		
		Saturated liquid	Evaporation	Saturated vapor	Saturated liquid	Evaporation	Saturated vapor	Saturated liquid	Evaporation	Saturated vapor
300	67.013	0.01745	6.449	6.466	269.59	910.1	1179.7	0.4369	1.1980	1.6350
310	77.68	0.01755	5.609	5.626	279.92	902.6	1182.5	0.4504	1.1727	1.6231
320	89.66	0.01765	4.896	4.914	290.28	894.9	1185.2	0.4637	1.1478	1.6115
330	103.06	0.01776	4.289	4.307	300.68	887.0	1187.7	0.4769	1.1233	1.6002
340	118.01	0.01787	3.770	3.788	311.13	879.0	1190.1	0.4900	1.0992	1.5891
350	134.63	0.01799	3.324	3.342	321.63	870.7	1192.3	0.5029	1.0754	1.5783
360	153.04	0.01811	2.939	2.957	332.18	862.2	1194.4	0.5158	1.0519	1.5677
370	173.37	0.01823	2.606	2.625	342.79	853.5	1196.3	0.5286	1.0287	1.5573
380	195.77	0.01836	2.317	2.335	353.45	844.6	1198.1	0.5413	1.0059	1.5471
390	220.37	0.01850	2.0651	2.0836	364.17	835.4	1199.6	0.5539	0.9832	1.5371
400	247.31	0.01864	1.8447	1.8633	374.97	826.0	1201.0	0.5664	0.9608	1.5272
410	276.75	0.01878	1.6512	1.6700	385.83	816.3	1202.1	0.5788	0.9386	1.5174
420	308.83	0.01894	1.4811	1.5000	396.77	806.3	1203.1	0.5912	0.9166	1.5078
430	343.72	0.01910	1.3308	1.3499	407.79	796.0	1203.8	0.6035	0.8947	1.4982
440	381.59	0.01926	1.1979	1.2171	418.90	785.4	1204.3	0.6158	0.8730	1.4887
450	422.6	0.0194	1.0799	1.0993	430.1	774.5	1204.6	0.6280	0.8513	1.4793
460	466.9	0.0196	0.9748	0.9944	441.4	763.2	1204.6	0.6402	0.8298	1.4700
470	514.7	0.0198	0.8811	0.9009	452.8	751.5	1204.3	0.6523	0.8083	1.4606
480	566.1	0.0200	0.7972	0.8172	464.4	739.4	1203.7	0.6645	0.7868	1.4513
490	621.4	0.0202	0.7221	0.7423	476.0	726.8	1202.8	0.6766	0.7653	1.4419
500	680.8	0.0204	0.6545	0.6749	487.8	713.9	1201.7	0.6887	0.7438	1.4325
520	812.4	0.0209	0.5385	0.5594	511.9	686.4	1198.2	0.7130	0.7006	1.4136
540	962.5	0.0215	0.4434	0.4649	536.6	656.6	1193.2	0.7374	0.6568	1.3942
560	1133.1	0.0221	0.3647	0.3868	562.2	624.2	1186.4	0.7621	0.6121	1.3742
580	1325.8	0.0228	0.2989	0.3217	588.9	588.4	1177.3	0.7872	0.5659	1.3532
600	1542.9	0.0236	0.2432	0.2668	617.0	548.5	1165.5	0.8131	0.5176	1.3307
620	1786.6	0.0247	0.1955	0.2201	646.7	503.6	1150.3	0.8398	0.4664	1.3062
640	2059.7	0.0260	0.1538	0.1798	678.6	452.0	1130.5	0.8679	0.4110	1.2789
660	2365.4	0.0278	0.1165	0.1442	714.2	390.2	1104.4	0.8987	0.3485	1.2472
680	2708.1	0.0305	0.0810	0.1115	757.3	309.9	1067.2	0.9351	0.2719	1.2071
700	3093.7	0.0369	0.0392	0.0761	823.3	172.1	995.4	0.9905	0.1484	1.1389
705.4	3206.2	0.0503	0	0.0503	902.7	0	902.7	1.0580	0	1.0580

Abridged from "Thermodynamic Properties of Steam" by Joseph H. Keenan and Frederick G. Keyes. John Wiley & Sons. New York. 1937

APPENDIX F

Metric (SI) Units; English and Metric Conversions

Metric notation and terminology employs a base term, preceded by a prefix, which indicates a multiple of that base term. For example, mass is measured in grams. To indicate 1000 grams, the prefix *kilo* is placed in front of gram, and thus, we obtain kilogram. The multiples of the base are always expressed as powers of ten, as follows:

micro	1/1,000,000
milli	1/1,000
centi	1/100
deci	1/10
deca	10
hecta	100
kilo	1,000
mega	1,000,000

An example of pressure vessel thickness calculations is demonstrated for both English and SI metric units as follows:

English units		Metric units
Pressure *P*	250 psi	1723.7 kPa
Diameter *D*	24 in	609.6 mm
Radius *R*	12 in	304.8 mm
Allowable stress *S*	13,800 psi	95,148.2 kPa
Joint efficiency *E*	1.0	1.0
Crown radius *L*	24 in	609.6 mm

English	Metric
Shell Thickness	
$T = \frac{P(R)}{SE - 0.6P}$	$T = \frac{P(R)}{SE - 0.6P}$
$= \frac{250(12)}{13{,}800(1) - 0.6(250)}$	$= \frac{1723.7(304.8)}{95{,}148.2(1) - 0.6(1723.7)}$
$= 0.220$ in	$= 5.582$ mm
Ellipsoidal Head Thickness	
$T = \frac{P(D)}{2SE - 0.2P}$	$T = \frac{P(D)}{2SE - 0.2P}$
$= \frac{250(24)}{2(13{,}800)(1) - 0.2(250)}$	$= \frac{1723.7(609.6)}{2(95{,}148.2)(1) - 0.2(1723.7)}$
$= 0.218$ in	$= 5.532$ mm
Torispherical Head Thickness	
$T = \frac{0.885\ PL}{SE - 0.1P}$	$T = \frac{0.885\ PL}{SE - 0.1P}$
$= \frac{0.885(250)(24)}{13{,}800(1) - 0.1(250)}$	$= \frac{0.885(1723.7)(609.6)}{95{,}148(1) - 0.1(1723.7)}$
$= 0.385$ in	$= 9.791$ mm

TABLE F.1 Sample Calculations

Multiply	By	To obtain
	Length	
Inches	25.4	Millimeters
Inches	0.0254	Meters
Inches	2.540	Centimeters
Feet	304.8	Millimeters
Feet	30.48	Centimeters
Feet	0.3048	Meters
Millimeters	0.03937	Inches
Centimeters	0.3937	Inches
Meters	39.37	Inches
Meters	3.2808	Feet
	Area	
Square inches	645.2	Square millimeters
Square inches	0.0006452	Square meters
Square feet	0.09290	Square meters
Square inches	6.4516	Square centimeters
Square millimeters	0.00155	Square inches
Square centimeters	0.1550	Square inches
Square meters	10.764	Square feet
	Volume and volumetric flow rates	
Cubic inches	0.01639	Liters
Cubic inches	16390.0	Cubic millimeters
Gallons	3.785	Liters
Liters	0.2642	Gallons
Gallons/minute	0.06308	Liters/second
Gallons	0.003785	Cubic meters
Gallons	3785000.0	Cubic millimeters
Cubic feet	0.02832	Cubic meters
Cubic feet	28.32	Liters
Liters	0.03531	Cubic feet
Liters	61.02	Cubic inches
Cubic meters	264.2	Gallons
Cubic millimeters	0.00006102	Cubic inches
Cubic meters	35.31	Cubic feet
Liters/second	15.8529	Gallons/minute
	Weight or mass	
Pounds	0.4536	Kilograms
Pounds	453.6	Grams
Kilograms	2.2046	Pounds
Grams	0.0022046	Pounds

TABLE F.2 English and Metric Conversions

Multiply	By	To obtain
	Force	
Pounds force	4.448	Newtons
Newtons	0.2248	Pound force
	Bending (torque)	
Pound-foot	1.3558	Newton-meters
Pound-inch	0.11298	Newton-meters
Newton-meters	0.73757	Pound-foot
Newton-meters	8.8511	Pound-foot
	Pressure, stress	
Pound-square inch (psi)	6894.757	Pascals
Kips/square inch	6894757	Pascals
Pound/square inch	6.8948	Kilopascals
Kips/square inch	6.8948	Megapascals
Kilopascals	0.14504	Pounds/square inch
Kilogram/square centimeter	14.22	Pounds/square inch
Pounds/square inch	0.0703	Kilograms/square inch
Pound/square foot	47.8803	Pascals
Kilogram/square meter	9.8067	Pascals
	Fracture toughness	
Thousand pounds-$\sqrt{\text{inch}}$	1.0988×10^6	Pascal-$\sqrt{\text{meter}}$

TABLE F.2 English and Metric Conversions (*Continued*)

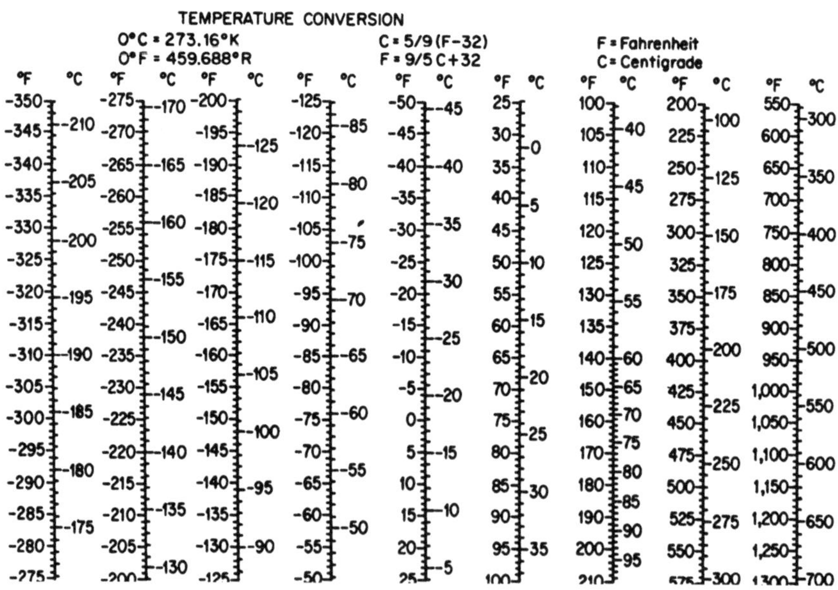

FIGURE F.1 Temperature conversion chart.

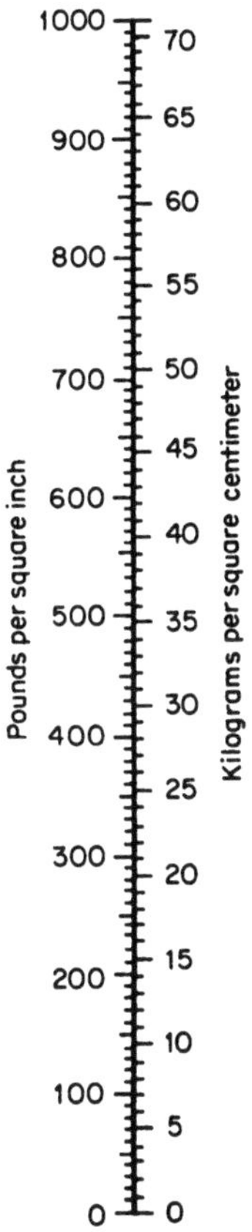

FIGURE F.2 English and metric conversion table.

Index

Page numbers followed by "f" are to figures; page numbers followed by "t" are to tables.

A

Accidents
- basic causes of, 101–102
- in data reports, 143
- design precautions against, 102–109
- in history of ASME code, 1–3
- during hydrostatic tests, 155–156, 157
- nondestructive examination to prevent, 194
- poor fit-up and, 183
- quick-opening closures and, 106–108

Acoustic emission testing, 203
Agitators, 104
AIA (Authorized Inspection Agency)
- for nuclear vessels, 225
- overview of, 6

Alloy steels, 188, 190
Alterations, 16–31
American National Standards Institute (ANSI), 5
American Society for Nondestructive Testing, 195–198, 200
American Society of Mechanical Engineers (ASME)
- Boiler and Pressure Vessel Committee of, 1, 4–5, 9
- website for, 6

American Welding Society, 177–178
Ammonium persulfate, 154
Annealing, 184, 187
ANSI (American National Standards Institute), 5
Appendices, of quality control manual, 135–136
ASME Boiler and Pressure Vessel Code
- Addenda to, 38–39
- additions and updates to, 3–5, 38–39
- Code Cases for, 10
- descriptive guide to, 33
- enforcement of, 6–7
- history of, 1–3
- interpretations of, 9–10
- parts of, 205
- sections of, 3–4, 211, 222–223
- visual reference for, 34f

ASME Certificate of Authorization
- for field fabrications, 145
- National Board Review in, 7–8
- overview of, 6
- in quality control manual creation, 112, 123
- for welded repairs or alterations, 16–31

Assemblers, field, 145
Atomic arrangements, of metals, 183, 184
Audits, 135, 195, 234
Austenitic materials
- tests of, 142, 169
- welding of, 184, 187

Authorized Inspection Agency (AIA)
- for nuclear vessels, 225
- overview of, 6

Authorized Inspectors. *See also* Inspections
- of Canadian shops, 11–15
- Code interpretations by, 9
- of Division 2 vessels, 209
- education requirements of, 7
- for field fabrications, 145–146

Authorized Inspectors (*continued*)
most important functions of, 161
of nuclear vessels, 224
origin of, 7
of power boilers, 217
providing desk/chair for, 159
quality control manual and, 112–136
service restrictions and, 95
shop inspections by, 137
shop organization and, 31
welded repairs and alterations, 16–17
Authorized Nuclear Inspectors, 231–233

B

Backing strips
in quality control tests, 130
radiograph interpretation and, 149, 151
shop inspections and, 139, 142
Base metals, 165
Bend tests, 167
Blumberg, H. S., 184
Boiler explosions, 1–3
Boilers and Pressure Vessels, 10
Boiling water reactors (BWRs), 221–222
Bolted flat cover plates, 73–74
Braced surfaces, 75, 78t, 142
Brazing, 163
Breech-lock doors, 106–108
Breeder reactors, 221–222
Brittle fracture, 82, 84, 92–93, 120
Burn-through, 151
Butt joints
preparation for welding of, 183
quality control of, 130
radiography requirements of, 147, 148t
service restrictions and, 93–94
in shop inspections, 42
Buttons, corrosion resistant, 101–102
BWRs (boiling water reactors), 221–222

C

Calculations, design
for circumference, 251, 252t–261t, 262
decimal equivalents in, 264
of Division 2 vessels, 207
pipe dimensions/weights for, 266t–271t
in quality control manual, 112–114, 115f
theoretical weights in, 264
volume tables/diagrams for, 245–246, 247t–249t, 250f
for weld inspections, 158
Calibrations of measurement, 132
Canada, 9, 10f, 11t–15t
Canadian Registration Number (CRN), 12–15
Cargo tanks, 238–242
Castings, 147
Certification Mark, 209
Checklists, 120, 121f–122f, 133
Circumference, calculating, 251, 252t–261t, 262
Circumferential joints
radiographic examination, 39–42
service restrictions and, 93–94
Clad vessels, 147, 164
Code Cases, ASME, 10
Code Inspectors. *See* Authorized Inspectors
Code symbol, 6. *See also specific symbols*
Code tolerances, 139–142
Compressed gas transport vessel, 243
Concrete reactor vessels, 223
Concrete saddles, 102
Conical heads, 75, 76t
Conical shells, 43f, 45f
Construction, 235
Continued service, 235
Controlled manuals, 226
Coolant systems, 220–224
Cooling process, 184, 188
Corrective actions, 233, 234
Corrosion
allowances for, 53t, 70, 82t, 103t
causes of, 101
design precautions for, 103–104
effects of, 101
inspection for, 102, 104–106
resistance data for, 103
in stress failures, 101–102
Court proceedings, 143
Cracks, 153, 183
CRN (Canadian Registration Number), 12–15
Cryogenic fluids, 242
Cryogenic vessels
materials for, 81–83
procedure qualification tests of, 169
service restrictions of, 93–95
Currents, electrical, 166
Cylindrical shells
circumference of, 251, 252t–261t, 262
external pressure and thickness of, 53–56, 56t–57t
internal pressure and thickness of, 44–46, 53t
joint efficiencies of, 43f, 45f
volume tables/diagrams for, 245–246, 247t–249t, 250f

D

Data reports
of accidents, 143
for field fabrications, 146
impact test exemptions, 90
importance of, 143
incomplete, 144
joint efficiencies in, 43
overpressure protection statement in, 109
in quality control manual preparation, 123, 133, 135
in quality control of nuclear vessels, 232–233
record disposal and, 133
service restrictions and, 95
for transport tanks, 238
Decimal equivalents, 264t
Department of Transportation (DOT) regulations, 15, 240
Design-by-analysis requirements, 207, 208
Design-by-rule requirements, 207, 208
Detailed process flowcharts, 119
Dimpled jacketed vessels, 95, 97t
Dished heads, 71t–72t
Division 1 Code, 205–206, 209, 223
Division 2 Code
versus Division 1 Code, 205–206, 209
inspectors in, 209
for nuclear vessels, 223
parts of, 206–210
Division 3 Code, 210
Document control, 228–229
DOT (Department of Transportation) regulations, 15, 240
Drawings
for field fabrication, 124, 145
in nuclear vessel quality assurance manuals, 228
in quality control systems, 112–114
of welding technique, 166
Dry saturated steam temperatures, 273f, 274t–275t
Ductility, 187, 189

E

E symbol, 39
Eddy current testing, 200t
Efficiency. *See* Joint efficiency
Eisenberg, G. M., 39
Electrical specifications, 166
Electrodes, 128, 168
Ellipsoidal heads
internal pressure thickness tables for, 57, 58t–69t, 70
joint design for, 70, 71t
Employee injuries, 143
Engineering
for Division 3 vessels, 210
for Division 2 vessels, 206, 207
of nuclear vessels, 224, 227–228
quality control responsibilities of, 113–119, 128
English conversions, 273f, 277, 278f, 279t–280t, 280f–281f
Erection companies, 123–124, 145, 215
Escape pipe discharge, 109
Etching of sectioned specimens, 154–155
Explosions, 1–3
External piping, 124, 145
External pressure thickness charts, 53–56, 56t–57t

F

Fatigue, 102, 206, 241, 242
Field fabrication
inspections of, 144–146
of power boilers, 215–217
quality control manuals for, 123–124
Filler metals, 165
Fillet welds, 130, 142, 177
Fit-ups, 127, 142, 180–183
Flange testing, 87–88
Flat cover plates, 73–74
Flat heads, 73–74
Fluorescent penetrant inspection, 198t
F-number grouping, 168
Forgings, 147
Fossil-fuel power plants, 219
Freon gas, 200
Full radiography, 41t
Furnace checks, 142, 189

G

Gages
calibration of, 132
in hydrostatic tests, 143, 157
in safety design, 107
Gamma rays, 148–149, 196t, 201f
Gas, 166
Gas and bubble formation tests, 201
Grain size, 184, 187

H

Halide torch, 200
Halogen diode detector testing method, 200

Handhole sizes, 105–106
H&D (Hurter and Driffield) curve, 150, 154
Heads. *See also specific types*
internal pressure thickness tables for, 57, 58t–69t, 70
joint design for, 70–81
joint efficiencies of, 44f, 45f
mill orders, 29
quality control of, 117
warehouse orders, 30
welding details/symbols for, 180f
Heat treatments, 131–132
Heat-affected zone, 187, 188
Helium mass spectrometer leak detector, 200–201
Hemispherical heads, 75, 77t
Hold tag, 125f
Hurter and Driffield (H&D) curve, 150, 154
Hydrochloric acid, 154
Hydrostatic tests
blocking prior to, 142–143
of Division 2 vessels, 209
mishaps/accidents involving, 155–156, 157
of nuclear vessels, 232
for quality control manuals, 120, 123, 134–135
in shop inspections, 138t
test gage requirements of, 157

I

Impact tests
for cargo tanks, 241
for cryogenic vessels, 83, 169, 170f
exemptions from, 84–93
rationale for, 84
Indicating gages, 157
Injuries, employee, 143
Inspections. *See also* Authorized Inspectors
of Division 2 vessels, 208–209
of field fabrications, 144–146
of hydrostatic testing materials, 157
of nuclear vessels, 231–233
of openings, 102, 104–106
quality control systems in, 112–136
of received materials, 116–119
of shops, 137–142
simplified steps for, 157–159
vessels exempt from, 95, 98t
Inspectors. *See* Authorized Inspectors
Interior/internal corrosion. *See* Corrosion
Internal pressure thickness tables
for cylindrical shells, 44–46, 53t
for ellipsoidal and torispherical heads, 57, 58t–69t, 70, 71t–72t
Interpretation Submittal Form, 9

J

Jacketed vessels, 95
Jigs, 167–168, 168f–169f
Joint efficiency
in data reports, 43
degree of radiography in, 39–43, 43f–45f
radiography requirements regarding, 147
in thickness tables, 47f–52f, 58t–69t
Joints, welded. *See* Welded joints

L

Laboratory testing, 131
Laminated plates, 153
Leak testing, 200–201
Leg sizes, 177
Legal documents, 143
Lined vessels, 147, 164
Liquid penetrant tests, 138t–139t, 197t
Loadings
of cargo tanks, 240, 241
Division 2 vessels, 206, 207
weather concerns, 104
Longitudinal cracks, 153
Longitudinal joints
radiographic examination, 39–42
service restrictions and, 93–94
Low-temperature vessels
impact test exemptions for, 84, 92–93
materials for, 81–83
service restrictions of, 93–95

M

Magnetic particle tests, 138t–139t, 197t, 202f
Manhole sizes, 105–106
Manufacturers
benefits of ASME Code to, 6
business challenges of, 19–21
of Division 2 vessels, 206–208
efficient business practices for, 22
field fabrication and, 123–124
litigation against, 143
of nuclear vessels, 224, 226
of power boilers, 215–217
quality control systems of, 112–136

Manufacturers (*continued*)
responsibilities of, 108–109, 144–145, 206–208
Manufacturer's Design Report, 207–208
Material control, 114, 116
Materials
classes of, 35t–38t
defects in, 102
of Division 2 vessels, 206, 208
in field fabrications, 124
impact test exemptions of, 91–93
life cycle of, 183–184
in low-temperature vessels, 82–83
mill orders of, 27–29
in nuclear vessel quality control system, 229
in procedure specification forms, 165
procurement control of, 114, 116–119
requiring postweld heat treatment, 190–191
in shop inspections, 137, 139
shop organization and, 31–32
stress values of, 35t–38t, 39
for transport tanks, 236
warehouse orders of, 29–30
Maximum allowable working pressure (MAWP), 39
MDMT (minimum design metal temperature), 84–92
Metallurgy, 183–187
Metric conversions, 273f
Mill orders, 27–29
Mill test reports, 116, 119, 137, 158
Minimum design metal temperature (MDMT), 84–92

N

Narrative welding procedures, 163
National Board Bulletin, 39
National Board Inspection Code (NBIC), 16, 17
National Board of Boiler and Pressure Vessel Inspectors
Canadian registrations, 9, 10f, 11t–15t
contact information for, 8
laboratory tests by, 8
nuclear vessel quality and, 225
overview of, 6–7
radiographic examinations and, 39
registration with, 8, 123
requirements of, 8–15
safety valve inspections by, 109
shop reviews by, 7–8, 11–12
stamping and, 7, 8, 11
National Board R stamp
forms related to, 18–31f
issuance of, 16–17, 18–19f
nameplate/stamping, 17, 32f
purpose of, 8
National Institute of Standards Testing (NIST), 132
NBIC (National Board Inspection Code), 16, 17
Neutrons, 219
Noncircular-cross-section-type vessels, 95, 96t–97t
Nonconformance
definition of, 125
handling of, 125
of nuclear vessels, 233
in quality control manual preparation, 124–127
of received materials, 119
types of, 124
Nonconformity reports, 126–127
Noncontrolled manuals, 226
Nondestructive examinations. *See also specific tests*
of Division 2 vessels, 208–209
methods of, 193–194, 196f–200f
of nuclear vessels, 226, 231
qualified/certified personnel for, 195–200
for quality control, 130–131, 195
rationale for, 193, 194
in shop inspections, 137–139
written procedures for, 195–196
Nonpermanent records, 233
Normalizing process, 187
Notch toughness, 84, 167
Nozzles
impact test exemptions for, 87–88
quality control of, 116–117
radiography requirements of, 147
reinforced openings for, 75–81, 82t
service restrictions and, 93–94
Nuclear Components, 10
Nuclear fission, 219, 222
Nuclear fusion, 222
Nuclear vessels
Code sections related to, 222–223
coolant systems in, 220–224
inspectors of, 224, 231–233
manufacturers of, 224, 225
nonconformance of, 233
nondestructive examinations for, 226, 231
overview of, 219
quality control manual for, 226–234
quality control systems for, 136, 224–234

O

Openings
  design precautions for, 103–104
  inspection of, 102, 104–106
  minimum sizes of, 105–106
  radiography requirements of, 147
  reinforcement of, 75, 77–81, 82t
  in transport tank design, 236
  in vessel failures, 102
Organizational charts, 112, 226, 227
Out-of-roundness measures, 139, 140t
Overpressure protection, 108–109, 207, 209
Overwelding, 176–177

P

Parent metals, 187
Partial data reports, 119
Partial radiography, 39
Peening, 166
Penetrameters, 149, 154
Permanent records, 233
Piezoelectric transducers, 203
Pipe walls
  dimensions/weights of, 266t–271t
  in joint preparation, 183
Plates. *See also specific types*
  decimal equivalents/theoretical weights for, 264t
  design precautions for, 104, 105t
  mill orders, 27–29
  quality control of, 116–117, 119
  warehouse orders, 29–30
  welding details/symbols for, 180f
Pneumatic tests
  of Division 2 vessels, 209
  of nuclear vessels, 232
  in shop inspections, 138t, 142
P-number grouping, 167
Porosity, 152, 154
Portable tanks, 242–243
Positions, welding, 165
Postweld heat treatment (PWHT)
  furnace checks in, 142
  impact test exemptions and, 87
  in procedure specification forms, 166
  in quality control manual preparation, 131–132
  rationale for, 188–189
  requirements for, 189–192
  in shop inspections, 137
Power Boiler Code, 211–217
PQR (procedure qualification records), 162–165, 167
Preheating
  definition of, 187
  rationale for, 188, 189f
  specifications for, 165–166
Pressure gages, 107, 132
Pressure relief devices
  for cargo tanks, 242
  in Division 2 vessels, 209
  in field fabrications, 144
  importance of, 108
  manufacturers' responsibilities regarding, 108–109
  overview of, 109
  for transport tanks, 237–238
Pressure vessels
  definition of, 235
  visual reference for, 34f
Pressurized water reactors (PWRs), 220–221
Probe saws, 154
Procedure qualification records (PQR), 162–165, 167
Purchase requisitions, 114, 116
PWHT. *See* Postweld heat treatment

Q

Quality control managers, 133
Quality control manuals
  Code requirements for, 112
  definition of, 6
  importance of, 111
  for nuclear vessels, 226–234
  revisions to, 133
  writing/implementation of, 112–136, 195
Quality control systems
  audits of, 135
  importance of, 111
  for nuclear vessels, 136, 224–234
  for power boilers, 217
  requirements for, 111–112
  shop inspections in, 137–142
  tests in, 142–143
  written description of, 112–136
Quick-opening closures, 106–108

R

R stamp. *See* National Board R stamp
Radiation, 148–149
Radiography
  benefits of, 196t
  conditions requiring, 146–147, 148t
  definition of, 146
  design thickness based on, 33–43, 40t, 41–42t, 43–44f
  of Division 2 vessels, 206
  interpretation of, 148–154

Radiography (*continued*)
for quality control, 114, 130–131, 142
in quality control manual preparation, 114, 130–131
in shop inspections, 137
Rail tank cars, 242
Receiving Inspection Report, 118f, 119
Recommended Practice for Nondestructive Examining Personnel Qualification and Certification (American Society for Nondestructive Testing), 195–198, 200
Recording gages, 157
Records
of nuclear vessel quality assurance, 233–234
of qualified welding procedures, 162
retention of, 132–133, 143, 234
Reflective surfaces, 240
Reinforced openings, 75, 77–81, 82t
Rupture disk devices, 109

S

Safety interlocking devices
in breech-lock doors, 106–107
failure of, 108
Safety valves
installation and settings, 109
National Board certification of, 8
in vessel failures, 102
Sectioning, 154–155
Service restrictions, 93–95
Shielding gas, 166
Shop supervisors, 176
Shops
inspection of, 137–142
National Board reviews of, 7–8, 11–12
organization of, 31–32
welded repairs and alterations, 16–17
work-flow schedules in, 24–27
Slag inclusions, 152, 153
SNT-TC-1A (American Society for Nondestructive Testing), 195–198, 200
Solutions, etching, 154–155
Specification cargo tanks, 15
Specification control, 112–114
Spherical shells, 53–56
Spot radiography, 39, 41t–42t, 152–153
Stamping
in Canadian provinces, 11–12
for efficient inspections, 158
of field fabrications, 145
in formation of National Board, 7
in hydrostatic testing, 120, 123
location of, 97
of pressure relief devices, 109
of qualified welders, 128
of received materials, 119
requirements for, 8, 11, 16–31, 97, 99f–100f
in shop inspections, 137, 139
of transport tanks, 238
welded repairs and alterations, 16–31
Standard welding procedure specifications (SWPSs), 163–164
Standard Welding Symbols (American Welding Society), 177–178
Statement of authority/responsibility, 112, 227
Stayed surfaces, 75, 78t
Stereoscopic principle, 151–152
Stiffening rings, 56, 57t
Stress failures, 101–102
Stress multipliers, 39
Stress values of materials
Code tables for, 35t–38t
in cylindrical shell thicknesses, 45–46
in ellipsoidal and torispherical heads, 70
in radiography, 39
Subcontracted examinations, 130–131, 195
Supplementary essential variables, 167
Survey teams, 225–226
SWPSs (standard welding procedure specifications), 163–164
Symbols, code. *See specific symbols*
Symbols, welding, 176–179, 180f–182f
Synopsis of Boiler and Pressure Vessel Laws, Rules, and Regulations, 9

T

Tack welds, 127, 128, 130
Tanks, 235
Telltale holes, 104
Tests. *See also specific tests*
of field fabrications, 146
of joint efficiency, 42
of nuclear vessels, 231
for quality control, 120, 123, 130–133, 142–143
in quality control manuals, 120, 123, 130–133
in shop inspections, 137–139
of transport tanks, 237
Throat dimensions, 177
Ton containers, 243
Toriconical heads, 75, 76t

Torispherical heads
internal pressure thickness tables for, 57, 58t–69t, 70
joint design for, 70, 72t
Training, of personnel, 195–200
The Transactions of the ASME Series J Journal of Pressure Vessel Technology, 39
Transport tanks
data reports for, 238
design and material requirements of, 236
examination and testing requirements for, 237
general Code requirements of, 236
pressure relief devices for, 237–238
repairs and alternations of, 238
specific types of, 238–243
stamping of, 238
terminology related to, 235
welding and fabrication requirements for, 237
Transverse cracks, 153
Trepanning, 154–155

U

Ultrasonic tests, 130–131, 199t, 202f
UM symbol, 95, 98t
Undercuts, 154
Underwelding, 176–177
Uniform Boiler and Pressure Vessel Laws Society, 9
Unstayed flat heads, 73t
Uranium, 221–222
U.S. customary measures, 273f, 277, 278f, 279t–280t, 280f–281f
User's Design Specification, 206–207

V

Vacuum insulated cargo tanks, 241–242
Vents, 157
Verified designs, 227
View box, radiography, 153
Voltage, 166
Volume tables/diagrams, 245–246, 247t–249t, 250f

W

Warehouse orders, 29–30
Warning devices, 107
Water pressure, 156f
Water temperature, 120, 143, 156
Weld metal zone, 187
Weld overlay, 153
Weld sizes, 176–177
Welded joints
categories of, 93–94
efficiencies of, 38–43, 43f–45f
fit-up of, 127, 142, 180–183
impact test exemptions for, 90–91
postweld heat treatment of, 190–192
preparation of, 180–183
in procedure specification forms, 164–165
quality control of, 127–131
radiographic examination of, 148–154
radiography requirements of, 146–147, 148t
sectioning for examination of, 154–155
service restrictions and, 93–95
specification drawings of, 166
Welded repairs
National Board R stamp for, 16–31
postweld heat treatment of, 191
radiography requirements of, 147
Welders
activity logs of, 128–129
of nuclear vessels, 230
qualifications of, 127, 128, 145, 163, 170–176, 230
stamps of, 128
written welding procedures for, 163
Welding
to close holes left by trepanning, 155
definition of, 184
details and symbols for, 176–179, 180f–182f
of Division 2 vessels, 208
metallurgical changes in, 183–187
of nuclear vessels, 229–230
preheating prior to, 187–188
procedure specification forms for, 164–166
qualified procedures for, 162–170
in quality control programs, 127–131, 133, 229–230
radiography interpretation of, 148–154
in shop inspections, 139, 141t, 142
of transport tanks, 237
variables, 164, 167–170
written procedures for, 159
Welding metallurgy, 183–187
Welding procedure specifications (WPSs), 162–165, 167
Wire penetrameters, 150–151
Work-flow schedules, 24–27

X

X-rays, 148–149, 196t, 201f

9 781260 455410

Printed by Libri Plureos GmbH in Hamburg,
Germany